Soil Mechanics

Soil Mechanics

A. Verruijt

Revised by S. van Baars

VSSD

First edition 2007

Published by VSSD
Leeghwaterstraat 42, 2628 CA Delft, The Netherlands
tel. +31 15 27 82124, telefax +31 15 27 87585, e-mail: hlf@vssd.nl
internet: http://www.vssd.nl/hlf
URL about this book: **http://www.vssd.nl/hlf/f033.htm**

Dutch hardback edition
ISBN-13 978-90-71301-45-2
Dutch electronic version
ISBN-13 978-90-71301-85-8
English language version:
ISBN-13 978-90-6562-058-3
English electronic version
ISBN-13 978-90-6562-163-4

NUR 956

Key words.: soil mechanics

Preface

This book is part of the introductory course of Soil Mechanics in the Department of Civil Engineering of the Delft University of Technology. It contains an introduction into the major principles and methods of soil mechanics, such as the analysis of stresses, deformations, and stability. The most important methods of determining soil parameters, in the laboratory and in situ, are also described. Some basic principles of applied mechanics that are frequently used are presented in Appendices. The subdivision into chapters is such that one chapter can be treated in a single lecture of 45 minutes, approximately.
Comments of students and other users on the material in earlier versions of this book have been implemented in the present version, and errors have been corrected. Remaining errors are the author's responsibility, of course, and all comments will be appreciated.
Several users, from all over the world, have been kind enough to send me their comments or their suggestions for corrections or improvements.
The "double sliding" logo was produced by Professor G. de Josselin de Jong, who played an important role in developing soil mechanics as a branch of science, and who taught me Soil Mechanics.

Zoetermeer, November 2002 Arnold Verruijt

In this new edition several small, but also some major adjustments have been carried through. Not only pressure, instead of tension, but also settlement and volume decrease are defined positive. Besides this, more attention is given to the difference between the failure criterion of Coulomb (based on one normal stress) and the failure criterion of Mohr-Coulomb (based on two principal stresses). Especially the shear tests reflect this difference. My special thanks goes out to Prof.dr.ir. A. Verruijt, who taught me the fundamentals of Soil Mechanics and gave me the opportunity to revise this book to the developments the Geotechnical Engineering is undergoing.

Delft, May 2007 Stefan Van Baars

Contents

Part I
Introduction

1 Introduction

1.1 The discipline

Soil mechanics is the science of equilibrium and motion of soil bodies. Here soil is understood to be the weathered material in the upper layers of the earth's crust. The non-weathered material in this crust is denoted as *rock*, and its mechanics is the discipline of *rock mechanics*. In general the difference between soil and rock is roughly that in soils it is possible to dig a trench with simple tools such as a spade or even by hand. In rock this is impossible, it must first be splintered with heavy equipment such as a chisel, a hammer or a mechanical drilling device. The natural weathering process of rock is that in the long run the influence of sun, rain and wind it degenerates into stones. This process is stimulated by fracturing of rock bodies by freezing and thawing of the water in small crevices in the rock. The coarse stones that are created in mountainous areas are transported downstream by gravity, often together with water in rivers. By internal friction the stones are gradually reduced in size, so that the material becomes gradually finer: gravel, sand and eventually silt. In flowing rivers the material may be deposited, the coarsest material at high velocities, but the finer material only at very small velocities. This means that gravel will be found in the upper reaches of a river bed, and finer material such as sand and silt in the lower reaches.

The Netherlands is located in the lower reaches of the rivers Rhine and Meuse. In general the soil consists of weathered material, mainly sand and clay. This material has been deposited in earlier times in the delta formed by the rivers. Much fine material has also been deposited by flooding of the land by the sea and the rivers. This process of sedimentation occurs in many areas in the world, such as the deltas of the Nile and the rivers in India and China. In the Netherlands it has come to an end by preventing the rivers and the sea from flooding by building dikes. The process of land forming has thus been stopped, but subsidence continues, by slow tectonic movements. In order to compensate for the subsidence of the land, and sea water level rise, the dikes must gradually be raised, so that they become heavier and cause more subsidence. This process will probably continue forever if the country is to be maintained.

People use the land to live on, and build all sort of structures: houses, roads, bridges, etcetera. It is the task of the geotechnical engineer to predict the behaviour of the soil as a result of these human activities. The problems that arise are, for instance, the settlement of a road or a railway under the influence of its own weight and the traffic

load, the margin of safety of an earth retaining structure (a dike, a quay wall or a sheet pile wall), the earth pressure acting upon a tunnel or a sluice, or the allowable loads and the settlements of the foundation of a building.
For all these problems soil mechanics should provide the basic knowledge.

1.2 History

Soil mechanics has been developed in the beginning of the 20th century. The need for the analysis of the behaviour of soils arose in many countries, often as a result of spectacular accidents, such as landslides and failures of foundations. In the Netherlands the slide of a railway embankment near Weesp, in 1918 (see Figure 1-1) gave rise to the first systematic investigation in the field of soil mechanics, by a special commission set up by the government. Many of the basic principles of soil mechanics were well known at that time, but their combination to an engineering discipline had not yet been completed. The first important contributions to soil mechanics are due to Coulomb, who published an important treatise on the failure of soils in 1776, and to Rankine, who published an article on the possible states of stress in soils in 1857. In 1856 Darcy published his famous work on the permeability of soils, for the water supply of the city of Dijon. The principles of the mechanics of continua, including statics and strength of materials, were also well known in the 19th century, due to the work of Newton, Cauchy, Navier and Boussinesq.
The union of all these fundamentals to a coherent discipline had to wait until the 20th century. It may be mentioned that the committee to investigate the disaster near Weesp came to the conclusion that the water levels in the railway embankment had risen by sustained rainfall, and that the embankment's strength was insufficient to withstand these high water pressures.
Important pioneering contributions to the development of soil mechanics were made by Karl Terzaghi, who, among many other things, has described how to deal with the influence of the pressures of the pore water on the behaviour of soils. This is an essential element of soil mechanics theory. Mistakes on this aspect often lead to large disasters, such as the slides near Weesp, Aberfan (Wales) and the Teton Valley Dam disaster. In the Netherlands much pioneering work was done by Keverling Buisman, especially on the deformation rates of clay. A stimulating factor has been the establishment of the Delft Soil Mechanics Laboratory in 1934, now known as GeoDelft and soon as Deltaris. In many countries of the world there are similar institutes and consulting companies that specialize on soil mechanics. Usually they also deal with *Foundation engineering*, which is concerned with the application of soil mechanics principle to the design and the construction of foundations in engineering practice. Soil mechanics and Foundation engineering together are often denoted as *Geotechnics* or *Geo-Engineering*.

De Prins

Figure 1-1. Landslide near Weesp, 1918.

1.3 Why Soil Mechanics?

Soil mechanics has become a distinct and separate branch of engineering mechanics because soils have a number of special properties, which distinguish the material from other materials. Its development has also been stimulated, of course, by the wide range of applications of soil engineering in civil engineering, as all structures require a sound foundation and should transfer its loads to the soil. The most important special properties of soils will be described briefly in this chapter. In further chapters they will be treated in greater detail, concentrating on quantitative methods of analysis.

1.3.1 Stiffness dependent upon stress level

Many engineering materials, such as metals, but also concrete and wood, exhibit linear stress-strain-behaviour, at least up to a certain stress level. This means that the deformations will be twice as large if the stresses are twice as large. This property is described by Hooke's law, and the materials are called *linear elastic*. Soils do not satisfy this law. For instance, in compression soil becomes gradually stiffer. At the surface sand will slip easily through the fingers, but under a certain compressive stress it gains an ever increasing stiffness and strength. This is mainly caused by the increase of the forces between the individual particles, which gives the structure of particles an increasing strength. This property is used in daily life by the packaging of coffee and other granular materials by a plastic envelope, and the application of vacuum inside the package. The package becomes very hard when the air is evacuated from it. In civil engineering the non-linear property is used to great advantage in a pile foundation for buildings on very soft soil, underlain by a layer of sand. In the sand below a thick deposit of soft clay the stress level is high, due to the weight of the clay. This makes the sand very hard and strong, and it is possible to apply large compressive forces to the piles, provided that they reach into the sand.

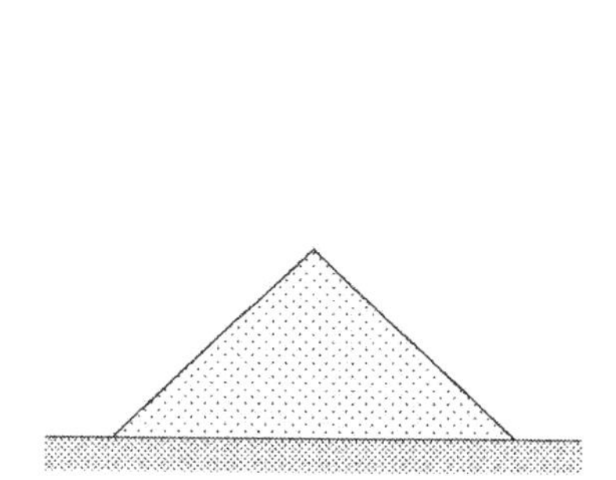

Figure 1-2. A heap of sand.

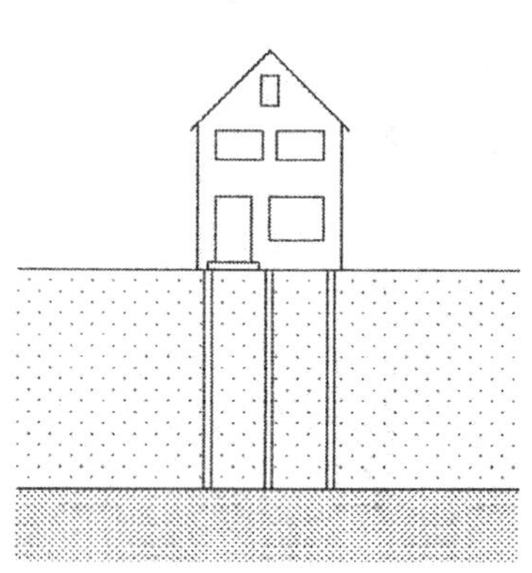

Figure 1-3. Pile foundation.

1.3.2 Shear

In compression soils become gradually stiffer. In shear, however, soils become gradually softer, and if the shear stresses reach a certain level, with respect to the normal stresses, it is even possible that failure of the soil mass occurs. This means that the slope of a sand heap, for instance in a depot or in a dam, can not be larger than about 30 or 40 degrees. The reason for this is that particles would slide over each other at greater slopes. As a consequence of this phenomenon many countries in deltas of large rivers are very flat. It has also caused the failure of dams and embankments all over the world, sometimes with very serious consequences for the local population. Especially dangerous is that in very fine materials, such as clay, a steep slope is often possible for some time, due to capillary pressures in the water, but after some time these capillary pressures may vanish (perhaps because of rain), and the slope will fail.

A positive application of the failure of soils in shear is the construction of guard rails along highways. After a collision by a vehicle the foundation of the guard rail will rotate in the soil due to the large shear stresses between this foundation and the soil body around it. This will dissipate large amounts of energy (into heat), creating a permanent deformation of the foundation of the rail, but the passengers, and the car, may be unharmed. Of course, the guard rail must be repaired after the collision.

1.3.3 Dilatancy

Shear deformations of soils often are accompanied by volume changes. Loose sand has a tendency to contract to a smaller volume, and densely packed sand can practically deform only when the volume expands somewhat, making the sand looser. This is called *dilatancy*, a phenomenon discovered by Reynolds, in 1885. This property causes the soil around a human foot on the beach near the water line to be drawn dry during walking. The densely packed sand is loaded by the weight of the foot, which causes a shear deformation, which in turn causes a volume expansion, which sucks in some water from the surrounding soil. The expansion of a dense soil during shear is shown in Figure 1-4. The space between the particles increases.

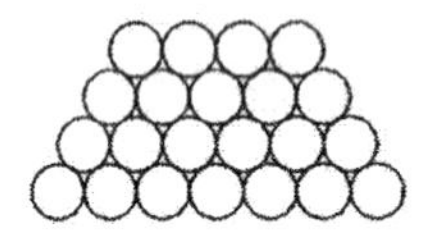
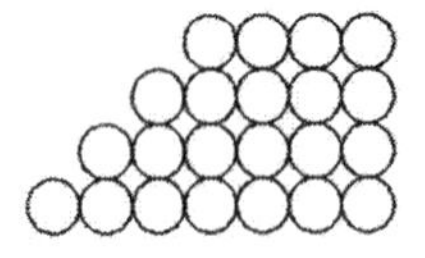

Figure 1-4. Dilatancy.

On the other hand a very loose assembly of sand particles will have a tendency to collapse when it is sheared, with a decrease of the volume. Such volume deformations may be especially dangerous when the soil is saturated with water. The tendency for volume decrease then may lead to a large increase in the pore water pressures. Many geotechnical accidents have been caused by increasing pore water pressures. During earth quakes in Japan, for instance, saturated sand is sometimes densified in a short time, which causes large pore pressures to develop, so that the sand particles may start to float in the water. This phenomenon is called *liquefaction*. In the Netherlands the sand in the channels in the Eastern Scheldt estuary was very loose, which required large densification works before the construction of the storm surge barrier. The sand used to create the airport Tjek Lap Kok in Hongkong was densified before the construction of the runways and the facilities of the airport.

1.3.4 Creep

The deformations of a soil often depend upon time, even under a constant load. This is called *creep*. Clay in particular shows this phenomenon. It causes structures founded on clay to settlements that practically continue forever. A new road, built on a soft soil, will continue to settle for many years. For buildings such settlements are particular damaging when they are not uniform, as this may lead to cracks in the building.

The building of dikes in the Netherlands, on compressible layers of clay and peat, results in settlements of these layers that continue for many decades. In order to maintain the level of the crest of the dikes, they must be raised after a number of

years. This results in increasing stresses in the subsoil, and therefore causes additional settlements. This process will continue forever. Before the construction of the dikes the land was flooded now and then, with sediment being deposited on the land. This process has been stopped by man building dikes. Safety has an ever increasing price.
Sand and rock show practically no creep, except at very high stress levels. This may be relevant when predicting the deformation of porous layers form which gas or oil are extracted.

1.3.5 Groundwater

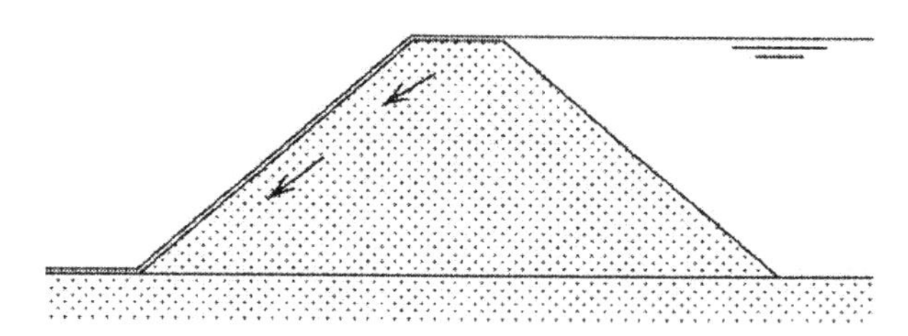

Figure 1-5. Overflowing dike.

A special characteristic of soil is that water may be present in the pores of the soil. This water contributes to the stress transfer in the soil. It may also be flowing with respect to the granular particles, which creates friction stresses between the fluid and the solid material. In many cases soil must be considered as a two phase material. As it takes some time before water can be expelled from a soil mass, the presence of water usually prevents rapid volume changes.
In many cases the influence of the groundwater has been very large. In 1953 in the Netherlands many dikes in the south-west of the country failed because water flowed over them, penetrated the soil, and then flowed through the dike, with a friction force acting upon the dike material. see Figure 1-5. The force of the water on and inside the dike made the slope slide down, so that the dike lost its water retaining capacity, and the low lying land was flooded in a short time.

Figure 1-6. Pisa.

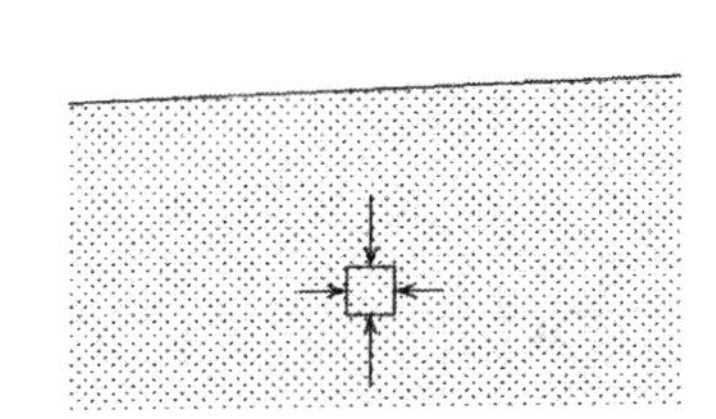

Figure 1-7. Stresses.

In other countries of the world large dams have sometimes failed also because of rising water tables in the interior of the dam (for example, the Teton Valley Dam in the USA, in which water could enter the coarse dam material because of a leaky clay core). Even excessive rainfall may fill up a dam, as happened near Aberfan in Wales in 1966, when a dam of mine tailings collapsed onto the village.
It is also very important that lowering the water pressures in a soil, for instance by the production of groundwater for drinking purposes, leads to an increase of the stresses between the particles, which results in settlements of the soil. This happens in many big cities, such as Venice and Bangkok, that may be threatened to be swallowed by the sea. It also occurs when a groundwater table is temporarily lowered for the construction of a dry excavation. Buildings in the vicinity of the excavation may be damaged by lowering the groundwater table. On a different scale the same phenomenon occurs in gas or oil fields, where the production of gas or oil leads to a volume decrease of the reservoir, and thus to subsidence of the soil. The production of natural gas from the large reservoir in Groningen is estimated to result in a subsidence of about 50 cm.

1.3.6 Unknown initial stresses

Soil is a natural material, created in historical times by various geological processes. Therefore the initial state of stress is often not uniform, and often even partly unknown. Because of the non-linear behaviour of the material, mentioned above, the initial stresses in the soil are of great importance for the determination of soil behaviour under additional loads. These initial stresses depend upon geological history, which is never exactly known, and this causes considerable uncertainty. In particular, the initial horizontal stresses in a soil mass are usually unknown. The initial vertical stresses may be determined by the weight of the overlying layers. This means that the stresses increase with depth, and therefore stiffness and strength also increase with depth. The horizontal stresses, however, usually remain unknown. When the soil has been compressed horizontally in earlier times, it can be expected that the horizontal stress is high, but when the soil is known to have spread out, the horizontal stresses may be very low. Together with the stress dependency of the soil behaviour all this means that there may be considerable uncertainty about the initial behaviour of a soil mass. It may also be noted that further theoretical study can not provide much help in this matter. Studying field history, or visiting the site, and talking to local people, may be more helpful.

1.3.7 Variability

The creation of soil by ancient geological processes also means that soil properties may be rather different on different locations. Even in two very close locations the soil properties may be completely different, for instance when an ancient river channel has been filled with sand deposits. Sometimes the course of an ancient river can be traced on the surface of a soil, but often it can not be seen at the surface. When an embankment is built on such a soil, it can be expected that the settlements

will vary, depending upon the local material in the subsoil. The variability of soil properties may also be the result of a heavy local load in the past.
A global impression of the soil composition can be obtained from geological maps. These indicate in the first place the geological history of the soils. Together with geological knowledge and experience this may give a first indication of the soil properties. Other geological information may also be helpful. Large areas of Western Europe have, for instance, been covered by thick layers of ice in earlier ice ages, and this means that the soils in these areas have been subject to a preload of considerable magnitude.
An accurate determination of soil properties can not be made from desk studies. It requires testing of the actual soils in the laboratory, using samples taken from the field, or testing of the soil in the field (*in situ*). This will be elaborated in later chapters.

Problems

1.1 In times of high water in the rivers in the Netherlands, when the water table rises practically to the crest of the dikes, local authorities sometimes put sand bags on top of the dike. Is that useful?

1.2 Another measure to prevent failure of a dike during high floods, is to place large sheets of plastic on the slope of the dike. On which side?

1.3 Will the horizontal stress in the soil mass near a deep river be relatively large or small?

1.4 The soil at the bottom of the sea is often much stiffer in the Northern parts than it is in the Southern parts. What can be the reason?

1.5 A possible explanation of the leaning of the Pisa tower is that the subsoil contains a compressible clay layer of variable thickness. On what side of the tower would that clay layer be thickest?

1.6 Another explanation for the leaning of the Pisa tower is that in earlier ages (before the start of the building of the tower, in 1400) a heavy structure stood near that location. On which side of the tower would that building have been?

1.7 The tower of the Old Church of Delft in Holland, along the canal Old Delft, is also leaning. What is the probable cause, and is there a possible simple technical solution to prevent further leaning?

Part II
Soil and stresses

2 Classification

2.1 Grain size

Soils are usually classified into various types. In many cases these various types also have different mechanical properties. A simple subdivision of soils is on the basis of the *grain size* of the particles that constitute the soil. Coarse granular material is often denoted as gravel and finer material as sand. In order to have a uniformly applicable terminology it has been agreed internationally to consider particles larger than 2 mm, but smaller than 63 mm as *gravel*. Larger particles are denoted as *stones*. *Sand* is the material consisting of particles smaller than 2 mm, but larger than 0.063 mm. Particles smaller than 0.063 mm and larger than 0.002 mm are denoted as *silt*. Soil consisting of even smaller particles, smaller than 0.002 mm, is denoted as *clay* or *luthum*, see Table 2-1. In some countries, such as the Netherlands, the soil may also contain layers of *peat*, consisting of organic material such as decayed plants. Particles of peat usually are rather small, but it may also contain pieces of wood. It is then not so much the grain size that is characteristic, but rather the chemical composition, with large amounts of carbon. The amount of carbon in a soil can easily be determined by measuring how much is lost when burning the material.

Soil type	min.	max.
Clay		0.002 mm
Silt	0.002 mm	0.063 mm
Sand	0.063 mm	2 mm
Gravel	2 mm	63 mm

Table 2-1: Grain sizes.

The mechanical behaviour of the main types of soil, sand, clay and peat, is rather different. Clay usually is much less permeable for water than sand, but it usually is also much softer. Peat is usually is very light (some times hardly heavier than water), and strongly anisotropic because of the presence of fibers of organic material. Peat usually is also very compressible. Sand is rather permeable, and rather stiff, especially under a certain preloading. It is also very characteristic of granular soils such as sand and gravel, that they can not transfer tensile stresses. The particles can only transfer compressive forces, no tensile forces. Only when the particles are very small and the soil contains some water, can a tensile stress be transmitted, by capillary forces in the contact points.

The grain size may be useful as a first distinguishing property of soils, but it is not very useful for the mechanical properties. The quantitative data that an engineer needs depend upon the mechanical properties such as stiffness and strength, and these must be determined from mechanical tests. Soils of the same grain size may have different mechanical properties. Sand consisting of round particles, for instance, can have a strength that is much smaller than sand consisting of particles with sharp points. Also, a soil sample consisting of a mixture of various grain sizes can have a very small permeability if the small particles just fit in the pores between the larger particles.
The global character of a classification according to grain size is well illustrated by the characterization sometimes used in Germany, saying that gravel particles are smaller than a chicken's egg and larger than the head of a match, and that sand particles are smaller than a match head, but should be visible to the naked eye.

2.2 Grain size diagram

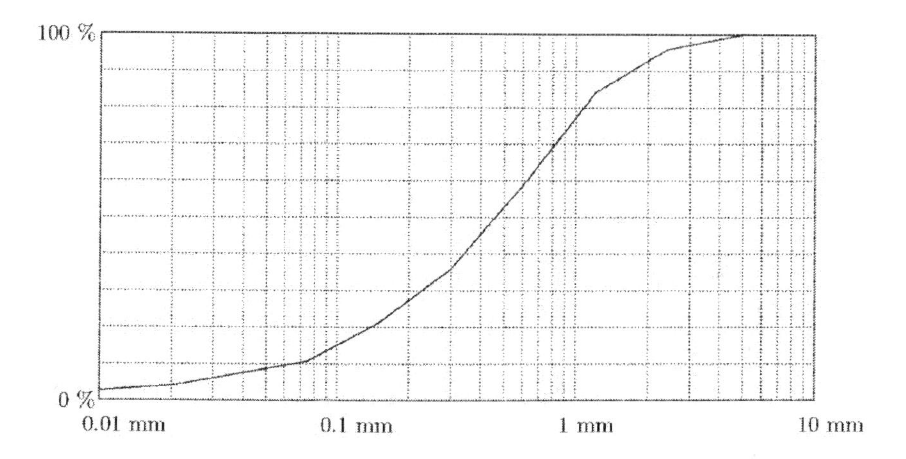

Figure 2-1. Grain size diagram.

The size of the particles in a certain soil can be represented graphically in a *grain size diagram*, see Figure 2-1. Such a diagram indicates the percentage of the particles smaller than a certain diameter, measured as a percentage of the weight. A steep slope of the curve in the diagram indicates a uniform soil, a shallow slope of the diagram indicates that the soil contains particles of strongly different grain sizes. For rather coarse particles, say larger than 0.05 mm, the grain size distribution can be determined by sieving. The usual procedure is to use a system of sieves having different mesh sizes, stacked on top of each other, with the coarsest mesh on top and the finest mesh at the bottom. After shaking the assembly of sieves, by hand or by a shaking machine, each sieve will contain the particles larger than its mesh size, and smaller than the mesh size of all the sieves above it. In this way the grain size diagram can be determined. Special standardized sets of sieves are available, as well

as convenient shaking machines. The example shown in Figure 2-1 illustrates normal sand. In this case there appear to be no grains larger than 5 mm.
The grain size distribution can be characterized by the quantities D_{60} and D_{10}. These indicate that 60 %, respectively 10 % of the particles (expressed as weights) is smaller than that diameter. In the case illustrated in Figure 2-1 it appears that $D_{60} \approx 0.6$ mm, and $D_{10} \approx 0.07$ mm. The ratio of these two numbers is denoted as the *uniformity coefficient* C_u,

$$C_u = \frac{D_{60}}{D_{10}}. \qquad (2.1)$$

In the case of Figure 2-1 this is about 8.5. This indicates that the soil is not uniform. This is sometimes denoted as a *well graded soil.* In a *poorly graded soil* the particles all have about the same size. The uniformity coefficient is than only slightly larger than 1, say $C_u = 2$. For particles smaller than about 0.05 mm the grain size can not be determined by sieving, because the size of the holes in the mesh would become unrealistically small, and also because during shaking the small particles might fly up in the air, as dust. The amount of particles of a particular size can then be determined much better by measuring the velocity of deposition in a glass of water. This method is based upon a formula derived by Stokes. This formula expresses that the force on a small sphere, sinking in a viscous fluid, depends upon the viscosity of the fluid, the size of the sphere and the velocity. Because the force acting upon the particle is determined by the weight of the particle under water, the velocity of sinking of a particle in a fluid can be derived. The formula is

$$v = \frac{(\gamma_k - \gamma_w)D^2}{18\mu}, \qquad (2.2)$$

where γ_k is the volumetric weight of the particle, γ_w is the volumetric weight of the fluid, D is the grain size, and γ is the dynamic viscosity of the fluid. Because for very small particles the velocity may be very small, the test may take rather long.

2.3 Chemical composition

Besides the difference in grain size, the chemical composition of soil can also be helpful in distinguishing between various types of soils. Sand and gravel usually consist of the same minerals as the original rock from which they were created by the erosion process. This can be quartz, feldspar or glimmer. In Western Europe sand mostly consist of quartz. The chemical formula of this mineral is SiO_2.
Fine-grained soils may contain the same minerals, but they also contain the so-called *clay minerals*, which have been created by chemical erosion. The main clay minerals are kaolinite, montmorillonite and illite. In the Netherlands the most frequent clay mineral is illite. These minerals consist of compounds of aluminium with hydrogen, oxygen and silicates. They differ from each other in chemical composition, but also in geometrical structure, at the microscopic level. The microstructure of clay usually

resembles thin plates. On the microscale there are forces between these very small elements, and ions of water may be bonded. Because of the small magnitude of the elements and their distances, these forces include electrical forces and the Van der Waals forces.
Although the interaction of clay particles is of a different nature than the interaction between the much larger grains of sand or gravel, there are many similarities in the global behaviour of these soils. There are some essential differences, however. The deformations of clay are time dependent, for instance. When a sandy soil is loaded it will deform immediately, and then remain at rest if the load remains constant. Under such conditions a clay soil will continue to deform, however. This is called *creep*. It is very much dependent upon the actual chemical and mineralogical constitution of the clay. Also, some clays, especially clays containing large amounts of montmorillonite, may show a considerable swelling when they are getting wetter.
As mentioned before, peat contains the remains of decayed trees and plants. Chemically it therefore consists partly of carbon compounds. It may even be combustible, or it may be produce gas. As a foundation material it is not very suitable, also because it is often very light and compressible. It may be mentioned that some clays may also contain considerable amounts of organic material.
For a civil engineer the chemical and mineralogical composition of a soil may be useful as a warning of its characteristics, and as an indication of its difference from other materials, especially in combination with data from earlier projects. A chemical analysis does not give much quantitative information on the mechanical properties of a soil, however. For the determination of these properties mechanical tests, in which the deformations and stresses are measured, are necessary. These will be described in later chapters.

2.4 Consistency limits

For very fine soils, such as silt and clay, the consistency is an important property. It determines whether the soil can easily be handled, by soil moving equipment, or by hand. The consistency is often very much dependent on the amount of water in the soil. This is expressed by the *water content* w (see also Chapter 4). It is defined as the weight of the water per unit weight of solid material,

$$w = W_w / W_k \,. \tag{2.3}$$

When the water content is very low (as in a very dry clay) the soil can be very stiff, almost like a stone. It is then said to be in the *solid state*. Adding water, for instance if the clay is flooded by rain, may make the clay plastic, and for higher water contents the clay may even become almost liquid. In order to distinguish between these states (solid, plastic and liquid) two standard tests have been agreed upon, that indicate the *consistency limits*. They are sometimes denoted as the *Atterberg limits*, after the Swedish engineer who introduced them.
The transition from the *liquid state* to the *plastic state* is denoted as the *liquid limit*, w_L. It represents the lowest water content at which the soil behaviour is still mainly

liquid. As this limit is not absolute, it has been defined as the value determined in a certain test, due to Casagrande, see Figure 2-2. In the test a hollow container with a soil sample may be raised and dropped by rotating an axis. The liquid limit is the value of the water content for which a standard V-shaped groove cut in the soil, will just close after 25 drops. When the groove closes after less than 25 drops, the soil is too wet, and some water must be allowed to evaporate. By waiting for some time, and perhaps mixing the clay some more, the water content will have decreased, and the test may be repeated, until the groove is closed after precisely 25 drops. Then the water content must immediately be determined, before any more water evaporates, of course.

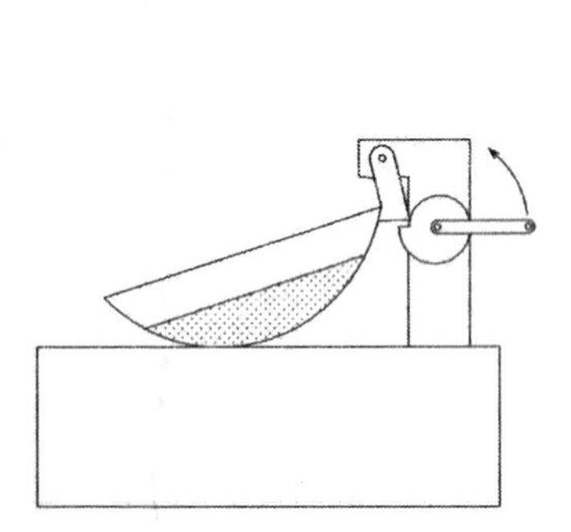

Figure 2-2. Liquid limit.

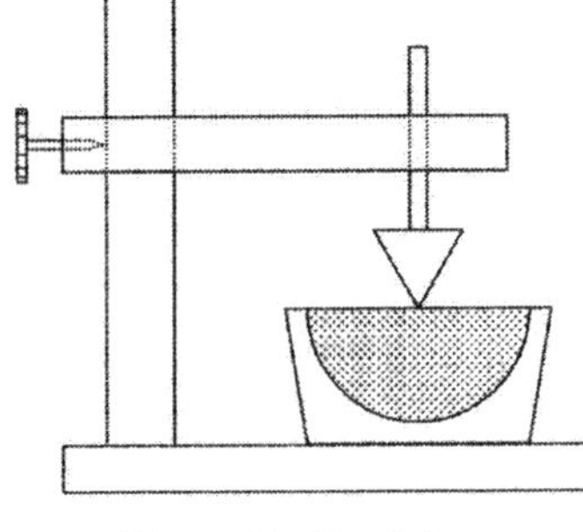

Figure 2-3. The fall cone.

An alternative for Casagrande's test is the *fall cone*, see Figure 2-3. In this test a steel cone, of 60 grams weight, and having a point angle of 60°, is placed upon a clay sample, with the point just at the surface of the clay. The cone is then dropped and its penetration depth is measured. The liquid limit has been defined as the water content corresponding to a penetration of exactly 10 mm. Again the liquid limit can be determined by doing the test at various water contents. It has also been observed, however, that the penetration depth, when plotted on a logarithmic scale, is an approximately linear function of the water content. This means that the liquid limit may be determined from a single test, which is much faster, although less accurate.
The transition from the plastic state to the solid state is called the *plastic limit*, and denoted as w_P. It is defined as the water content at which the clay can just be rolled to threads of 3 mm diameter. Very wet clay can be rolled into very thin threads, but dry clay will break when rolling thick threads. The (arbitrary) limit of 3 mm is supposed to indicate the plastic limit. In the laboratory the test is performed by starting with a rather wet clay sample, from which it is simple to roll threads of 3 mm. By continuous rolling the clay will gradually become drier, by evaporation of the water, until the threads start to break.
For many applications (potteries, dike construction) it is especially important that the range of the plastic state is large. This is described by the *plasticity index PI*. It is defined as the difference of the liquid limit and the plastic limit,

$$PI = w_L - w_P.$$

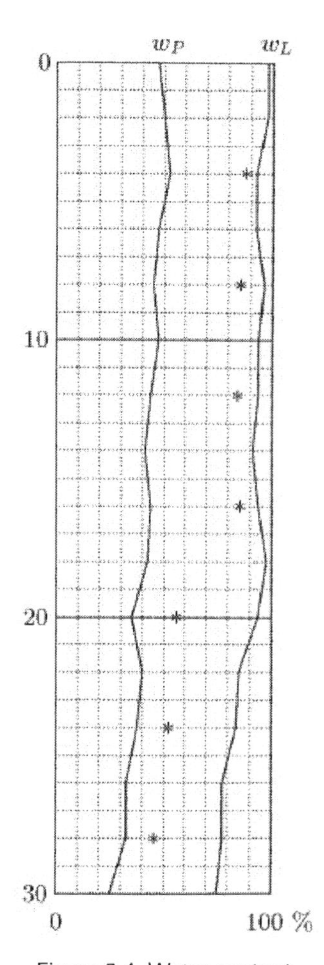

Figure 2-4. Water content.

The plasticity index is a useful measure for the possibility to process the clay. It is important for potteries, for the construction of the clay core in a high dam, and for the construction of a layer of low permeability covering a deposit of polluted material. In all these cases a high plasticity index indicates that the clay can easily be used without too much fear of it turning into a liquid or a solid.
In countries with very thick clay deposits (England, Japan, Scandinavia) it is often useful to determine a profile of the plastic limit and the liquid limit as a function of depth, see Figure 2-4. In this diagram the natural water content, as determined by taking samples and immediately determining the water content, can also be indicated.

2.5 An international classification system

The large variability of soil types, even in small countries such as the Netherlands, leads to large variations in soil properties in soils that may resemble each other very much at first sight. This is enhanced by confusion between terms such as *sandy clay* and *clayey sand* that may be used by local firms. In some areas tradition may have also lead to the use of terms such as *blue clay* or *brown clay*, that may be very clear to experienced local engineers, but have little meaning to others.

Character 1		Character 2	
G	gravel	W	well graded
S	sand	P	poorly graded
M	silt	M	silty
C	clay	C	clayey
O	organic	L	low plasticity
Pt	peat	H	high plasticity

Table 2-2: Unified Classification System (USA).

Uniform criteria for the classification of soils do not exist, especially because of local variations and characteristics. The soil in a plane of Tibet may be quite different from the soil in Bolivia or Canada, as their geological history may be quite different. The engineer should be aware of such differences and remain open to characterizations that are used in other countries. Nevertheless, a classification system that has been developed by the United States Bureau of Reclamation, is widely used all over the world. This system consists of two characters to indicate a soil type, see Table 2-2. A soil of type SM, for instance, is a silty sand, which indicates that it is a sand, but containing considerable amounts of non-organic fine silty particles. This type of soil is found in the Eastern Scheldt in the Netherlands. The sand on the beaches of the Netherlands usually is of the type SW. A clay of very low plasticity, that is a clay with a relatively small plasticity index is denoted as CL. The clay in a polder in Holland will often be of the type CH. It has a reasonably large range of plastic behaviour.

The characterization *well graded* indicates that a granular material consists of particles that together form a good framework for stress transfer. It usually is relatively stiff and strong, because the smaller particles fill well in the pores between the larger particles. A material consisting of large gravel particles and fine sand is called *poorly graded*, because it has little coherence. A well graded material is suitable for creating a road foundation, and is also suitable for the production of concrete.

Global classifications as described above usually have only little meaning for the determination of mechanical properties of soils, such as stiffness and strength. There may be some correlation between the classification and the strength, but this is merely indicative. For engineering calculations mechanical tests should be performed, in which stresses and deformations are measured. Such tests are described in later chapters.

3 Soil exploration

In this chapter some of the most effective or popular methods for soil exploration, or soil investigations in the field will be described.

3.1 Cone Penetration Test

A simple, but very effective method of soil investigation consists of pushing a steel rod into the soil, and then measuring the force during the penetration, as a function of depth. This force consists of the reaction of the soil at the point (the cone resistance), and the friction along the circumference of the rods. The method was developed in the 1930's in the Netherlands. It was mainly intended as an exploration tool, to give an indication of the soil structure, and as a modelling tool for the design of a pile foundation. This *sounding test*, *cone penetration test*, or simply CPT, has been developed from a simple tool, that was pushed into the ground by hand or a manual pressure device, into a sophisticated electronic measuring device, with an advanced hydraulic loading system. The load is often provided by the weight of a heavy truck.
Originally the CPT was a purely mechanical test, as shown schematically in Figure 3-1. The instrument consists of three movable parts, with a common central axis. The upper part is connected, by a screw thread, to a hollow rod, that reaches to the soil surface, using extension rods of 1 meter length. The procedure was that pressure was alternately exerted upon the central axis or the outer rods. When pushing on the internal axis at first only the cone is pushed into the ground, over a distance of 35 mm. The other two parts do not move with respect to the soil (by the friction of the soil), so that the force represents the cone resistance only. When pushing the instrument beyond a distance of 35 mm the second part, the *friction sleeve*, moves with the cone, so that in this stage the force consists of the cone resistance plus the friction along the friction sleeve. The upper part of the instrument is still stationary in this stage. If it is assumed that the cone resistance is still the same as before, the sleeve friction can be determined by subtraction. If in the next step the force is exerted on the outer rods, the cone remains stationary and the system is compressed to its original state, but at a greater depth (10 cm). The diameter of the lowest part of the sleeve, which is attached to the cone and moves with it, was sometimes reduced, to ensure that in the first stage only point resistance is measured.
Modern versions of the CPT use an electrical cone, see Figure 3-2. Both the cone resistance and the friction are measured continuously, using a system of strain gauges in the interior of the cone. The instrument again consists of three parts, that

are separated by thin rings of rubber. The very sensitive strain gauges can measure the forces on the lower two parts of the instrument independently.

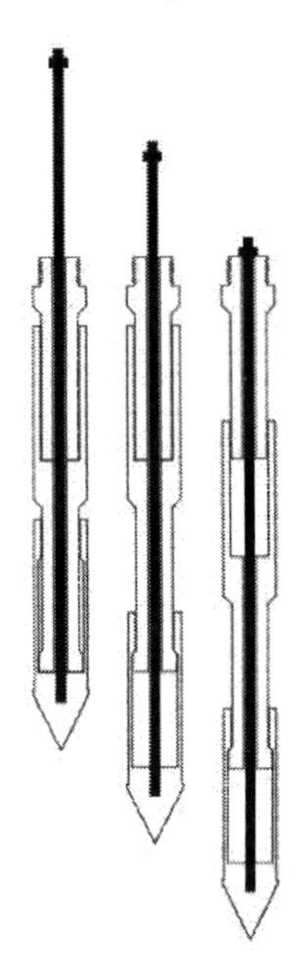

Figure 3-1. Mechanical CPT.

Figure 3-2. Electrical cone.

The results of a cone penetration test give a good insight into the layered structure of the soil. Clay layers have a much smaller cone resistance than sand. A typical cone resistance q_c for a sand layer is 5 MPa or 10 MPa, or even higher, whereas the cone resistance of soft clay layers is below 1 MPa. If the local friction is also measured the difference is even more pronounced. The ratio of friction to cone resistance for clays is much higher than for sand. In sands the friction usually is only about 1 % of the cone resistance, whereas in clays this ratio usually is 3 % to 5 %. Higher values (8 % – 10 %) may suggest a layer of peat. In peat the friction usually is substantial, but it has a very small cone resistance.

Recent developments are to install additional measuring devices in the cone, such a pore pressure meter. This type of cone is denoted as a *piezocone*. A small chamber inside the cone is connected to the pores in the soil by a number of tiny holes in the cone. This enables to measure the local pore water pressure. This pressure is determined by the actual pore water pressure in the soil, but also by the penetration of the cone in the soil, at least in materials of low permeability. In a very dense clay the material may have a tendency to expand, which will lead to and under pressure in the water, with respect to the hydrostatic pressure. This enables to distinguish very thin layers of clay. In measuring the cone resistance or the friction such thin layers are not observed, because of the averaging procedure in measuring forces.
An example of the results of as cone penetration test is shown in Figure 3-3. At a depth of 7 meter a sand layer of about 2 meter thickness can be observed. At a depth

of 18 meter the top of a thick sand layer is found. The low values above the first sand layer, and between the two sand layers indicate soft soil, probably clay. A simple building (a house) can be founded on the top sand layer, provided that the presence of this layer is general. A single CPT is insufficient to conclude the existence of this layer everywhere, having it observed in 3 CPT's at practically the same depth (and at about the same thickness) usually is sufficient evidence of its general existence. A heavy foundation, for a large building, usually requires a foundation reaching into the deep sand.

Soil type	Friction ratio	Cone resistance q_c
sand, medium – coarse	0.4%	
sand, fine – medium	0.6%	5 - 30 MPa
sand, fine	0.8%	
sand, silty	1.1%	
sand, clayey	1.4%	5 - 10 MPa
sandy clay or loam	1.8%	
silt	2.2%	
clay, silty	2.5%	
clay	3.3%	0.5 - 2 MPa
clay, peaty	5.0%	
peat	8.1%	0.1 - 1 MPa

Table 3-3. Friction ratio and cone resistance

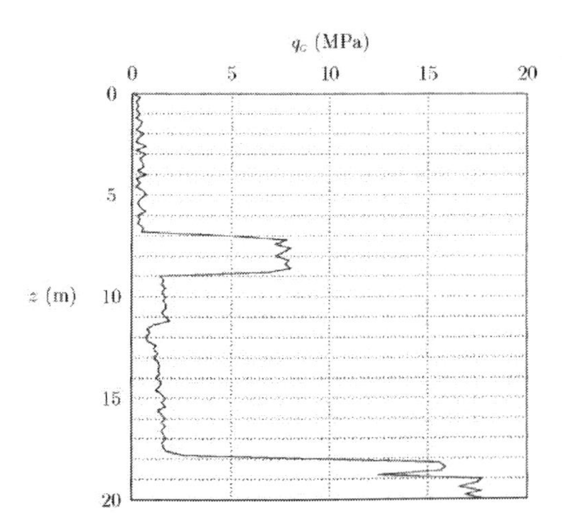

Figure 3-3. Result of CPT.

The cone penetration test is in the Netherlands also used as a model test for estimating the bearing capacity of pile foundations. In the west of the country generally about 10 m to 20 m of soft soil layers lay on top of a stiff sand layer, which is excellent for using pile foundations. The bearing capacity of a pile depends mostly on the capacity of the sand layer. The capacity of the tip of the pile with area A can be estimated by:

$$F_{tip} = q_c A \tag{3.1}$$

3.2 Vane test

The shear strength of soils can be measured reasonably accurately in situ using the *vane test.* In this test a small instrument in the shape of a vane is pushed into the ground, through a system of rods, just as in the cone penetration test. The vane is connected, by a central steel axis, to a screw at the top of the rods. This screw can be rotated, so that the soil in a cylindrical element of soil is sheared along its surface, against the soil outside the cylinder. Measuring the moment necessary for the rotation enables to determine the average shear stress along the boundary, which is about equal to the (undrained) shear strength of the soil. The vane test is very popular in Scandinavian countries, where the soil very often consists of thick layers of clay of reasonable strength.

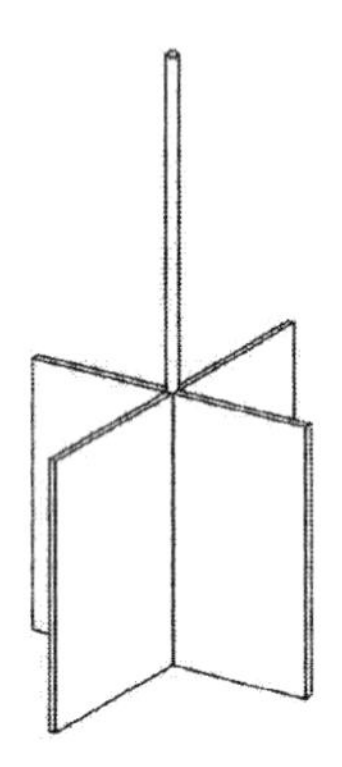

Figure 3-4. Vane test.

3.3 Standard Penetration Test

In many parts of the world, especially in Anglo-Saxon countries, the properties of the soil are often determined by using a *Standard Penetration Test*, or SPT. In this test a sampling tube is driven into a borehole in the ground using a standardized hammering weight. The actual test consists of measuring the number of blows needed to achieve a penetration of 300 mm (1 foot) into the ground. This is denoted as N, the *blow count*, the number of blows per foot. An advantage of the SPT is that no heavy equipment is needed, as for instance in the CPT, which has to be pushed into the ground statically, and thus requires a large counter weight. Another advantage of the SPT is that immediately provides a soil sample. The sample is not of the best quality, but at least there is a sample. The reproducibility of the SPT usually is not so very good, and the difference between sand and clay is not so pronounced as it is in the CPT. It is also not possible to immediately derive the shear

strength from the blow count.
For many projects the initial soil data often may be restricted to a series of SPT-results. Then it is useful to know that a characteristic blow count for sand is $N = 20$, and that for soft clay the value may be $N = 5$, or even lower, down to $N = 1$. A first indication can be obtained from Table 3-4, derived from Terzaghi & Peck. Many researchers have tried to obtain a correlation with the CPT, but their results are not very consistent.

Sand		Clay	
N	Density	N	Consistency
<4	Very Loose	<2	Very soft
4-10	Loose	2-4	Soft
10-30	Normal	4-8	Normal
30-50	Dense	8-15	Stiff
>50	Very dense	15-30	Very stiff
		>30	Hard

Table 3-4: Interpretation of SPT according to Terzaghi &Peck.

3.4 Soil sampling

For many engineering projects it is very useful to take a sample of the soil, and to investigate its properties in the laboratory. The investigation may be a visual inspection (which indicates the type of materials: sand, clay or peat), a chemical analysis, or a mechanical test, such as a compression test or a triaxial test.

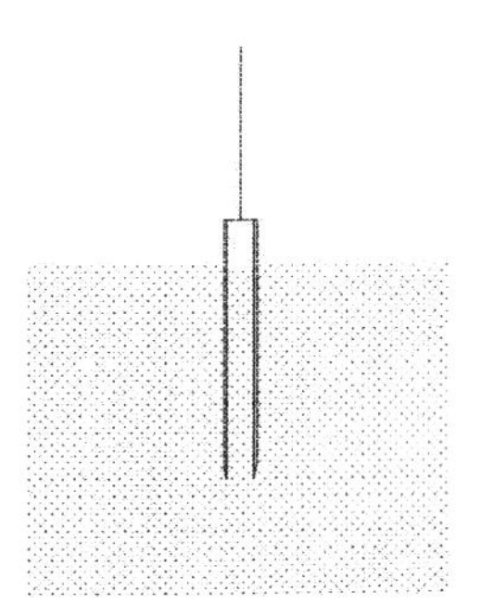

Figure 3-5. SPT.

A simple method to take a sample is to drive a tube into the ground, and then recovering the tube with the soil in it. The tube may be about 1 meter long, see Figure 3-5, and may have a valve at its bottom, to prevent loosing the sample. The tube may be brought into the soil by driving it into the ground using a falling weight, or a hammer. An advantage of this method is that it does not require heavy equipment. It is possible to take a sample in a terrain that is inaccessible to heavy vehicles. The sample is somewhat disturbed, of course, during the sampling process, but even so, a good impression of the composition of the soil can be obtained. The

sample is not very well suited for a refined test, however, as the initial state of stress is disturbed, and perhaps also the density. To take a deep sample the sampling tube may be of smaller diameter than the borehole, which is supported and deepened by a special boring tube.
An alternative method is to push the sampler into the ground, by using hydraulic equipment, mounted on a heavy truck. In this case the sampling process is somewhat more careful, and the disturbance of the sample is less. Due to friction of the sample with the wall of the sampling tube, however, the samples are not undisturbed.

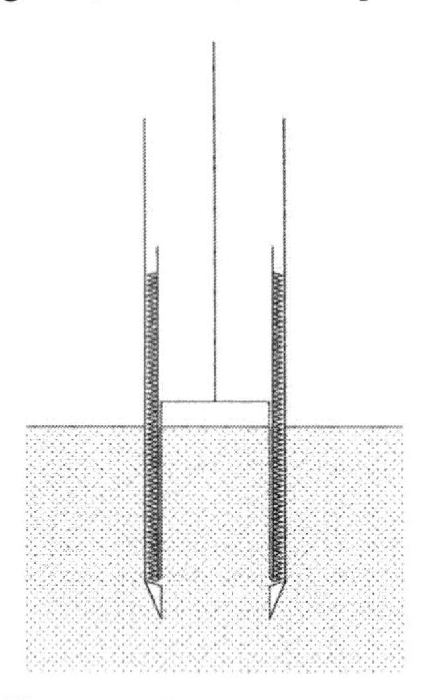

Figure 3-6. Begemann sampler.

Various institutes have developed systems in which the sample to be taken is almost undisturbed. A completely undisturbed sample is impossible, but some procedures come very close. Some methods are, for instance, to take a very large block of soil, and use the inner part only, or freezing a block of sand, and then cutting a sample from the frozen soil. Good quality samples can also be obtained using the Begemann sampler, developed at GeoDelft, see Figure 3-6. This sampler consists of two steel tubes, that are being pushed into the soil together. The sample is cut by the outer tube, which immediately widens behind the cutting edge, and the sample is surrounded by a nylon stocking, that initially is rolled up on the inner tube. The end of the stocking is attached to a plate at the top of the future sample, so that, when the tubes are pushed down, the stocking gradually displace downward the stocking is gradually stripped off the inner tube. The final result is a very long soil sample (for instance 20 meter long), enclosed by a nylon stocking. Around the stocking the sample is supported by a heavy fluid (of unit weight $\gamma \approx 15$ kN/m^3), that simulates the original lateral support of the soil. This fluid also reduces the friction along the circumference of the sample. The samples produced by this sampler are of high quality. Very thin layers of all sorts of materials can be identified, including loose sand. The quality of the samples is good enough to be used for accurate laboratory

testing, in compression tests or triaxial tests. The results of a boring may be presented in the form of a color photograph of one half of the sample, cut along its length. That the thin layers are not disturbed near the boundary confirms that there is very little friction.
It may be interesting to note that samples can also be taken from the bottom of the sea. One possible method is by using a diving bell, in which the air pressure is kept at the same level as the water pressure. From this diving bell a sample can be taken by the operators, or they can make a cone penetration test. Another method is to use a heavy frame, that is submerged in the water from a ship. Using a remote control system a cone can be made to penetrate the soil, or a sample can be taken. This method can even be used in water depths of 1000 meter, or more. An example of a continuous Begemann boring is shown in Figure 3-7.

Figure 3-7. Begemann sample.

Investigating the sea bottom is of special interest in offshore engineering, of course. For the production of oil and gas from the sea bottom large platforms are constructed, which usually need a pile foundation to withstand the extreme wave load conditions during a storm. The piles usually are steel tubular piles, of large diameter (one meter or more), and very large length (50 meter or more). These piles derive their bearing capacity mostly from the friction along the shaft, and not from the point resistance (as most piles in Western Netherlands). It is of great importance to predict the maximum shearing resistance along the pile shaft. This can be measured very well by a cone penetration test, from the bottom of the sea. Even though this is a costly operation, it gives very valuable information about the soil structure, and it gives numerical values for the cone resistance and the friction, as a function of depth.

4 Particles, water, air

4.1 Porosity

Soils usually consist of particles, water and air. In order to describe a soil various parameters are used to describe the distribution of these three components, and their relative contribution to the volume of a soil. These are also useful to determine other parameters, such as the weight of the soil. They are defined in this chapter. An important basic parameter is the *porosity* n, defined as the ratio of the volume of the pore space and the total volume of the soil,

$$n = V_p / V_t . \qquad (4.1)$$

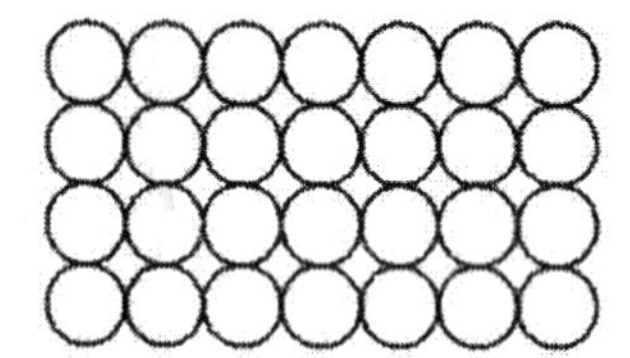

Figure 4-1. Cubic array.

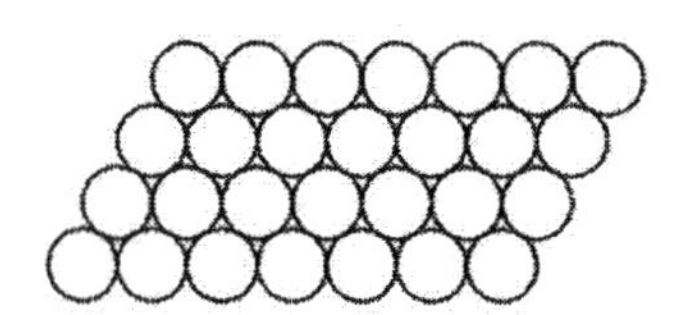

Figure 4-2. Densest array.

For most soils the porosity is a number between 0.30 and 0.45 (or, as it is usually expressed as a percentage, between 30 % and 45 %). When the porosity is small the soil is called densely packed, when the porosity is large it is loosely packed.
It may be interesting to calculate the porosities for two particular cases. The first case is a very loose packing of spherical particles, in which the contacts between the spheres occur in three mutually orthogonal directions only. This is called a *cubic array* of particles, see Figure 4-1. If the diameter of the spheres is D, each sphere occupies a volume $\pi D^3/6$ in space. The ratio of the volume of the solids to the total volume then is $V_p/V_t = \pi/6 = 0.5236$, and the porosity of this assembly thus is $n = 0.4764$. This is the loosest packing of spherical particles that seems possible. Of course, it is not stable: any small disturbance will make the assembly collapse.
A very dense packing of spheres can be constructed by starting from layers in which the spheres form a pattern of equilateral triangles, see Figure 4-2. The packing is constructed by packing such layers such that the spheres of the next layer just fit in the hollow space between three spheres of the previous layer. The axial lines from a sphere with the three spheres that support it from below form an regular tetrahedron,

having sides of magnitude D. The height of each tetrahedron is $D\sqrt{2/3}$. Each sphere of the assembly, with its neighbouring part of the voids, occupies a volume in space of magnitude $D\times(D\sqrt{3/4})\times(D\sqrt{2/3})=D^3\sqrt{1/2}$. Because the volume of the sphere itself is $\pi D^3/6$ is, the porosity of this assembly is $n=1-\pi/\sqrt{18}=0.2595$. This seems to be the most dense packing of a set of spherical particles.
Although soils never consist of spherical particles, and the values calculated above have no real meaning for actual soils, they may give a certain indication of what the porosity of real soils may be. It can thus be expected that the porosity n of a granular material may have a value somewhere in the range from 0.25 to 0.45. Practical experience confirms this statement.
The amount of pores can also be expressed by the *void ratio e*, defined as the ratio of the volume of the pores to the volume of the solids,

$$e=V_p/V_s. \tag{4.2}$$

In many countries this quantity is preferred to the porosity, because it expresses the pore volume with respect to a fixed volume (the volume of the solids). Because the total volume of the soil is the sum of the volume of the pores and the volume of the solids, $V_t=V_p+V_s$, the porosity and the void ratio can easily be related,

$$e=\frac{n}{1-n}, \quad n=\frac{e}{1+e} \tag{4.3}$$

The porosity can not be smaller than 0, and can not be greater than 1. The void ratio can be greater than 1.
The void ratio is also used in combination with the *relative density*. This quantity is defined as

$$RD=\frac{e_{\max}-e}{e_{\max}-e_{\min}}. \tag{4.4}$$

Here e_{max} is the maximum possible void ratio, and e_{min} the minimum possible value. These values may be determined in the laboratory. The densest packing of the soil can be obtained by strong vibration of a sample, which then gives e_{min}. The loosest packing can be achieved by carefully pouring the soil into a container, or by letting the material subside under water, avoiding all disturbances, which gives e_{max}. The accuracy of the determination of these two values is not very large. After some more vibration the sample may become even denser, and the slightest disturbance may influence a loose packing. It follows from eq. (4.4) that the relative density varies between 0 and 1. A small value, say $RD<0.5$, means that the soil can easily be densified. Such a densification can occur in the field rather unexpectedly, for instance in case of a sudden shock (an earthquake), with dire consequences.
Of course, the relative density can also be expressed in terms of the porosity, using eqs. (4.3), but this leads to an inconvenient formula, and therefore this is unusual.

4.2 Degree of saturation

The pores of a soil may contain water and air. To describe the ratio of these two the *degree of saturation* S is introduced as

$$S = V_w / V_p . \tag{4.5}$$

Here V_w is the volume of the water, and V_p is the total volume of the pore space. The volume of air (or any other gas) per unit pore space then is $1 - S$. If $S = 1$ the soil is completely saturated, if $S = 0$ the soil is perfectly dry.

4.3 Density

For the description of the density and the volumetric weight of a soil, the densities of the various components are needed. The *density* of a substance is the mass per unit volume of that substance. For water this is denoted by ρ_w, and its value is about $1000\ \mathrm{kg/m^3}$. Small deviations from this value may occur due to temperature differences or variations in salt content. In soil mechanics these are often of minor importance, and it is often considered accurate enough to assume that

$$\rho_w = 1000\, kg/m^3 . \tag{4.6}$$

For the analysis of soil mechanics problems the density of air can usually be disregarded.
The density of the solid particles depends upon the actual composition of the solid material. In many cases, especially for quartz sands, its value is about

$$\rho_p = 2650\, kg/m^3 . \tag{4.7}$$

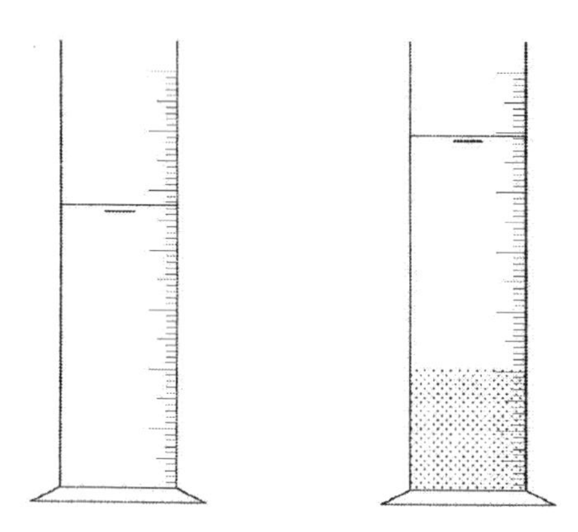

Figure 4-3. Measuring the density of solid particles.

This value can be determined by carefully dropping a certain mass of particles (say W_p) in a container partially filled with water, see Figure 4-3. The precise volume of the particles can be measured by observing the rise of the water table in the glass. This is particularly easy when using a graduated measuring glass. The rising of the water table indicates the volume of the particles, V_p. Their mass W_p can be measured most easily by measuring the weight of the glass before and after dropping the particles into it. The density of the particle material then follows immediately from its definition,

$$\rho_p = W_p / V_p. \tag{4.8}$$

For sand the value of ρ_p usually is about $2650\ \text{kg/m}^3$.

The principle of this simple test, in which the volume of a body having a very irregular shape (a number of sand particles) is measured, is due to Archimedes. He had been asked to check the composition of a golden crown, of which it was suspected that it contained silver (which is cheaper). He realized that this could be achieved by comparing the density of the crown with the density of a piece of pure gold, but then he had to determine the precise volume of the crown. The legend has it that when stepping into his bath he discovered that the volume of a body submerged in water equals the volume of water above the original water table. While shouting "Eureka!" he ran into the street, according to the legend.

4.4 Volumetric weight

In soil mechanics it is often required to determine the total weight of a soil body. This can be calculated if the porosity, the degree of saturation and the densities are known. The weight of the water in a volume *V* of soil is $Sn\rho_w gV$, and the weight of the particles in that volume is $(1-n)\rho_p gV$, where g is the strength of the gravity field, or the acceleration of gravity. The value of that constant is about $g = 9.8\ \text{N/kg}$, or, approximately, $g = 10\ \text{N/kg}$. Thus the total weight *W* is

$$W = \left[Sn\rho_w g + (1-n)\rho_p g\right]V. \tag{4.9}$$

This means that the *volumetric weight* γ, defined as the weight per unit volume, is

$$\gamma = W/V = Sn\rho_w g + (1-n)\rho_p g. \tag{4.10}$$

This formula indicates that the volumetric weight is determined by a large number of soil parameters: the degree of saturation, the porosity, the densities of water and soil particles, and the gravity constant. In reality it is often much simpler to determine the volumetric weight (often also denoted as the *unit weight*) directly by measuring the weight *W* of a volume *V* of soil. It is then not necessary to determine the contribution of each of the components.
If the soil is completely dry the dry volumetric weight is

$$\gamma_d = W_d/V = (1-n)\rho_p g. \tag{4.11}$$

This value can also be determined directly by weighing a volume of dry soil. In order to dry the soil a sample may be placed in an oven. The temperature in such an oven is usually close to 100 degrees, so that the water will evaporate quickly. At a much

higher temperature there would be a risk that organic parts of the soil would be burned.
From the dry volumetric weight the porosity n can be determined, see eq. (4.11), provided that the density of the particle material is known. This is a common method to determine the porosity in a laboratory.
If both the original volumetric weight γ and the dry volumetric weight γ_d are known, by measuring the weight and volumes both in the original state and after drying, the porosity n may be determined from eq. (4.11), and then the degree of saturation S may be determined using eq. (4.10). Unfortunately, this procedure is not very accurate for soils that are almost completely saturated, because a small error in the measurements may cause that one obtains, for example, $S = 0.97$ rather than the true value $S = 0.99$. In itself this is rather accurate, but the error in the air volume is then 300 %. In some cases, this may lead to large errors, for instance when the compressibility of the water-air-mixture in the pores must be determined.

4.5 Water content

The *water content* is another useful parameter, especially for clays. It has been used in Chapter 2. By definition the water content w is the ratio of the weight (or mass) of the water and the solids,

$$w = W_w/W_p. \tag{4.12}$$

It may be noted that this is not a new independent parameter, because

$$w = S\frac{n}{1-n}\frac{\rho_w}{\rho_p} = Se\frac{\rho_w}{\rho_p}. \tag{4.13}$$

For a completely saturated soil ($S = 1$) and assuming that $\rho_p/\rho_w = 2.65$, it follows that the void ratio e is about 2.65 times the water content.
A normal value for the porosity is $n = 0.40$. Assuming that $\rho_p = 2650\ \text{kg/m}^3$ it then follows from eq. (4.11) that $\gamma_d = 15900\ \text{N/m}^3$, or $\gamma_d = 15.9\ \text{kN/m}^3$. Values of the order of magnitude of $16\ \text{kN/m}^3$ are indeed common for dry sand. If the material is completely saturated it follows from eq. (4.10) that $\gamma \approx 20\ \text{kN/m}^3$. For saturated sand this is a common value. The volumetric weight of clay soils may also be about $20\ \text{kN/m}^3$, but smaller values are very well possible, especially when the water content is high or the degree of saturation is low. Peat is often much lighter, sometimes hardly heavier than water.

Problems

4.1 A truck loaded with 2 m^3 dry sand appears to weigh "3 tons" more than the weight of the empty truck. What is the meaning of the term "3 tons", and what is the volumetric weight of the sand?

4.2 If it is known that the density of the sand particles in the material of the previous problem is ,2600 kg/m^3 then what is the porosity n? And the void ratio e?

4.3 It would be possible to fill the pores of the dry sand of the previous problems with water. What is the volume of the water that the sand could contain, and then what is the volumetric weight of the saturated sand?

4.4 The soil in a polder consists of a clay layer of 5 meter thickness, with a porosity of 50 %, on top of a deep layer of stiff sand. The water level in the clay is lowered by 1.5 meter. Experience indicates that then the porosity of the clay is reduced to 40 %. What is the subsidence of the soil?

4.5 The particle size of sand is about 1 mm. Gravel particles are much larger, of the order of magnitude of 1 cm, a factor 10 larger. The shape of gravel particles is about the same as that of sand particles. What is the influence of the particle size on the porosity?

4.6 Using the data indicated in Figure 4-3, determine the volume of the soil on the bottom of the measuring glass, and also read the increment of the total volume from the rise of the water table. What is the porosity of this soil?

4.7 A container is partially filled with water. A scale on the wall indicates that the volume of water is 312 cm^3. The weight of water and container is 568 gram. Some sand is carefully poured into the water. The water level in the container rises to a level that it contains 400 cm^3 of material (sand and water). The weight of the container now is 800 gram. Determine the density of the particle material, in kg/m^3.

5 Stresses in soils

5.1 Stresses

As in other materials, stresses may act in soils as a result of an external load and the volumetric weight of the material itself. Soils, however, have a number of properties that distinguish it from other materials. Firstly, a special property is that soils can only transfer compressive normal stresses, and no tensile stresses. Secondly, shear stresses can only be transmitted if they are relatively small, compared to the normal stresses. Furthermore it is characteristic of soils that part of the stresses is transferred by the water in the pores. This will be considered in detail in this chapter.
Because the normal stresses in soils usually are compressive stresses only, it is standard practice to use a sign convention for the stresses that is just opposite to the sign convention of classical continuum mechanics, namely such that compressive stresses are considered positive, and tensile stresses are negative. The stress tensor will be denoted by σ. The sign convention for the stress components is illustrated in Figure 5-1. Its definition is that a stress component when it acts in a positive coordinate direction on a plane with its outward normal in a negative coordinate direction, or when it acts in negative direction on a plane in positive direction. This means that the sign of all stress components is just opposite to the sign that they would have in most books on continuum mechanics or in applied mechanics.

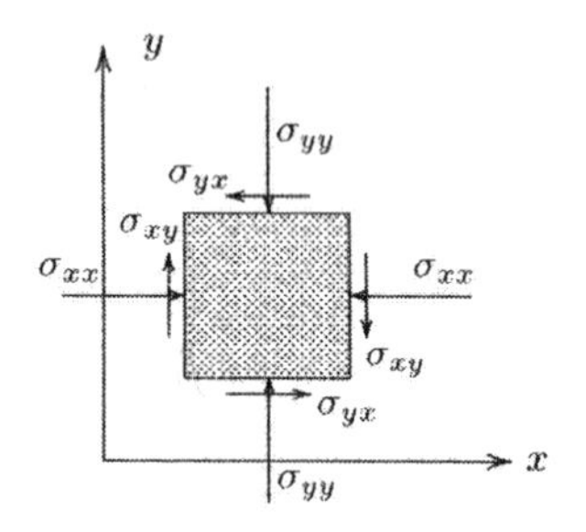

Figure 5-1. Stresses.

It is assumed that in indicating a stress component σ_{ij} the first index denotes the plane on which the stress is acting, and the second index denotes the direction of the pressure. This means, for instance, that the stress component σ_{xy} indicates that the force opposite to the y-direction, acting upon a plane having its normal in the x-direction is $F_y = \sigma_{xy} A_x$, where A_x denotes the area of the plane surface. This means that the sign of the forces is opposite to the sign convention of classical continuum mechanics.

5.2 Pore pressures

Soil is a porous material, consisting of particles that together constitute the grain skeleton. In the pores of the grain skeleton a fluid may be present: usually water. The pore structure of all normal soils is such that the pores are mutually connected. The water fills a space of very complex form, but it constitutes a single continuous body. In this water body a pressure may be transmitted, and the water may also flow through the pores. The pressure in the pore water is denoted as the *pore pressure*.

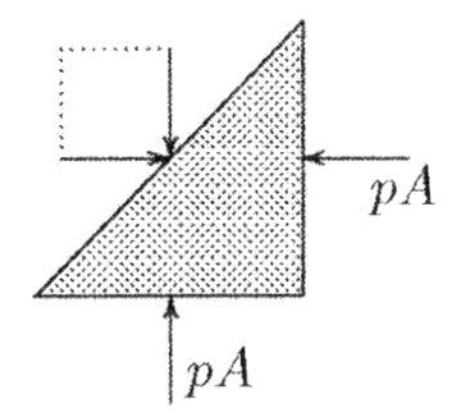

Figure 5-2.Pascal.

In a fluid at rest no shear stresses can be transmitted. This means that the pressure is the same in all directions. This can be proved by considering the equilibrium conditions of a small triangular element, see Figure 5-2, bounded by a vertical plane, a horizontal plane and a sloping plane at an angle of 45°. If the pressure on the vertical plane at the right is p, the force on that plane is pA, where A is the area of that plane. Because there is no shear stress on the lower horizontal plane, the horizontal force pA must be equilibrated by a force component on the sloping plane. That component must therefore also be pA. Because on this plane also the shear stress is zero, it follows that there must also be a vertical force pA, so that the resulting force on the plane is perpendicular to it. This vertical force must be in equilibrium with the vertical force on the lower horizontal plane of the element. Because the area of that element is also A, the pressure on that plane is p, equal to the pressure on the vertical plane. Using a little geometry it can be shown that this pressure p acts on every plane through the same point. This is often denoted as *Pascal's principle*.

If the water is at rest (i.e. when there is no flow of the water), the pressure in the water is determined by the location of the point considered with respect to the water surface. As shown by Stevin the magnitude of the water pressure on the bottom of a container filled with water, depends only upon the height of the column of water and the volumetric weight of the water, and not upon the shape of the container, see Figure 5-3. The pressure at the bottom in each case is

$$p = \gamma_w d, \tag{5.1}$$

where γ_w is the volumetric weight of the water, and d is the depth below the water surface. The total vertical force on the bottom is $\gamma_w dA$. Only in case of a container with vertical sides this is equal to the total weight of the water in the container. Stevin showed that for the other types of containers illustrated in Figure 5-3 the total force on the bottom is also $\gamma_w dA$ is. That can be demonstrated by considering equilibrium of the water body, taking into account that the pressure in every point on the walls must always be perpendicular to the wall. The container at the extreme right in Figure 5-3 resembles a soil body, with its pore space. It can be concluded that the water in a soil satisfies the principles of hydrostatics, provided that the water in the pore space forms a continuous body.

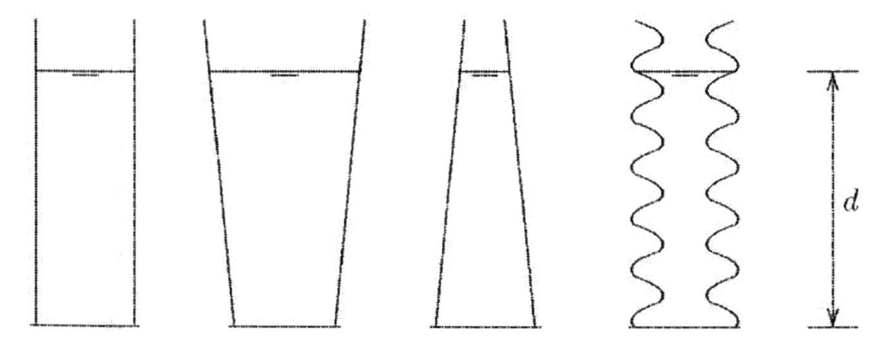

Figure 5-3. Hydrostatic water pressure depends upon depth only.

5.3 Effective stress

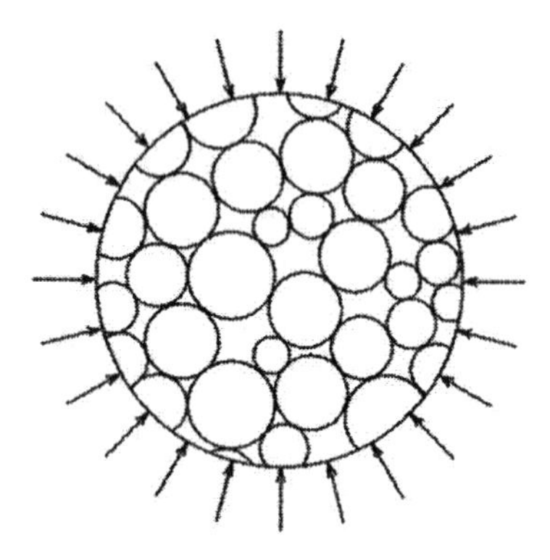

Figure 5-4. Isotropic stress.

On an element of soil normal stresses as well as shear stresses may act. The simplest case, however, is the case of an isotropic normal stress, see Figure 5-4. It is assumed that the magnitude of this stress, acting in all directions, is σ. In the interior of the soil, for instance at a cross section in the centre, this stress is transmitted by a pore pressure p in the water, and by stresses in the particles. The stresses in the particles are generated partly by the concentrated forces acting in the contact points between the particles, and partly by the pressure in the water, that almost completely surrounds the particles. It can be expected that the deformations of the particle skeleton are almost completely determined by the concentrated forces in the contact points, because the structure can deform only by sliding and rolling in these contact points. The pressure in the water results in an equal pressure in all the grains. It follows that this pressure acts on the entire surface of a cross section, and that by subtracting p from the total stress σ a measure for the contact forces is obtained. It can also be argued that when there are no contact forces between the particles, and a pressure p acts in the pore water, this same pressure p will also act in all the particles, because they are completely surrounded by the pore fluid. The deformations in this case are the compression of the particles and the water caused by this pressure p. Quartz and water are very stiff materials, having an elastic modulus about 1/10 of the elastic modulus of steel, so that the deformations in this case are very small (say 10^{-6}), and can be disregarded with respect to the large deformations that are usually observed in a soil (10^{-3} to 10^{-2}). These considerations indicate that it seems meaningful to introduce the difference of the total stress σ and the pore pressure p,

$$\sigma' = \sigma - p. \tag{5.2}$$

The quantity σ' is denoted as the *effective stress*. The effective stress is a measure for the concentrated forces acting in the contact points of a granular material. If $p = \sigma$ it follows that $\sigma' = 0$, which means that then there are no concentrated forces in the contact points. This does not mean that the stresses in the grains are zero in that case, because there will always be a stress in the particles equal to the pressure in the surrounding water. The basic idea is, as stated above, that the deformations of a granular material are almost completely determined by changes of the concentrated forces in the contact points of the grains, which cause rolling and sliding in the contact points. These are described (on the average) by the effective stress, a concept introduced by Terzaghi. Eq. (5.2) can, of course, also be written as

$$\sigma = \sigma' + p. \tag{5.3}$$

Terzaghi's effective stress principle is often quoted as "total stress equals effective stress plus pore pressure", but it should be noted that this applies only to the normal stresses. Shear stresses can be transmitted by the grain skeleton only.

It may be noted that the concept is based upon the assumption that the particles are very stiff compared to the soil as a whole, and also upon the assumption that the contact areas of the particles are very small. These are reasonable assumptions for a normal soil, but for porous rock they may not be valid. For rock the compressibility of the rock must be taken into account, which leads to a small correction in the formula.

To generalize the subdivision of total stress into effective stress and pore pressure it may be noted that the water in the pores can not contribute to the transmission of shear stresses, as the pore pressure is mainly isotropic. Even though in a flowing fluid viscous shear stresses may be developed, these are several orders of magnitude smaller than the pore pressure, and than the shear stresses than may occur in a soil. This suggests that the generalization of (5.3) is

$$\begin{aligned} \sigma_{xx} &= \sigma'_{xx} + p, \ \sigma_{yz} = \sigma'_{yz}, \\ \sigma_{yy} &= \sigma'_{yy} + p, \ \sigma_{zx} = \sigma'_{zx}, \\ \sigma_{zz} &= \sigma'_{zz} + p, \ \sigma_{xy} = \sigma'_{xy}. \end{aligned} \tag{5.4}$$

This is usually called the principle of effective stress. It is one of the basic principles of soil mechanics. The notation, with the effective stresses being denoted by an accent, σ', is standard practice. The total stresses are denoted by σ, without accent. Even though the equations (5.4) are very simple, and may seem almost trivial, different expressions may be found in some publications especially relations of the form $\sigma = \sigma' + np$, in which n is the porosity. The idea behind this is that the pore water pressure acts in the pores only, and that therefore a quantity np must be subtracted from the total stress σ to obtain a measure for the stresses in the particle skeleton. That seems to make sense, and it may even give a correct value for the average stress in the particles, but it ignores that soil deformations are not in the first place determined by deformations of the individual particles, but mainly by changes in the geometry of the *grain skeleton*. This average granular stress might be useful if one wishes to study the effect of stresses on the properties of the grains themselves (for instance a photo-elastic or a piezo-electric effect),

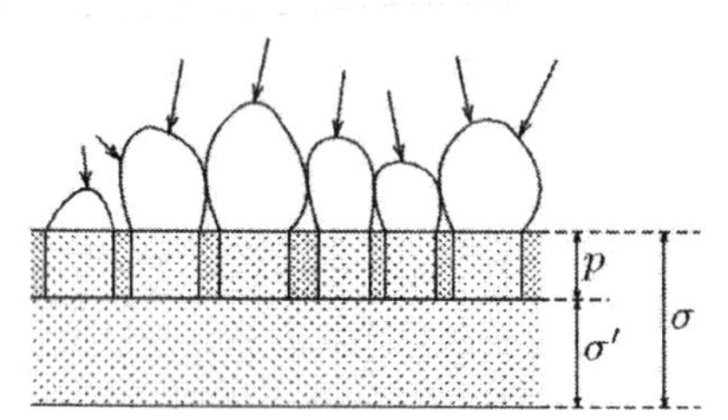

Figure 5-5. Effective stress.

but in order to study the deformation of soils it is not useful. Terzaghi's notion that the soil deformations are mainly determined by the contact forces only leads directly to the concept of effective stress, because only if one writes $\sigma' = \sigma - p$ do the effective stresses vanish when there are no contact forces. The pore pressure must be considered to act over the entire surface to obtain a good measure for the contact forces, see Figure 5-5.

The equations (5.4) can be written in matrix notation as

$$\sigma_{ij} = \sigma'_{ij} + p\delta_{ij}, \tag{5.5}$$

in which δ_{ij} is the Kronecker delta, or the unit matrix. Its definition is

$$\delta_{ij} = \begin{cases} 1 & if\ i = j, \\ 0 & if\ i \neq j. \end{cases} \tag{5.6}$$

Calculating the effective stresses in soils is one of the main problems of soil mechanics. The effective stresses are important because they determine the deformations. In the next chapter the procedure for the determination of the effective stress will be illustrated for the simplest case, of one-dimensional deformation. In later chapters more general cases will be considered, including the effect of flowing groundwater.

5.4 Archimedes and Terzaghi

The concept of effective stress is so important for soil mechanics that it deserves careful consideration. It may be illuminating, for instance, to note that the concept of effective stress is in agreement with the principle of Archimedes for the upward force on a submerged body.

Consider a volume of soil of magnitude V, having a porosity n. The total weight of the particles in that volume is $(1-n)\gamma_p V$, in which γ_p is the volumetric weight of the particle material, which is about 26.5 kN/m^3. Following Archimedes, the upward force under water is equal to the weight of the water that is being displaced by the particles, that is $(1-n)\gamma_w V$, in which γ_w the volumetric weight of water, about $10\ \text{kN/m}^3$. The remaining force is

$$F = (1-n)\gamma_p V - (1-n)\gamma_w V,$$

which must be transmitted to the bottom on which the particles rest. If the area of the volume is denoted by A, and the height by h, then the average stress is, with $\sigma' = F/A$,

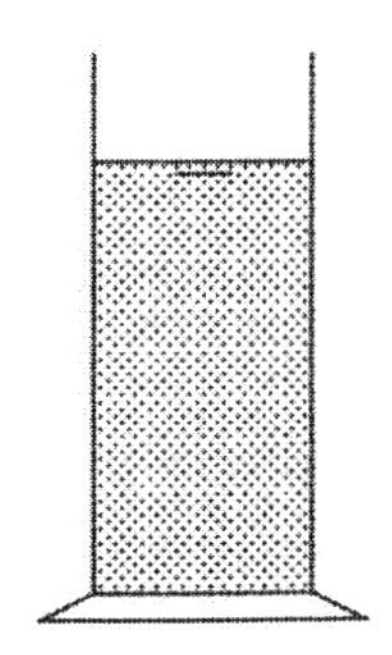

Figure 5-6. Archimedes.

$$\sigma' = (1-n)\gamma_p h - (1-n)\gamma_w h = (1-n)(\gamma_p - \gamma_w)h. \tag{5.7}$$

The quantity $(\gamma_p - \gamma_w)$ is sometimes denoted as the *submerged volumetric weight*. Following Terzaghi the effective stresses must be determined as the difference of the total stress and the pore pressure. The total stress is generated by the weight of the soil, whatever its constitution, i.e. $\sigma = \gamma_s h$, in which γ_s is the volumetric weight of the soil. If the ground water is at rest the pore pressure is determined by the depth below the water table, i.e. $p = \gamma_w h$. This means that the effective stress is

$$\sigma' = \gamma_s h - \gamma_w h. \tag{5.8}$$

Because for a saturated soil the volumetric weight is

$$\gamma_s = n\gamma_w + (1-n)\gamma_p,$$

this can also be written as

$$\sigma' = (1-n)\gamma_p h - (1-n)\gamma_w h = (1-n)(\gamma_p - \gamma_w)h. \tag{5.9}$$

This is identical to the expression (5.7). Terzaghi's principle of effective stress appears to be in agreement with the principle of Archimedes, which is a fundamental principle of physics. It may be noted that in the two methods it has been assumed that the determining factor is the force transmitted between the particles and an eventual rigid surface, or the force transmittance between the grains. This is another basic aspect of the concept of effective stress, and it cannot be concluded that Archimedes' principle automatically leads to the principle of effective stress.
Terzaghi's approach, leading to the expression (5.8), is somewhat more direct, and especially more easy to generalize. In this method the porosity n is not needed, and hence it is not necessary to determine the porosity to calculate the effective stress. On the other hand, the porosity is hidden in the volumetric weight γ_s. The generalization of Terzaghi's approach to more complicated cases, such as non-saturated soils, or flowing groundwater, is relatively simple. For a non-saturated soil the total stresses will be smaller, because the soil is lighter. The pore pressure remains hydrostatic, and hence the effective stresses will be smaller, even though there are just as many particles as in the saturated case. The effective principle can also be applied in cases involving different fluids (oil and water, or fresh water and salt water). In the case of flowing groundwater the pore pressures must be calculated separately, using the basic laws of groundwater flow. Once these pore pressures are known they can be subtracted from the total stresses to obtain the effective stresses.
The procedure for the determination of the effective stresses usually is that first the total stresses are determined, on the basis of the total weight of the soil and all possible loads. Then the pore pressures are determined, from the conditions on the groundwater. Then finally the effective stresses are determined by subtracting the pore pressures from the total stresses.

Problems

5.1 A rubber balloon is filled with dry sand. The pressure in the pores is reduced by 5 kPa with the aid of a vacuum pump. Then what is the change of the total stress, and the change of the effective stress?

5.2 An astronaut carries a packet of vacuum packed coffee into space. What is the rigidity of the pack (as determined by the effective stresses) in a spaceship? And after landing on the moon, where gravity is about one sixth of gravity on earth?

5.3 A packet of vacuum coffee is dropped in water, and it sinks to a depth of 10 meter. Is it harder now?

5.4 A treasure hunter wants to remove a collection of antique Chinese plates from a sunken ship. Under water the divers must lift the plates very carefully, of course, to avoid damage. Is it important for this damage to know the depth below water of the ship?

5.5 The bottom of a lake consists of sand. The water level in the lake rises, so that the water pressure at the bottom is increased. Will the bottom of the lake subside by deformation of the sand?

6 Stresses in a layer

6.1 Vertical stresses

In many places on earth the soil consists of practically horizontal layers. If such a soil does not carry a local surface load, and if the groundwater is at rest, the vertical stresses can be determined directly from a consideration of vertical equilibrium. The procedure is illustrated in this chapter.

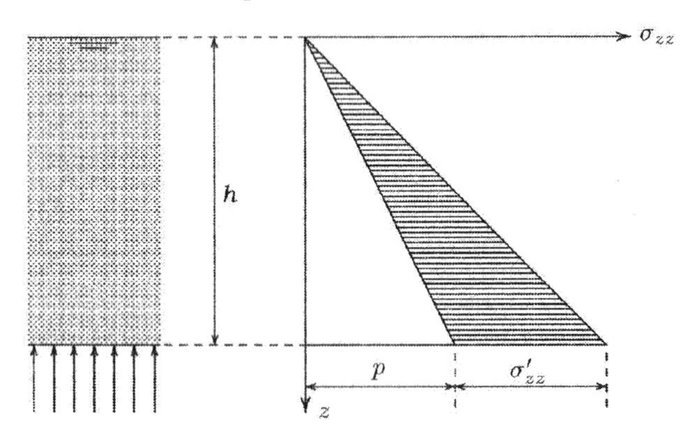

Figure 6-1. Stresses in a homogeneous layer.

A simple case is a homogeneous layer, completely saturated with water, see Figure 6-1. The pressure in the water is determined by the location of the *phreatic surface*. This is defined as the plane where the pressure in the groundwater is equal to the atmospheric pressure. If the atmospheric pressure is taken as the zero level of pressures, as is usual, it follows that $p = 0$ at the phreatic surface. If there are no capillary effects in the soil, this is also the upper boundary of the water, which is denoted as the *groundwater table*. It is assumed that in the example the phreatic surface coincides with the soil surface, see Figure 6-1. The volumetric weight of the saturated soil is supposed to be $\gamma = 20\ \text{kN/m}^3$. The vertical normal stress in the soil now increases linearly with depth,

$$\sigma_{zz} = \gamma h. \qquad (6.1)$$

This is a consequence of vertical equilibrium of a column of soil of height h. It has been assumed that there are no shear stresses on the vertical planes bounding the column in horizontal direction. That seems to be a reasonable assumption if the terrain is homogeneous and very large, with a single geological history. Often this is assumed, even when there are no data.

At a depth of 10 m, for instance, the vertical total stress is $200\ \text{kN/m}^2 = 200\ \text{kPa}$. Because the groundwater is at rest, the pressures in the water will be hydrostatic. The soil can be considered to be a container of water of very complex shape, bounded by all the particles, but that is irrelevant for the actual pressure in the water. This means that the pressure in the water at a depth h will be equal to the weight of the water in a column of unit area, see also Figure 5-3,

$$p = \gamma_w h, \tag{6.2}$$

Where γ_w is the volumetric weight of water, usually $\gamma_w = 10\ \text{kN/m}^3$. It now follows that a depth of 10 m the effective stress is 200 kPa - 100 kPa = 100 kPa.

Formally, the distribution of the effective stress can also be found from the basic equation $\sigma'_{zz} = \sigma_{zz} - p$, or, with (6.1) and (6.2),

$$\sigma'_{zz} = (\gamma - \gamma_w)h. \tag{6.3}$$

The vertical effective stresses appear to be linear with depth. That is a consequence of the linear distribution of the total stresses and the pore pressures, with both of them being zero at the same level, the soil surface.

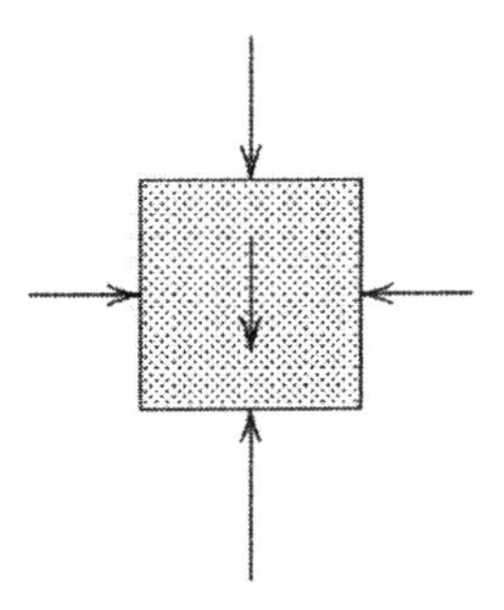

Figure 6-2. Equilibrium.

It should be noted that the vertical stress components, both the total stress and the effective stress, can be found using the condition of vertical equilibrium only, together with the assumption that the shear stresses are zero on vertical planes. The horizontal normal stresses remain undetermined at this stage. Even by also considering horizontal equilibrium these horizontal stresses can not be determined. A consideration of horizontal equilibrium, see Figure 6-2, does give some additional information, namely that the horizontal normal stresses on the two vertical planes at the left and at the right must be equal, but their magnitude remains unknown. The determination of horizontal (or lateral) stresses is one of the essential difficulties of soil mechanics. Because the horizontal stresses can not be determined from equilibrium conditions they often remain unknown. It will be shown later that even when also considering the deformations, the determination of the horizontal stresses remains very difficult, as this requires detailed knowledge of the geological history, which is usually not available. Perhaps the best way to determine the horizontal stresses is by direct or indirect measurement in the field. The problem will be discussed further in later chapters.

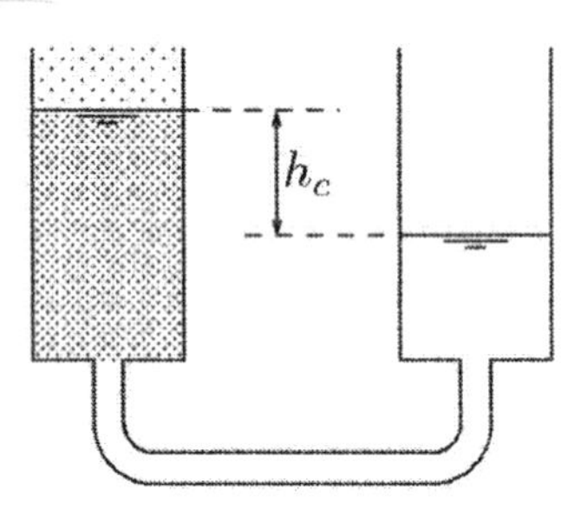

Figure 6-3. Capillary rise.

The simple example of Figure 6-1 may be used as the starting point for more complex cases. As a second example the situation of a somewhat lower phreatic surface is considered, say when it is lowered by 2 m. This may be caused by the action of a pumping station in the area, such that the water level in the canals and the ditches in a polder is to be kept at a level of 2 m below the soil surface. In this case there are two possibilities, depending upon the size of the particles in the soil. If the soil consists of very coarse material, the groundwater level in the soil will coincide with the phreatic surface (the level where $p=0$), which will be equal to the water level in the open water, the ditches. However, when the soil is very fine (for instance clay), it is possible that the top of the groundwater in the soil (the groundwater level) is considerably higher than the phreatic level, because of the effect of *capillarity*. In the fine pores of the soil the water may rise to a level above the phreatic level due to the suction caused by the surface tension at the interface of particles, water and air. This surface tension may lead to pressures in the water below atmospheric pressure, i.e. negative water pressures. The zone above the phreatic level is denoted as the *capillary zone*. The maximum height of the groundwater above the phreatic level is denoted as h_c, the *capillary rise*.

If the capillary rise h_c in the example is larger than 2 meter, the soil in the polder will remain saturated when the water table is lowered by 2 meter. The total stresses will not change, because the weight of the soil remains the same, but the pore pressures throughout the soil are reduced by $\gamma_w \times 2\text{ m} = 20\text{ kN/m}^2$. This means that the effective stresses are increased everywhere by the same amount, see Figure 6-4.

Lowering the phreatic level appears to lead to an increase of the effective stresses. In practice this will cause deformations, which will be manifest by a subsidence of the ground level. This indeed occurs very often, wherever the groundwater table is lowered. Lowering the water table to construct a dry building pit, or lowering the groundwater table in a newly reclaimed polder, leads to higher effective stresses, and therefore settlements. This may be accompanied by severe damage to buildings and houses, especially if the settlements are not uniform. If the subsidence is uniform there is less risk for damage to structures founded on the soil in that area.

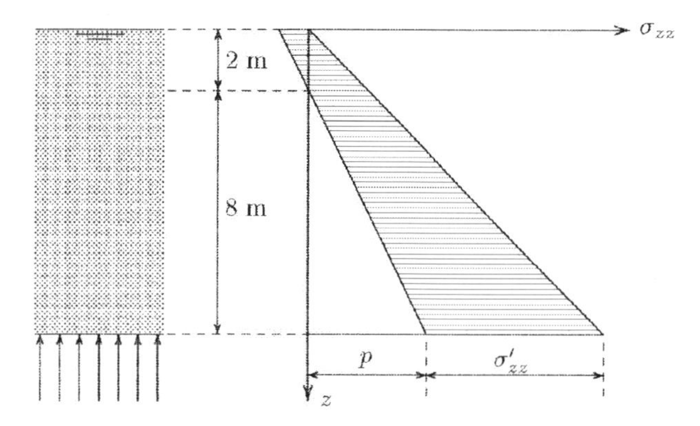

Figure 6-4. Lowering the freatic surface by 2 m, with capillary rise.

Lowering the phreatic level may also have some positive consequences. For instance, the increase of the effective stresses at the soil surface makes the soil much stiffer and stronger, so that heavier vehicles (tractors or other agricultural machines) can be supported. In case of a very high phreatic surface, coinciding with the soil surface, as illustrated in Figure 6-1, the effective stresses at the surface are zero, which means that there is no force between the soil particles. Man, animal and machine then can not find support on the soil, and they may sink into it. The soil is called soggy or swampy. It seems natural that in such cases people will be motivated to lower the water table. This will result in some subsidence, and thus part of the effect of the lower groundwater table is lost. This can be restored by a further lowering of the water table, which in turn will lead to further subsidence. In some places on earth the process has had almost catastrophic consequences (Venice, Bangkok). The subsidence of Venice, for instance, was found to be caused for a large part by the production of ever increasing amounts of drinking water from the soil in the immediate vicinity of the city. Further subsidence has been reduced by finding a water supply farther form the city.

When the soil consists of very coarse material, there will practically be no capillarity. In that case lowering the phreatic level by 2 meter will cause the top 2 meter of the soil to become dry, see Figure 6-5. The upper 2 meter of soil then will become lighter. A reasonable value for the dry volumetric weight is $\gamma_d = 16\ \text{kN/m}^3$. At a depth of 2 m the vertical effective stress now is $\sigma'_{zz} = 32\ \text{kPa}$, and at a depth of 10 m the effective stress is $\sigma'_{zz} = 112\ \text{kPa}$. It appears that in this case the effective stresses increase by 12 kPa, compared to the case of a water table coinciding with the ground surface. The distribution of total stresses, effective stresses and pore pressures is shown in Figure 6-5. Again there will be a tendency for settlement of the soil. In later chapters a procedure for the calculation of these settlements will be presented. For this purpose first the relation between effective stress and deformation must be considered.

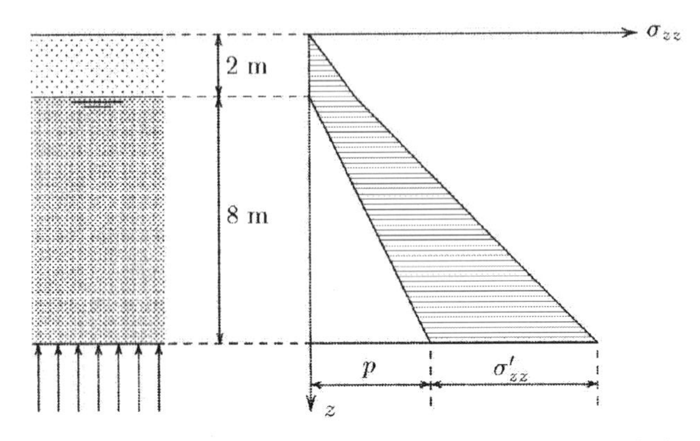

Figure 6-5. Lowering the freatic surface by 2 m, no capillarity.

Subsidence of the soil can also be caused by the extraction of gas or oil from soil layers. The reservoirs containing oil and gas are often located at substantial depth (in Groningen at 2000 m depth). These reservoirs usually consist of porous rock, that have been consolidated through the ages by the weight of the soil layers above it, but some porosity (say 10 % or 20 %) remains, filled with gas or oil. When the gas or oil is extracted from the reservoir, by reducing the pressure in the fluid, the effective stresses increase, and the thickness of the reservoir will be reduced. This will cause the soil layers above the reservoir to settle, and it will eventually give rise to subsidence of the soil surface. In Groningen the subsidence above the large gas reservoir is estimated to reach about 50 cm, over a very large area. All structures subside with the soil, with not very much risk of damage, as there are no large local variations to be expected. However, because the soil surface is below sea level, great care must be taken to maintain the drainage capacity of the hydraulic infrastructure. Sluices may have to be renewed because they subside, whereas water levels must be maintained. The dikes also have to be raised to balance the subsidence due to gas production.

In some parts of the world subsidence may have very serious consequences, for instance in areas of coal mining activities. In mining the entire soil is being removed, and sudden collapse of a mine gallery may cause great damage to the structures above it.

6.2 The general procedure

It has been indicated in the examples given above how the total stresses, the effective stresses and the pore pressures can be determined on a horizontal plane in a soil consisting of practically horizontal layers. In most cases the best general procedure is that first the total stresses are determined, from the vertical equilibrium of a column of soil. The total stress then is determined by the total weight of the column (particles and water), plus an eventual surcharge caused by a structure. In the next step the pore pressures are determined, from the hydraulic conditions. If the groundwater is at rest it is sufficient to determine the location of the phreatic surface. The pore pressures then are hydrostatic, starting from zero at the level of the phreatic surface, i.e. linear with the depth below the phreatic surface. When the soil is very fine a capillary zone may develop above the phreatic surface, in which the pore pressures are negative. The maximum negative pore pressure depends upon the size of the pores, and can be measured in the laboratory. Assuming that there are sufficient data to determine the pore pressures, the effective stresses can be determined as the difference of the total stresses and the pore pressures.

A final example is shown in Figure 6-6. This concerns a layer of 10 m thickness, carrying a surcharge of 50 kPa. The phreatic level is located at a depth of 5 m, and it has been measured that in this soil the capillary rise is 2 m. The volumetric weight of the soil when dry is 16 $\mathrm{kN/m^3}$, and when saturated it is 20 $\mathrm{kN/m^3}$. Using these data it can be concluded that the top 3 m of the soil will be dry, and that the lower 7 m will be saturated with water. The total stress at a depth of 10 m then is $50\ \mathrm{kPa} + 3\ \mathrm{m} \times 16\ \mathrm{kN/m^3} + 7\ \mathrm{m} \times 20\,\mathrm{kN/m^3} = 238\ \mathrm{kPa}$. At that depth the pore pressure is $5\ \mathrm{m} \times 10\ \mathrm{kN/m^3} = 50\ \mathrm{kPa}$. It follows that the effective stress at 10 m depth is 188 kPa. The distribution of total stresses, effective stresses and pore pressures is shown in Figure 6-6.

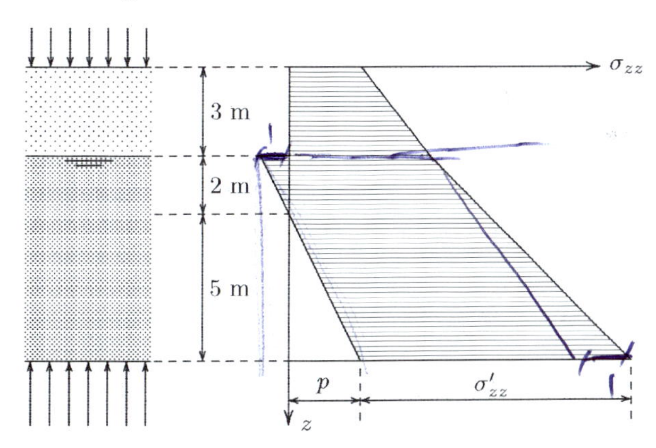

Figure 6-6. Example.

It should be noted that throughout this chapter it has been assumed that the groundwater is at rest, so that the pressure in the groundwater is hydrostatic. When the groundwater is flowing this is not so, and more data are needed to determine the pore pressures. For this purpose the flow of groundwater is considered in the next chapters.

Problems

6.1 A lake is being reclaimed. The soil consist of 10 meter of homogeneous clay, having a saturated volumetric weight of 18 $\mathrm{kN/m^3}$. Below the clay the soil is sand. After the reclamation the phreatic level is at 2 m below the ground surface, but the soil remains saturated. Construct a graph of total stresses, effective stresses and pore pressures before and after the reclamation.

6.2 A concrete caisson having a mass of 5000 ton, a foundation surface of $20\ \text{m} \times 20\ \text{m}$, and a height of 10 m, is being placed on dry sand. Calculate the average total stress and the average effective stress just below the caisson.

6.3 A similar caisson is placed in open water, on a bottom layer of sand. The water level is 5 m above the top of the sand, so that the top of the caisson is at 5 m above water. Again calculate the average total stress and the average effective stress just below the caisson.

6.4 Solve the same problem if the depth of the water is 15 m. And when it is 100 m.

6.5 A certain soil has a dry volumetric weight of 157 kN/m^3, and a saturated volumetric weight of 21.4 kN/m^3. The phreatic level is at 2.5 m below the soil surface, and the capillary rise is 1.3 m. Calculate the vertical effective stress at a depth of 6.0 m, in kPa.

6.6 A layer of saturated clay has a thickness of 4 m, and a volumetric weight of 18 kN/m^3. Above this layer a sand layer of 4 m is located, having a dry volumetric weight of 16 kN/m^3 and a saturated volumetric weight of 20 kN/m^3. The groundwater level is at a depth of 1 m below soil surface, which is the top of the sand layer. There is no capillary rise in the sand, and the pore pressures are hydrostatic. Calculate the average effective stress in the clay, in kPa.

6.7 The soil in the previous problem is loaded by a surcharge of 2 m of the same sand. The groundwater level is maintained. Calculate the increase of the average effective stress in the clay, in kPa.

Part III
Groundwater and flow

7 The law of Darcy

7.1 Hydrostatics

As already mentioned in earlier chapters, the stress distribution in groundwater at rest follows the rules of hydrostatics. More precise it can be stated that in the absence of flow the stresses in the fluid in a porous medium must satisfy the equations of equilibrium in the form

$$\begin{aligned} \frac{\partial p}{\partial x} &= 0, \\ \frac{\partial p}{\partial y} &= 0, \\ \frac{\partial p}{\partial z} + \gamma_w &= 0. \end{aligned} \tag{7.1}$$

Here it has been assumed that the z-axis is pointing in vertically upward. The quantity γ_w is the volumetric weight of the water, for which γ_w=10 kN/m^3. It has further been assumed that there are no shear stresses in the water. This is usually a very good approximation. Water is a viscous fluid, and shear stresses may occur in it, but only when the fluid is moving, and it has been assumed that the water is at rest. Furthermore, even when the fluid is moving the shear stresses are very small compared to the normal stress, the fluid pressure.

The first two equations in (7.1) mean that the pressure in the fluid can not change in horizontal direction. This is a consequence of horizontal equilibrium of a fluid element, see Figure 7-1. Equilibrium in vertical direction requires that the difference of the fluid pressures at the top and bottom of a small element balances the weight of the fluid in the element, i.e. $\Delta p = -\gamma_w \Delta z$. Here Δz represents the height of the element. By passing into the limit $\Delta z \rightarrow 0$ the third equation of the system (7.1) follows.

The value of the volumetric weight in the last of eqs. (7.1) need not be constant for the equations to be valid. If the volumetric weight is variable the equations are still valid. Such a variable density may be the result of variable salt contents in the water, or variable temperatures. It may even be that the density is discontinuous, for instance, in case of two different fluids, separated by a sharp interface. This may happen for oil and water, or fresh water and salt water. Even in those cases the equations (7.1) correctly express equilibrium of the fluid.

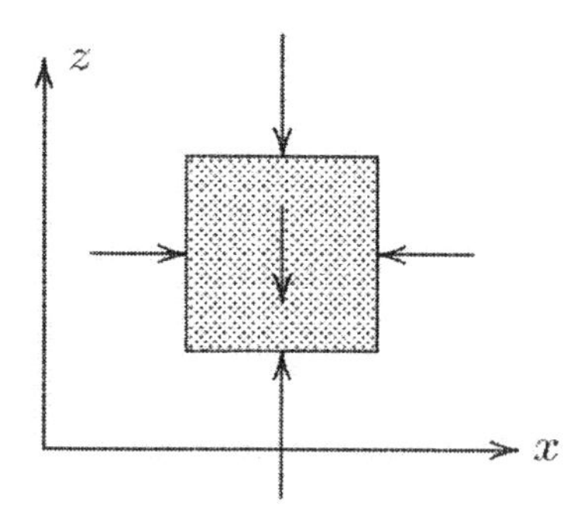

Figure 7-1. Equilibrium of water.

In soil mechanics the fluid in the soil usually is water, and it can often be assumed that the groundwater is homogeneous, so that the volumetric weight γ_w is a constant. In that case the system of equations (7.1) can be integrated to give

$$p = -\gamma_w z + C, \tag{7.2}$$

where C is an integration constant. Equation (7.2) means that the fluid pressure is completely known if the integration constant C can be found. For this it is necessary, and sufficient, to know the water pressure in a single point. This may be the case if the phreatic surface has been observed at some location. In that point the water pressure $p = 0$ for a given value of z.

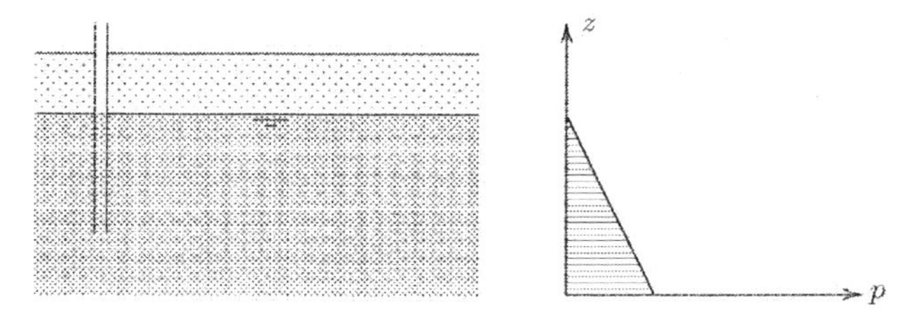

Figure 7-2. Standpipe.

The location of the phreatic surface in the soil can be determined from the water level in a ditch or pond, if it known that there is no, or practically no, groundwater flow. In principle the phreatic surface could be determined by digging a hole in the ground, and then wait until the water has come to rest. It is much more accurate, and easy, to determine the phreatic surface using an open *standpipe*, see Figure 7-2. A standpipe is a steel tube, having a diameter of for instance 2.5 cm, with small holes at the bottom, so that the water can rise in the pipe. Such a pipe can easily be installed into the ground, by pressing or eventually by hammering it into the ground. The diameter of the pipe is large enough that capillary effects can be disregarded. After some time, during which the water has to flow from the ground into the pipe, the level of the water in the standpipe indicates the location of the phreatic surface, for the point of the pipe. Because this water level usually is located below ground surface, it can be observed with the naked eye. The simplest method to measure the

water level in the standpipe is to drop a small iron or copper weight into the tube, attached to a flexible cord. As soon as the weight touches the water surface, a sound can be heard, especially by holding an ear close to the end of the pipe.
Of course, the measurement can also be made by accurate electronic measuring devices. Electronic pore pressure meters measure the pressure in a small cell, by a flexible membrane and a strain gauge, glued onto the membrane. The water presses against the membrane, and the strain gauge measures the small deflection of the membrane. This can be transformed into the value of the pressure if the device has been calibrated before.

7.2 Groundwater head

The concept of groundwater head can be illustrated by considering a standpipe in the soil, see Figure 7-3. The water level in the standpipe, measured with respect to a certain horizontal level where $z=0$, is the groundwater head h in the point indicated by the open end of the standpipe. In the standpipe the water is at rest, and therefore the pressure at the bottom end of the pipe is $p=(h-z)\gamma_w$.

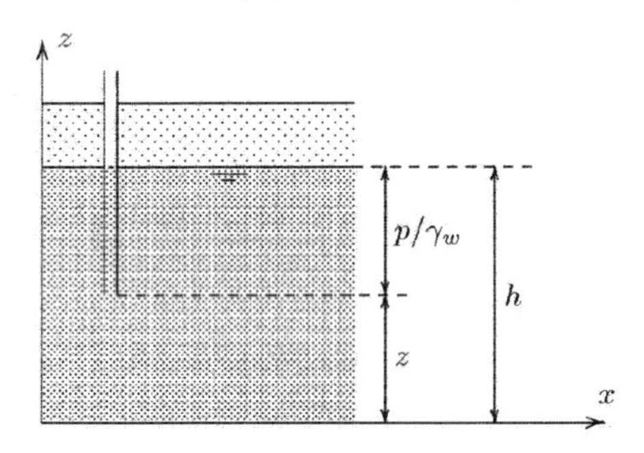

Figure 7-3. Groundwater head.

In civil engineering many problems are concerned with a single fluid, mostly fresh water, and the volumetric weight can then be considered as constant. In that case it is convenient to introduce the ***groundwater head h***, defined as

$$h=z+\frac{p}{\gamma_w}. \tag{7.3}$$

7.3 Groundwater flow

The hydrostatic distribution of pore pressures is valid when the groundwater is at rest. When the groundwater is flowing through the soil the pressure distribution will not be hydrostatic, because then the equations of equilibrium (7.1) are no longer complete. The flow of groundwater through the pore space is accompanied by a friction force between the flowing fluid and the soil skeleton, and this must be taken into account.

This friction force (per unit volume) is denoted by $\vec{f}$. Then the equations of equilibrium are

$$\begin{aligned} &\frac{\partial p}{\partial x} - f_x = 0, \\ &\frac{\partial p}{\partial y} - f_y = 0, \\ &\frac{\partial p}{\partial z} + \gamma_w - f_z = 0. \end{aligned} \tag{7.4}$$

Here f_x, f_y and f_z are the components of the force, per unit volume, exerted onto the soil skeleton by the flowing groundwater. The sign of these terms can be verified by considering the equilibrium in one of the directions, say the x-direction, see Figure 7-3. If the pressure increases in x-direction there must be a force in positive x-direction acting on the water to ensure equilibrium. Both terms in the equation of equilibrium then are positive, so that they cancel.

It may be mentioned that in the equations the accelerations of the groundwater might also be taken into account. This could be expressed by terms of the form ρa_x, ρa_y en ρa_z in the right hand sides of the equations. Such terms are usually very small, however. It may be noted that the velocity of flowing groundwater usually is of the order of magnitude of 1 m/d, or smaller. If such a velocity would be doubled in one hour the acceleration would be $(1/24)\times(1/3600)^2 \text{ m/s}^2$, which is extremely small with respect to the acceleration of gravity g, which also appears in the equations. In fact the acceleration terms would be a factor 3×10^8 smaller, and therefore may be neglected.

It seems probable that the friction force between the particles and the water depends upon the velocity of the water, and in particular such that the force will increase with increasing velocity, and acting in opposite direction. It can also be expected that the friction force will be larger, at the same velocity, if the viscosity of the fluid is larger (the fluid is then more sticky). From careful measurements it has been established that the relation between the velocity and the friction force is linear, at least as a very good first approximation. If the soil has the same properties in all directions (i.e. is isotropic) the relations are

$$\begin{aligned} f_x &= -\frac{\mu}{\kappa} q_x, \\ f_y &= -\frac{\mu}{\kappa} q_y, \\ f_z &= -\frac{\mu}{\kappa} q_z. \end{aligned} \tag{7.5}$$

Here q_x, q_y and q_z are the components of the specific discharge, that is the discharge per unit area. The precise definition of q_x is the discharge (a volume per unit time) Q

through a unit area A perpendicular to the x-direction, $q_x = Q/A$, see Figure 7-5. This quantity is expressed in m^3/s per m^2, a discharge per unit area. In the SI-system of units that reduces to m/s. It should be noted that this is not the average velocity of the groundwater, because for that quantity the discharge should be divided by the area of the pores only, and that area is a factor n smaller than the total area. The specific discharge is proportional to the average velocity, however,

$$\vec{v} = \vec{q}/n. \tag{7.6}$$

The fact that the specific discharge is expressed in m/s, and its definition as a discharge per unit area, may give rise to confusion with the velocity. This confusion is sometimes increased by denoting the specific discharge q as the *filter velocity*, the *seepage velocity* or the *Darcian velocity*. Such terms can better be avoided: it should be denoted as the specific discharge.

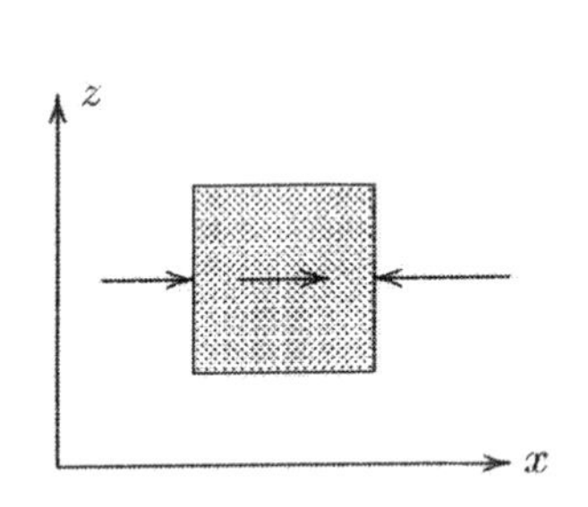

Figure 7-4. Groundwater flow and force equilibrium.

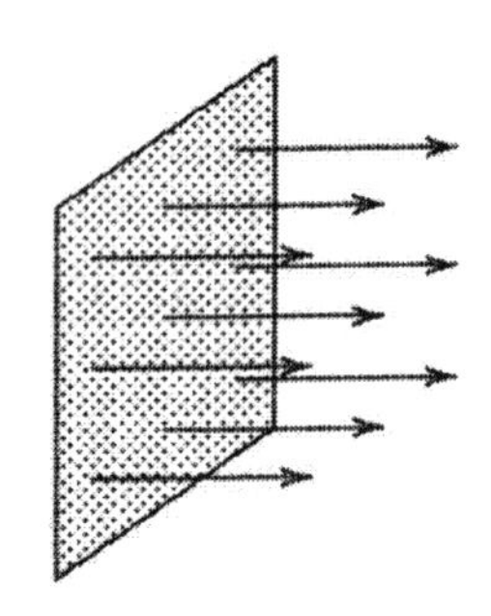

Figure 7-5. Specific discharge.

It may be interesting to note that in the USA the classical unit of volume of a fluid is the gallon (3.785 liter), so that a discharge of water is expressed in gallon per day, gpd. An area is expressed in square foot (1 foot = 30 cm), and therefore a specific discharge is expressed in gallons per day per square foot (gpd/sqft). That may seem an antique type of unit, but at least it has the advantage of expressing precisely what it is: a discharge per unit area. There is no possible confusion with a velocity, which in the USA is usually expressed in miles per hour, mph.

Equation (7.5) expresses that there is an additional force in the equations of equilibrium proportional to the specific discharge (and hence proportional to the velocity of the water with respect to the particles, as intended). The constant of proportionality has been denoted by μ/κ, where μ is the *dynamic viscosity* of the fluid, and κ is the *permeability* of the porous medium. The factor $1/\kappa$ is a measure for the resistance of the porous medium. In general it has been found that κ is larger if the size of the pores is larger. When the pores are very narrow the friction will be very large, and the value of κ will be small.

Substitution of equations (7.5) into (7.4) gives

$$\frac{\partial p}{\partial x}+\frac{\mu}{\kappa}q_x=0,$$
$$\frac{\partial p}{\partial y}+\frac{\mu}{\kappa}q_y=0, \quad (7.7)$$
$$\frac{\partial p}{\partial z}+\gamma_w+\frac{\mu}{\kappa}q_z=0.$$

In contrast with equations (7.1), which may be used for an infinitely small element, within a single pore, equations (7.7) represent the equations of equilibrium for an element containing a sufficiently large number of pores, so that the friction force can be represented with sufficient accuracy as a factor proportional to the average value of the specific discharge. It may be noted that the equations (7.7) are also valid when the volumetric weight γ_w is variable, for instance due to variations of salt content, or in the case of two fluids (e.g. oil and water) in the pores. That can easily be demonstrated by noting that these equations include the hydrostatic pressure distribution as the special case for zero specific discharge, i.e. for the no flow case.

The equations (7.7) can also be written as

$$q_x=-\frac{\kappa}{\mu}\left(\frac{\partial p}{\partial x}\right),$$
$$q_y=-\frac{\kappa}{\mu}\left(\frac{\partial p}{\partial y}\right), \quad (7.8)$$
$$q_z=-\frac{\kappa}{\mu}\left(\frac{\partial p}{\partial z}+\gamma_w\right).$$

These equations enable to determine the components of the specific discharge if the pressure distribution is known.

The equations (7.8) are a basic form of *Darcy's law*. They are named after the city engineer of the French town Dijon, who developed that law on the basis of experiments in 1856. Darcy designed the public water works of the town of Dijon, by producing water from the ground in the centre of town. He realized that this water could be supplied from the higher areas surrounding the town, by flowing through the ground. In order to assess the quantity that could be produced he needed the permeability of the soil, and therefore measured it. The grateful citizens of Dijon honoured him by erecting a statue, and by naming the central square of the town the Place Henri Darcy.

The equations (7.8) are generally valid, also if the volumetric weight γ_w of the fluid is not constant. If the volumetric weight γ_w is constant it follows from the derivatives of the groundwater head equation (7.3) that

$$\frac{\partial h}{\partial x}=\frac{1}{\gamma_w}\left(\frac{\partial p}{\partial x}\right),$$
$$\frac{\partial h}{\partial y}=\frac{1}{\gamma_w}\left(\frac{\partial p}{\partial y}\right), \tag{7.9}$$
$$\frac{\partial h}{\partial z}=\frac{1}{\gamma_w}\left(\frac{\partial p}{\partial z}+\gamma_w\right).$$

Using these relations Darcy's law, eqs. (7.8), can also be written as

$$q_x=-k\frac{\partial h}{\partial x},$$
$$q_y=-k\frac{\partial h}{\partial y}, \tag{7.10}$$
$$q_z=-k\frac{\partial h}{\partial z}.$$

The quantity k in these equations is the hydraulic conductivity, defined as

$$k=\frac{\kappa\gamma_w}{\mu}. \tag{7.11}$$

It is sometimes denoted as the *coefficient of permeability*. The permeability κ then should be denoted as the *intrinsic permeability* to avoid confusion.
Darcy himself wrote his equations in the simpler form of eq. (7.10). For engineering practice that is a convenient form of the equations, because the groundwater head h can often be measured rather simply, en because the equations are of a simple character, and are the same in all three directions. It should be remembered, however, that the form (7.8) is more fundamentally correct. If the volumetric weight γ_w is not constant, only the equations (7.8) can be used. The definition (7.3) then does not make sense.
When the groundwater head is the same in every point of a soil mass, the groundwater will be at rest. If the head is not constant, however, the groundwater flow, and according to eq. (7.10) it will flow from locations with a large head to locations where the head is low. If the groundwater flow is not maintained by some external influence (rainfall, or wells) the water will tend towards a situation of constant head.
Darcy's law can be written in an even simpler form if the direction of flow is known, for instance if the water is flowing through a narrow tube, filled with soil. The water is then forced to flow in the direction of the tube. If that directions is the s-direction, the specific discharge in that direction is, similar to (7.10),

$$q=-ki \quad \text{with} \quad i=\frac{dh}{ds} \quad \text{and} \quad q=\frac{Q}{A}. \tag{7.12}$$

This is the form of Darcy's law as it is often used in simple flow problems. The quantity dh/ds is the increase of the groundwater head per unit of length, in the direction of flow. The quantity dh/ds is called the *hydraulic gradient i* and represents the inclination of the freatic surface. The minus sign expresses that the water flows in the direction of decreasing head.

7.4 Seepage force

It has been seen that the flow of groundwater is accompanied by a friction between the water and the particles. According to (7.4) the friction force (per unit volume) that the particles exert on the water is

$$
\begin{aligned}
f_x &= \frac{\partial p}{\partial x}, \\
f_y &= \frac{\partial p}{\partial y}, \\
f_z &= \frac{\partial p}{\partial z} + \gamma_w.
\end{aligned}
\tag{7.13}
$$

With $h = z + p/\gamma_w$ this can be expressed into the groundwater head h, assuming that γ_w is constant,

$$
\begin{aligned}
f_x &= \gamma_w \frac{\partial h}{\partial x}, \\
f_y &= \gamma_w \frac{\partial h}{\partial y}, \\
f_z &= \gamma_w \frac{\partial h}{\partial z}.
\end{aligned}
\tag{7.14}
$$

The force that the water exerts on the soil skeleton is denoted by $\vec{j}$. Because of Newton's third law (the principle of equality of action and reaction), this is just the opposite of the $\vec{f}$. The vector quantity $\vec{j}$ is denoted as the *seepage force*, even though it is actually not a force, but a force per unit volume. It now follows that

$$
\begin{aligned}
j_x &= -\gamma_w \frac{\partial h}{\partial x}, \\
j_y &= -\gamma_w \frac{\partial h}{\partial y}, \\
j_z &= -\gamma_w \frac{\partial h}{\partial z}.
\end{aligned}
\tag{7.15}
$$

The seepage force is especially important when considering local equilibrium in a soil, for instance when investigating the conditions for internal erosion, when some particles may become locally unstable because of a high flow rate.

Problems

7.1 In geohydrology the unit m/d is often used to measure the hydraulic conductivity k. What is the relation with the SI-unit m/s?

7.2 In the USA the unit gpd/sqft (gallon per day per square foot) is sometimes used to measure the hydraulic conductivity k, and the specific discharge q. What is the relation with the SI-unit m/s?

7.3 A certain soil has a hydraulic conductivity $k = 5$ m/d. This value has been measured in summer. In winter the temperature is much lower, and if it supposed that the viscosity μ then is a factor 1.5 as large as in summer, determine the value of the hydraulic conductivity in winter.

8 Permeability

8.1 Permeability test

In the previous chapter Darcy's law for the flow of a fluid through a porous medium has been formulated, in its simplest form, as

$$q = -ki. \tag{8.1}$$

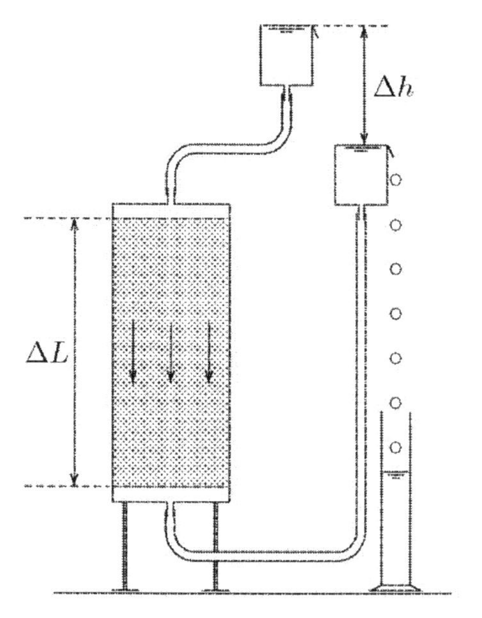

Figure 8-1. Permeability test.

This means that the hydraulic conductivity k can be determined if the specific discharge q can be measured in a test in which the gradient $i = dh/ds$ is known. An example of a test setup is shown in Figure 8-1.

It consists of a glass tube, filled with soil. The two ends are connected to small reservoirs of water, the height of which can be adjusted. In these reservoirs a constant water level can be maintained. Under the influence of a difference in head Δh between the two reservoirs, water will flow through the soil. The total discharge Q can be measured by collecting the volume of water in a certain time interval. If the area of the tube is A, and the length of the soil sample is ΔL, then Darcy's law gives

$$\frac{Q}{A} = k\frac{\Delta h}{\Delta L}. \tag{8.2}$$

Because $Q = qA$ this formula is in agreement with (8.1). Darcy performed tests as shown in Figure 8-1 to verify his formula (8.2). For this purpose he performed tests with various values of Δh, and indeed found a linear relation between Q and Δh. The same test is still used very often to determine the hydraulic conductivity (coefficient of permeability) k. For sand normal values of the hydraulic conductivity k range from 10^{-6} m/s to 10^{-3} m/s. For clay the hydraulic conductivity usually is several orders of magnitude smaller, for instance $k = 10^{-9}$ m/s, or even smaller. This is because the permeability is approximately proportional to the square of the grain size of the material, and the particles of clay are about 100 or 1000 times smaller than those of sand. An indication of the hydraulic conductivity of various soils is given in Table 8-1.

Type of soil	k (m/s)
gravel	$10^{-3} - 10^{-1}$
sand	$10^{-6} - 10^{-3}$
silt	$10^{-8} - 10^{-6}$
clay	$10^{-10} - 10^{-8}$

Table 8-1: Hydraulic conductivity k.

As mentioned before, the permeability also depends upon properties of the fluid. Water will flow more easily through the soil than a thick oil. This is expressed in the formula (7.11),

$$k = \frac{\kappa \gamma_w}{\mu}, \tag{8.3}$$

where μ is the dynamic viscosity of the fluid. The quantity κ (the intrinsic permeability) depends upon the geometry of the grains skeleton only. A useful relation is given by the formula of Kozeny-Carman,

$$\kappa = cd^2 \frac{n^3}{(1-n)^2}. \tag{8.4}$$

Here d is a measure for the grain size, and c is a coefficient, that now only depends upon the tortuosity of the pore system, as determined by the shape of the particles. Its value is about 1/200 or 1/100. Equation (8.4) is of little value for the actual determination of the value of the permeability κ, because the value of the coefficient c is still unknown, and because the hydraulic conductivity can easily be determined directly from a permeability test. The Kozeny-Carman formula (8.4) is of great value, however, because it indicates the dependence of the permeability on the grain size and on the porosity. The dependence on d^2 indicates, for instance, that two soils

for which the grain size differs by a factor 1000 (sand and clay) may have a difference in permeability of a factor 10^6. Such differences are indeed realistic.
The large variability of the permeability indicates that this may be a very important parameter. In constructing a large dam, for instance, the dam is often built from highly permeable material, with a core of clay. This clay core has the purpose to restrict water losses from the reservoir behind the dam. If the core is not very homogeneous, and contains thin layers of sand, the function of the clay core is disturbed to a high degree, and large amounts of water may be leaking through the dam.

8.2 Falling head test

For soils of low permeability, such as clay, the normal permeability test shown in Figure 8-1 is not suitable, because only very small quantities of fluid are flowing through the soil, and it would take very long to collect an appreciable volume of water. For such soils a test set up as illustrated in Figure 8-2, the *falling head test* is more suitable. In this apparatus a clay sample is enclosed by a circular ring, placed in a container filled with water. The lower end of the sample is in open connection with the water in the container, through a porous stone below the sample. At the top of the sample it is connected to a thin glass tube, in which the water level is higher than the constant water level in the container. Because of this difference in water level, water will flow through the sample, in very small quantities, but sufficient to observe by the lowering of the water level in the thin tube.

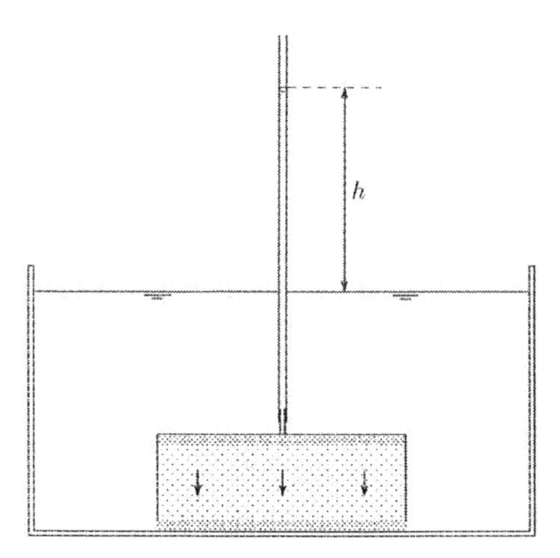

Figure 8-2. Falling head test.

In this case the head difference h is not constant, because no water is added to the system, and the level h is gradually reduced. This water level is observed as a function of time. On the basis of Darcy's law the discharge is

$$Q=\frac{kAh}{L}. \tag{8.5}$$

If the cross sectional area of the glass tube is a it follows that

$$Q = -a\frac{dh}{dt}. \tag{8.6}$$

Elimination of Q from these two equations gives

$$\frac{dh}{dt} = \frac{kA}{aL}h. \tag{8.7}$$

This is a differential equation for h, that can easily be solved,

$$h = h_0 \exp(-kAt/aL). \tag{8.8}$$

where h_0 is the value of the head difference h at time $t = 0$. If the head difference at time t is h, the hydraulic conductivity k can be calculated from the relation

$$k = \frac{aL}{At}\ln\left(\frac{h_0}{h}\right). \tag{8.9}$$

If the area of the tube a is very small compared to the area A of the sample, it is possible to measure relatively small values of k with sufficient accuracy. The advantage of this test is that very small quantities of flowing water can be measured. The Falling Head test is not often used in practice nowadays, because another test on clay or peat to determine the compressibility, determines the permeability at the same time (see Chapter 19).

It may be remarked that the determination of the hydraulic conductivity of a sample in a laboratory is relatively easy, and with great accuracy, but large errors may occur during sampling of the soil in the field, and perhaps during the transportation from the field to the laboratory. Furthermore, the measured value only applies to that particular sample, having small dimensions. This value may not be representative for the hydraulic conductivity in the field. In particular, if a thin layer of clay has been overlooked, the permeability of the soil for vertical flow may be much smaller than follows from the measurements. On the other hand, if it is not known that a clay layer contains holes, the flow in the field may be much larger than expected on the basis of the permeability test on the clay. It is often advisable to measure the permeability in the field (*in situ*), measuring the average permeability of a sufficiently large region (see Chapter 12).

Problems

8.1 In a permeability test (see Figure 8-1) a head difference of 20 cm is being maintained between the top and bottom ends of a sample of 40 cm height. The inner diameter of the circular tube is 10 cm. It has been measured that in 1 minute an amount of water of 35 cm^3 is collected in a measuring glass. What is the value of the hydraulic conductivity k?

8.2 A permeability apparatus (see Figure 8-1) is filled with 20 cm of sand, having a hydraulic conductivity of 10^{-5} m/s, and on top of that 20 cm sand having a hydraulic conductivity that is a factor 4 larger. The inner diameter of the circular tube is 10 cm. Calculate the discharge Q through this layered sample, if the head difference between the top and bottom of the sample is 20 cm.

8.3 In Figure 8-1 the fluid flows through the soil in vertical direction. In principle the tube can also be placed horizontally. The formulas then remain the same, and the measurement of the head difference is simpler. The test is usually not done in this way, however. Why not?

8.4 An engineer must give a quick estimate of the permeability of a certain sand. He remembers that the hydraulic conductivity of the sand in a previous project was 8 m/d. The sand in the current project seems to have particles that are about 1/4 times as large. What is his estimate?

9 Groundwater flow

In the previous chapters the relation of the flow of groundwater and the fluid pressure, or the groundwater head, has been discussed, in the form of Darcy's law. In principle the flow can be determined if the distribution of the pressure or the head is known. In order to predict or calculate this pressure distribution Darcy's law in itself is insufficient. A second principle is needed, which is provided by the principle of *conservation of mass.* This principle will be discussed in this chapter. Only the simplest cases will be considered, assuming isotropic properties of the soil, and complete saturation with a single homogeneous fluid (fresh water). It is also assumed that the flow is *steady*, which means that the flow is independent of time.

9.1 Flow in a vertical plane

Suppose that the flow is restricted to a vertical plane, with a cartesian coordinate system of axes x and z. The z-axis is supposed to be in upward vertical direction, or, in other words, gravity is supposed to act in negative z-direction. The two relevant components of Darcy's law now are

$$q_x = -k\frac{\partial h}{\partial x},\qquad q_z = -k\frac{\partial h}{\partial z}. \tag{9.1}$$

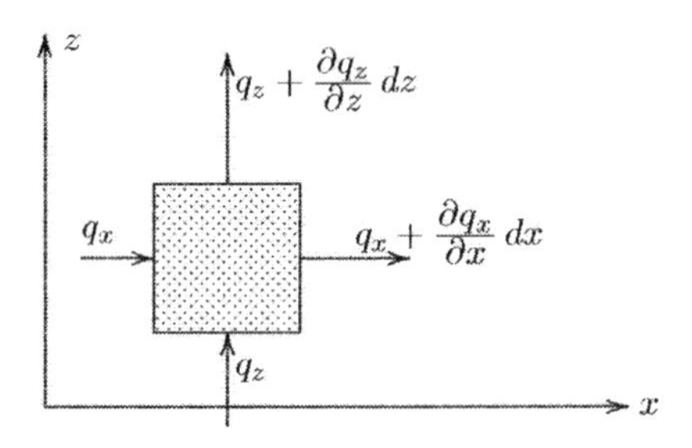

Figure 9-1. Continuity.

Conservation of mass now requires that no water can be lost or gained from a small element, having dimensions dx and dz in the x, z-plane, see Figure 9-1. In the x-direction water flows through a vertical area of magnitude $dydz$, where dy is the thickness of the element perpendicular to the plane of flow. The difference between the outflow from the element on the right end side and the inflow into the element on the left end side is the discharge

$$\frac{\partial q_x}{\partial x} dxdydz.$$

In the z-direction water flows through a horizontal area of magnitude $dxdy$. The difference of the outflow through the upper surface and the inflow through the lower surface is

$$\frac{\partial q_z}{\partial z} dxdydz.$$

The sum of these two quantities must be zero, and this gives, after division by $dxdydz$,

$$\frac{\partial q_x}{\partial x} + \frac{\partial q_z}{\partial z} = 0. \tag{9.2}$$

The validity of this equation, the *continuity equation*, requires that the density of the fluid is constant, so that conservation of mass means conservation of volume. Equation (9.2) expresses that the situation shown in Figure 9-1, in which both the flow in x-direction and the flow in z-direction increase in the direction of flow, is impossible. If the flow in x-direction increases, the element looses water, and this must be balanced by a decrease of the flow in z-direction.
Substitution of (9.1) into (9.2) leads to the differential equation

$$\frac{\partial^2 h}{\partial x^2} + \frac{\partial^2 h}{\partial z^2} = 0, \tag{9.3}$$

where it has been assumed that the hydraulic conductivity k is a constant. Eq. (9.3) is often denoted as the *Laplace equation*. This differential equation governs, together with the boundary conditions, the flow of groundwater in a plane, if the porous medium is isotropic and homogeneous, and if the fluid density is constant. It has also been assumed that no water can be stored. The absence of storage is valid only if the soil does not deform and is completely saturated.
The mathematical problem is to solve equation (9.3), together with the boundary conditions. For a thorough discussion of such problems many specialized books are available, both from a physical point of view (on groundwater flow) and from a mathematical point of view (on potential theory). Here only some particular solutions will be considered, and an approximate method using a *flow net*.

9.2 Upward flow

Figure 9-2 shows an example of a clay layer on a sand layer, with the groundwater level at the top of the clay layer coinciding with the soil surface, whereas in the deep sand the groundwater head is somewhat higher, as indicated in the figure by the water level in a standpipe, reaching into this sand layer. A case like this may occur in a polder, in case of a top layer of very low permeability, underlain by a very permeable layer in which the groundwater level is determined by the higher water levels in the canals surrounding the polder. It is assumed that the permeability of the sand is so large, compared to the permeability of the clay, that the water pressures in the sand layer are hydrostatic, even though there is a certain, small, velocity in the water. The upward flow through the clay layer is denoted as *seepage*. The drainage system of the polder must be designed so that the water entering the polder from above by rainfall, and the water entering the polder from below by seepage, can be drained away. The distribution of the pore water pressures in the sand layer can be sketched from the given water level, and the assumption that this distribution is practically hydrostatic. This leads to a certain value at the bottom of the clay layer. In this clay layer the pore pressures will be linear, between this value and the value $p=0$ at the top, assuming that the permeability of the clay layer is constant. Only then the flow rate through the clay layer is constant, and this is required by the continuity condition.

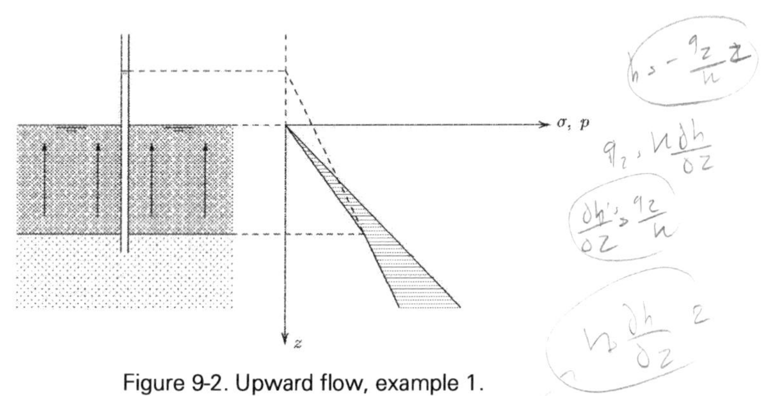

Figure 9-2. Upward flow, example 1.

A very simple special case of groundwater flow occurs when the water flows in vertical direction only. The solution for this case is $h=iz$, where i is a constant, a measure for the intensity of the flow. Actually i, that is dh/dz, is called the *gradient*. In this case $q_x=0$ and $q_z=-ki$. The equation of continuity (9.2) is now indeed satisfied. If the specific discharge is now denoted as q_z, the gradient appears to be $i=-q_z/k$, and $h=-q_z z/k$. Because the z- direction is chosen downwards, the groundwater head is now: $h=z-p/\gamma_w$ it now follows that the pore water pressure is

$$p=\gamma_w z(1-h/z)=\gamma_w z(1-i)=\gamma_w z(1+q_z\,/\,k). \tag{9.4}$$

The first term is the hydrostatic pressure, and the second term is due to the vertical flow. It appears that a vertical flow requires a pressure that increases with depth stronger than in the hydrostatic case.

In Figure 9-2 the total stresses (σ) have also been indicated, assuming that in the sand and the clay the volumetric weight is the same, and about twice as large as the volumetric weight of water. These total stresses are linear with depth, and at the surface $\sigma = 0$. The effective stresses are the difference of the total stresses and the pore water pressures ($\sigma' = \sigma - p$). They are indicated in the figure by horizontal hatching. It can be seen that the effective stresses in the clay are reduced by the upward flow, compared to the fully hydrostatic case, if the groundwater level in the sand were equal to the level of the soil surface. The upward flow appears to result in lower effective stresses.

It may be that the groundwater head in the deep sand is so high ($p = \sigma$) that the effective stresses in the clay layer reach the value $\sigma' = 0$. This is the smallest possible value, because tensile stresses can not be transmitted by the clay particles. The situation that the effective stresses become zero is a *critical* condition. In that case the effective stresses in the clay are zero, and no forces are transmitted between the particles. If the pressure in the water below the clay layer would become slightly larger, the clay layer will be lifted, and cracks will appear in it. If $\sigma' = 0$ the soil has no strength left. Even a small animal would sink into the soil. This situation is often indicated as *liquefaction*, because the soil (in this example the clay layer) has all the characteristics of a liquid : the pressure in it is linear with depth (although the apparent volumetric weight is about twice the volumetric weight of water), and shear stresses in it are impossible. The value of the gradient dh/dz for which this situation occurs is sometimes denoted as the *critical gradient*. In the case considered here the total stresses are

$$\sigma_{zz} = \gamma_s z, \tag{9.5}$$

where γ_s is the volumetric weight of the saturated soil (about 20 kN/m^3). In the case of a critical gradient the pore pressures, see (9.4), must be equal to the total stresses. This will be the case if $i = i_{cr}$, with

$$i_{cr} = -\frac{\gamma_s - \gamma_w}{\gamma_w}. \tag{9.6}$$

As the z-axis points in downward direction, this negative gradient indicates that the groundwater head increases in downward direction, which causes the upward flow. The order of magnitude of the absolute value of the critical gradient is about $|i_{cr}| = 1$, assuming that $\gamma_s = 2\gamma_w$.

In the critical condition the vertical velocity is so large that the upward friction of the water on the soil particles just balances the weight of the particles under water, so that they no longer are resting on each other. Such a situation, in which there is no more coherence in the particle skeleton, should be avoided by a responsible civil engineer. In engineering practice a sufficiently large margin of safety should be included. If the top layer is not homogeneous it is possible that an average gradient of 1 can easily lead to instabilities, because locally the thickness of the clay layer is somewhat smaller, for instance. Water has a very good capacity to find the weakest spot. In several cases this phenomenon has lead to large calamities and large costs, such as excavations of which the bottom layer has burst open, with flooding of the entire excavation as a result. Preventing such calamities may be costly, but is always much cheaper than the repair works that are necessary in case of collapse.

It may be interesting to note that the critical gradient can also be determined using the concept of *seepage force*, as introduced in the previous chapter. In this approach all the forces acting upon the particle skeleton are considered, and equilibrium of this skeleton is formulated. The force due to the weight of the material is a downward force caused by the volumetric weight under water, $\gamma_s - \gamma_w$. This leads to effective stresses of the form

$$\sigma'_{zz} = (\gamma_s - \gamma_w)z. \tag{9.7}$$

The particles have an apparent volumetric weight of $\gamma' = \gamma_s - \gamma_w$. The absolute value of the seepage force is, with (7.15), $j = -\gamma_w i$. The two forces can be balanced if the two values are equal $(\gamma' = j)$, i.e. if $i = i_{cr}$, with

$$i_{cr} = -\frac{\gamma_s - \gamma_w}{\gamma_w}. \tag{9.8}$$

This is in agreement with the value derived before, see (9.6).
Geotechnical engineers usually prefer the first approach, in which the effective stresses are derived as the difference of the total stresses and the pore pressures, and then the critical situation is generated if anywhere in the field the effective stress becomes zero. This is a much more generally applicable criterion than a criterion involving a critical gradient. As an illustration a somewhat more complex situation is shown in Figure 9-3, with two sand layers, above and below a clay layer. It has been assumed that in both sand layers the groundwater pressures are hydrostatic, with a higher zero level in the lower layer. Water will flow through the clay layer, in upward direction.

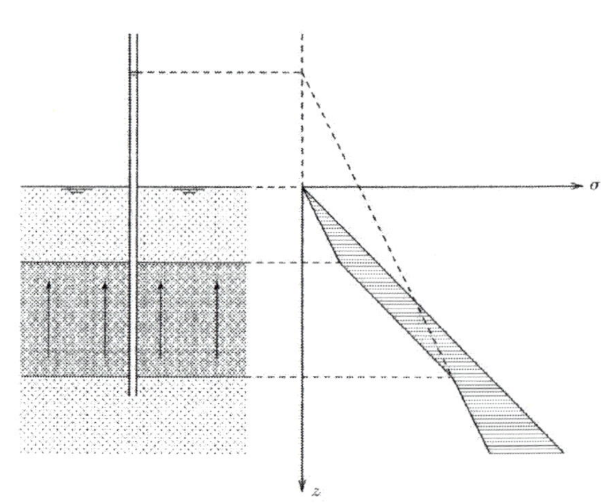

Figure 9-3. Upward flow, example 2.

The situation shown in Figure 9-3 is not yet critical, even though the upward gradient in the clay layer is $i = i_{cr}$, as can be seen by noting that the effective stresses in the clay layer do not increase with depth. Indeed, the upward seepage force in the clay layer is in equilibrium with the downward force due to the weight of the soil under water. However, at the top of the clay layer there is a non-zero effective stress at the top of the clay layer, due to the weight of the sand above it. Because of this surcharge the effective stresses are unequal to zero throughout the clay layer, and the situation is completely safe. The groundwater pressure below the clay layer could be considerably higher before the risk of loss of equilibrium by the effective stress becoming zero is reached, at the bottom of the clay layer. The concept of *critical gradient* appears to be irrelevant in this case, and its use should be discouraged.
It can be concluded that an upward groundwater flow may lead to loss of equilibrium, and this will occur as soon as the effective stress reaches zero, anywhere in the soil. Such a situation should be avoided, even if it seems to be costly.

9.3 Bursting of a clay layer

A common problem during excavation is the bursting of a clay layer. The excavation reduces the total stress, while the pore water pressure remains the same, which can lift a clay layer above. An easy method to prevent bursting of a clay layer is to lower the groundwater head below it, by a pumping well, such that $p < \sigma$. As an example Figure 9-4 shows an excavation for a building pit.

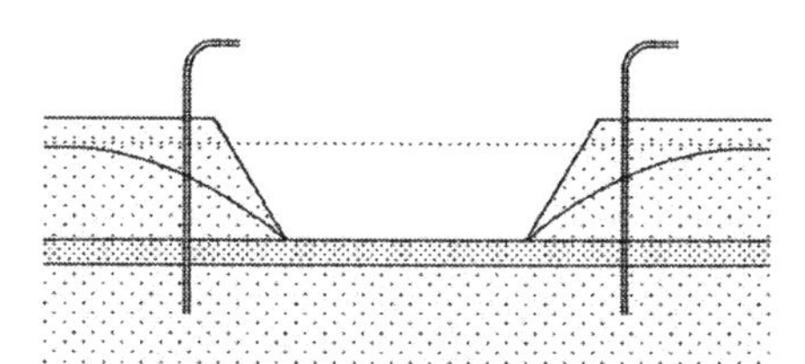

Figure 9-4. Draining an excavation.

If the groundwater level in the upper sand layered is lowered by a drainage system in the excavation, the shape of the phreatic level may be of the form sketched in the figure by the fully drawn curves. Water in the upper layer will flow into the excavation, and may be drained away by pumping at the bottom of the excavation. If the permeability of the clay layer is sufficiently small, the groundwater level in the lower layer will hardly be affected by this drainage system, and very little water will flow through the clay layer. The phreatic level in the lower sand layer is indicated in the figure by the dotted line. The situation drawn in the figure is very dangerous. Only a thin clay layer separates the deep sand from the excavation. The water pressures in the lower layer are far too high to be in equilibrium with the weight of the clay layer. This layer will certainly collapse, and the excavation will be flooded. To prevent this, the groundwater level in the lower layer may be lowered artificially, by pumping wells. These have also been indicated in the figure, but their influence has not yet been indicated. A disadvantage of this solution is that large amounts of water must be pumped to lower the groundwater level in the lower layer sufficiently, and this entails that over a large region the groundwater is affected. Another solution is that a layer of concrete is constructed, at the bottom of the excavation, before lowering the groundwater table.

9.4 Flow under a wall

A solution of the basic equations of groundwater flow, not so trivial as the previous one, in which the flow rate was constant, is the solution of the problem of flow in a very deep deposit, bounded by the horizontal surface $z = 0$, with a separation of two regions above that surface by a thin vertical wall at the location $x = 0$, see Figure 9-5. The water level at the right side of the wall is supposed to be at a height H above ground surface, and the water level at the left side of the wall is supposed to coincide with the ground surface. Under the influence of this water level difference groundwater will flow under the wall, from right to left. The solution of this problem can be obtained using the theory of functions of a complex variable. The actual solution procedure is not considered here. It is assumed, without any derivation, that in this case the solution of the problem is

$$h = \frac{H}{\pi} \arctan(z/x). \tag{9.9}$$

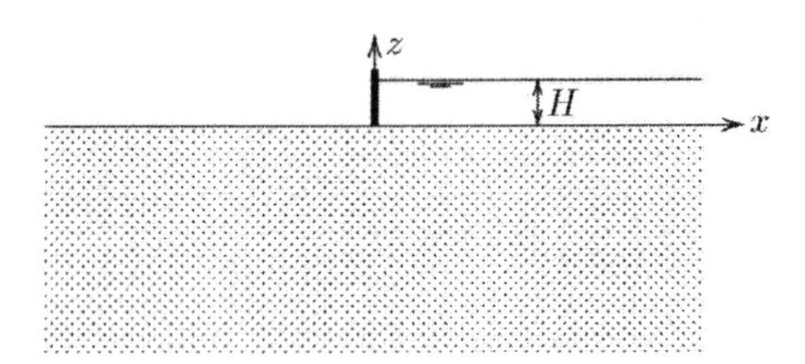

Figure 9-5. Flow under a wall.

In order to apply this solution, it should be verified first that it is indeed the correct solution. For this purpose it is sufficient to check that the solution satisfies the differential equation, and that it is in agreement with the boundary conditions.

That the solution (9.9) satisfies the differential equation (9.3) can easily be verified by substituting the solution into the differential equation. To verify the boundary conditions the behaviour of the solution for $z \uparrow 0$ must be investigated. The value of z/x then will approach 0 from below if $x > 0$, en it will approach 0 from above if $x < 0$. Let it now be assumed that the range of the function $\arctan(u)$ is from 0 to $\pi/2$ if the argument u goes from 0 to ∞, and from $\pi/2$ to π if the argument u goes from $-\infty$ to 0, see Figure 9-6.

In that case it indeed follows that $h = H$ if $x > 0$ and $z \uparrow 0$, and that $h = 0$ if $x < 0$ and $z \uparrow 0$. All this means that equation (9.9) is indeed the correct solution of the problem, as it satisfies all necessary conditions.

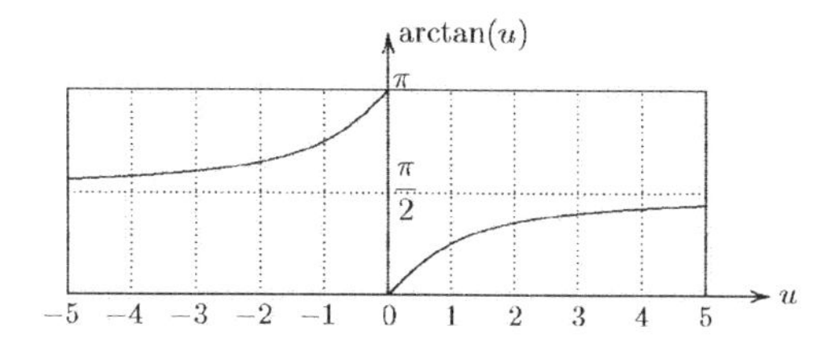

Figure 9-6. Function arctan(u).

The vertical component of the specific discharge can be obtained by differentiation of the solution (9.9) with respect to z. This gives

$$q_z = -\frac{kH}{\pi}\frac{x}{x^2+z^2}. \tag{9.10}$$

In particular, it follows that along the horizontal axis, where $z=0$,

$$z=0 \quad : \qquad q_z = -\frac{kH}{\pi x}. \tag{9.11}$$

If $x>0$ this is negative, so that the water flows in downward direction. This means that to the right of the wall the water flows in vertical direction into the soil, as was to be expected. If $x<0$, that means to the left of the wall, the specific discharge q_z is positive, i.e. the water flows in upward direction, as also was to be expected. Very close to the wall, i.e. for small values of x, the velocity will be very large. Locally that might result in erosion of the soil.
It also follows from the solution, because $\arctan(\infty)=\pi/2$, that on the vertical axis, i.e. for $x=0$, the groundwater head is $h=H/2$. That could have been expected, noting the symmetry of the problem.
The total discharge from the reservoir at the right side of the wall, between the two points $x=a$ and $x=b$ (with $b>a$) can be found by integration of eq. (9.11) from $x=a$ to $x=b$. The result is

$$Q=\frac{kHB}{\pi}\ln(b/a), \tag{9.12}$$

in which B is the thickness of the plane of flow, perpendicular to the figure. This formula indicates that the total discharge is infinitely large if $b\to\infty$ or if $a\to 0$. In reality such situations do not occur, fortunately.
Equation (9.12) can be used to obtain a first estimate for the discharge under a hydraulic
structure, such as a sluice, see Figure 9-7. If the length of the sluice is denoted by $2a$, and the thickness of the layer is d, it can be assumed that the water to the left and to the right of the sluice will mostly flow into the soil and out of it over a distance approximately equal to d. The flow then is somewhat similar to the flow in the problem of Figure 9-5 between $x=a$ and $x=b=a+d$. In Figure 9-7 it seems that

the values of a and d are approximately equal, so that $\ln(b/a) = 0.693$. This gives $Q = 0.22kHB$ as a first estimate
for the total discharge.

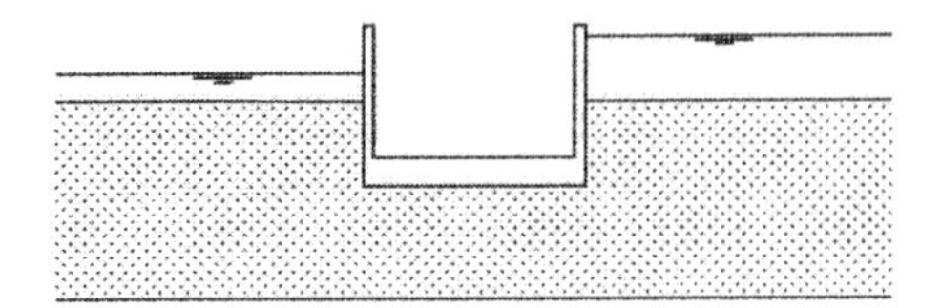

Figure 9-7. Flow under a sluice.

Problems

9.1 The thickness of a certain clay layer is 8 m, and its volumetric weight is 18 kN/m^3. It is covered by a layer of very permeable sand, having a thickness of 4 m, a saturated volumetric weight of 20 kN/m^3, and a dry volumetric weight o 16 kN/m^3. The phreatic surface coincides with the soil surface. In the sand layer directly below the clay layer the groundwater head is at a level 4 m above the soil surface. Sketch the distribution of total stresses, pore pressures and effective stresses in the three layers. In particular, calculate the effective stress in the centre of the clay layer.

9.2 Calculate the effective stress in the centre of the clay layer if the groundwater level in the upper sand layer is lowered to 2 m below the soil surface.

9.3 Next calculate the effective stress in the centre of the clay layer if the soil is loaded by a concrete plate, having a weight of 40 kN/m^2.

9.4 A clay layer has a thickness of 3 m, and a volumetric weight of 18 kN/m^3. Above the clay layer the soil consists of a sand layer, of thickness 3 m, a saturated volumetric weight of 20 kN/m^3, and a dry volumetric weight of 16 kN/m^3. The groundwater level in the sand is at 1 m below the soil surface. Below the clay layer, in another sand layer, the groundwater head is variable, due to a connection with a tidal river. What is the maximum head (above the soil surface) that may occur before the clay layer will fail?

10 Floatation

In the previous chapter it has been seen that under certain conditions the effective stresses in the soil may be reduced to zero, so that the soil looses its coherence, and a structure may fail. Even a small additional load, if it has to be supported by shear stresses, can lead to a calamity. Many examples of failures of this type can be given: the bursting of the bottom of excavation pits, and the floatation of basements, tunnels and pipelines. The floatation of structures is discussed in this chapter.

10.1 Archimedes

The basic principle of the uplift force on a body submerged in a fluid is due to Archimedes. This principle can best be explained by considering a small rectangular element, at rest in a fluid, see Figure 10-1. The material of the block is irrelevant, but it must be given to be at rest. The pressure in the fluid is a function of depth only, and in a homogeneous fluid the pressure distribution is

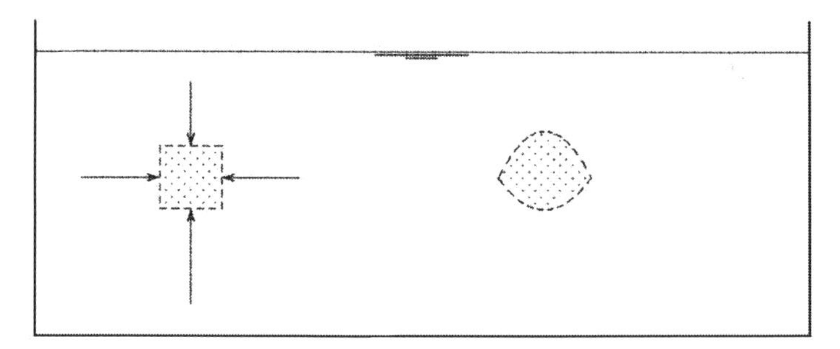

Figure 10-1. Archimedes' principle.

$$p = \rho g z, \tag{10.1}$$

where ρ is the density of the fluid, g the acceleration of gravity, and z the depth below the fluid surface.

The pressures on the left hand side and the right hand side are equal, but act in opposite direction, and therefore are in equilibrium. The pressure below the element is greater than the pressure above it. The resultant force is equal to the difference in pressure, multiplied by the area of the upper and lower surfaces. Because the pressure difference is just $\rho g h$, where h is the height of the element, the upward force equals ρg times the volume of the element. That is just the volumetric weight of the water multiplied by the volume of the element. Because any body can be

constructed from a number of such elementary blocks, the general applicability of Archimedes' principle follows.
A different argument, that immediately applies to a body of arbitrary shape, is that in a state of equilibrium the precise composition of a body is irrelevant for the force acting upon it. This means that the force on a body of water must be the same as the force on a body of some other substance, that then perhaps must be kept in equilibrium by some additional force. Because the body when composed of water is in equilibrium it follows that the upward force must be equal to the weight of the water in the volume. On a body of some other substance the resultant force of the water pressures must be the same, i.e. an upward force equal to the weight of the water in the volume. This is the proof that is given in most textbooks on elementary physics. The upward force is often denoted as the buoyant force, and the effect is denoted as *buoyancy*.
The buoyancy force on a body in a fluid may have as a result that the body floats on the water, if the weight of the body is smaller than the upward force. Floatation will happen if the body on the average is lighter than water. More generally, floatation may occur if the buoyancy force is larger than the sum of all downward forces together. This may happen in the case of basements, tunnels, or pipelines. In principle floatation can easily be prevented: the body must be heavy enough, and may have to be ballasted.
The problem of possible floatation of a foundation is that care must be taken that the effective stresses are always positive, taking into account a certain margin of safety. In practice this may be more difficult than imagined, because perhaps not all conditions have been foreseen. Some examples may illustrate the analysis.

10.2 A concrete floor under water

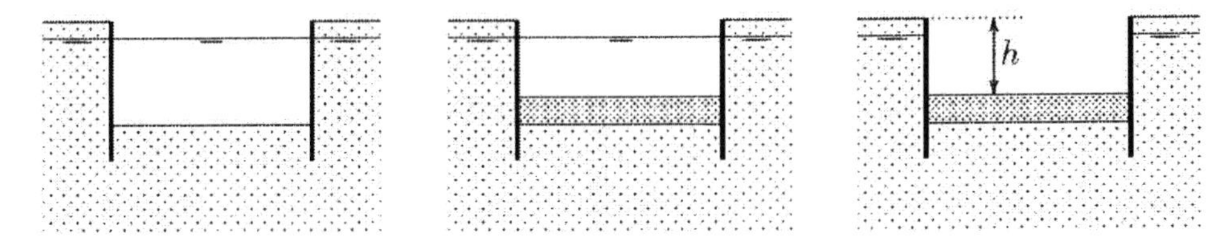

Figure 10-2. Excavation with concrete floor under water.

As a first example a concrete floor of an excavation is considered. Such structures are often used as foundations of basements, or as the pavement of the access road of a tunnel. One of the functions of the concrete plate is to give additional weight to the soil, so that it will not float. Care must be taken that the water table can only be lowered when the concrete plate is already present. Therefore a convenient procedure is to build the concrete plate under water, before the lowering of the water table, see Figure 10-2. After excavation of the pit, under water, perhaps using dredging equipment, the concrete floor must be constructed, taking great care of the continuity of the floor and the vertical walls of the excavation. When the concrete structure has

been finished, the water level can be lowered. In this stage the weight of the concrete is needed to prevent floatation.
There are two possible methods to perform the stability analysis. The best method is to determine the effective stresses just below the concrete floor. If these are always positive, in every stage of the building process, a compressive stress is being transferred in all stages, and the structure is safe. Whenever tensile stresses are obtained, even in a situation that is only temporary, the design must be modified. The structure will not always be in equilibrium, and will float or break. It is assumed that in the case shown in Figure 10-2 the groundwater level is at a depth d = 1 m below the soil surface, and that the depth of the top of the concrete floor should be located at a depth h = 5 m below the soil surface. Furthermore the thickness of the concrete layer (which is to be determined) is denoted as D. The total stress just below the concrete floor now is

$$\sigma = \gamma_c D, \tag{10.2}$$

where γ_c is the volumetric weight of the concrete, say $\gamma_c = 25\ \text{kN/m}^3$. The pore pressure just below the concrete floor is

$$p = (h - d + D)\gamma_w, \tag{10.3}$$

so that the effective stress is

$$\sigma'_{zz} = \sigma_{zz} - p = \gamma_c D - \gamma_w (h - d + D) = (\gamma_c - \gamma_w)D - \gamma_w (h - d). \tag{10.4}$$

The requirement that this must be positive gives

$$D > (h - d)\frac{\gamma_w}{\gamma_c - \gamma_w}. \tag{10.5}$$

The effective stress will be positive if the thickness of the concrete floor is larger than the critical value. In the example, with $h - d = 4$ m and the concrete being a factor 2.5 heavier than water, it follows that the thickness of the floor must be at least 2.67 m. It may be noted that the required thickness of the concrete floor should be somewhat larger, namely 3.33 m if the groundwater level would coincide with the soil surface. One must be very certain that this condition cannot occur if the concrete plate is taken thinner as 3.33 m. It may also be noted that in time of danger, perhaps when the groundwater pressures rises because of some emergency, the foundation can be saved by submerging with water.
The analysis can be done somewhat faster by directly requiring that the weight of the concrete must be sufficient to balance the upward force acting upon it from below. This leads to the same result. The analysis using the somewhat elaborate process of calculating the effective stresses may take some more time, but it can more easily be generalized, for instance in case of a groundwater flow, when the groundwater pressures are not hydrostatic.
The concrete floor in a structure as shown in Figure 10-2 may have to be rather thick, which requires a deep excavation and large amounts of concrete. In engineering

practice more advanced solutions have been developed, such as a thin concrete floor, combined with tension piles. It should be noted that this requires a careful (and safe) determination of the tensile capacity of the piles. A heavy concrete floor may be expensive, its weight is always acting.

10.3 Floatation of a pipe

The second example is concerned with a pipeline in the bottom of the sea (or a circular tunnel under a river), see Figure 10-3. The pipeline is supposed to consist of steel, with a concrete lining, having a diameter $2R$ and a total weight (above water) G, in kN/m. This weight consists of the weight of the steel and the concrete lining, per unit length of the pipe. For the risk of floatation the most dangerous situation will be when the pipe is empty.

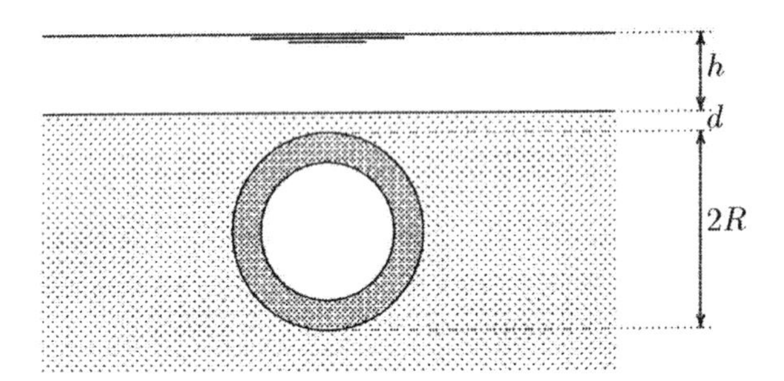

Figure 10-3. Pipe in the ground.

For the analysis of the stability of the pipeline it is convenient to express its weight as an average volumetric weight γ_p, defined as the total weight of the pipeline divided by its volume. In the most critical case of an empty pipeline this is

$$\gamma_p = G/\pi R^2. \tag{10.6}$$

The buoyant force F on the pipeline is, in accordance with Archimedes' principle,

$$F = \gamma_w \pi R^2, \tag{10.7}$$

where γ_w is the volumetric weight of water. If the upward force F is smaller than the weight G there will be no risk of floatation. The pipeline then sinks in open water. This will be the case if $\gamma_p > \gamma_w$. For a pipeline on the bottom of the sea this is a very practical criterion. If one would have to rely on the weight of the soil above the pipeline for its stability, floatation might occur if the soil above the pipeline is taken away by erosion, which is not unlikely. The pipeline then might float to the sea surface, and that should be avoided.
In case of a tunnel buried under a river there seems to be more certainty that the soil above the tunnel remains in place. Then the weight of the soil above the tunnel may prevent floatation even if the tunnel is lighter than water ($\gamma_p < \gamma_w$). The weight W of the soil above the tunnel is

$$W = \gamma_s \left[2Rd + (2 - \pi/2)R^2 \right], \tag{10.8}$$

Where γ_s is the volumetric weight of the saturated soil, and d is the cover thickness, the thickness of the soil at the top of the tunnel. It is now essential to realize, in accordance with Archimedes' principle that for the stability of the tunnel the soil above only contributes insofar as it is heavier than water. The water above the tunnel does not contribute. A block of wood will float in water, even if the water is very deep. This means that the effective downward force of the soil above the tunnel is

$$W' = (\gamma_s - \gamma_w) \left[2Rd + (2 - \pi/2)R^2 \right], \tag{10.9}$$

the difference of the weight of the soil and the weight of the water in the same volume. The amount of soil that is minimally needed now follows from the condition

$$W' + G - F > 0. \tag{10.10}$$

This gives

$$(\gamma_s - \gamma_w) \left[2Rd + (2 - \pi/2)R^2 \right] > (\gamma_w - \gamma_p)\pi R^2, \tag{10.11}$$

from which the ground cover d can be calculated. There still is some additional safety, because when the tunnel moves upward the soil above it must shear along the soil next to it, and the friction force along that plane has been disregarded. It is recommended to keep that as a hidden reserve, because floatation is such a serious calamity.

The analysis can, of course, also be performed in the more standard way of soil mechanics stress analysis: determine the effective stress as the difference of the total stress and the pore pressure. The procedure is as follows

The average total stress below the tunnel is (averaged over its width $2R$)

$$\sigma = \gamma_w h + W/2R + G/2R = \gamma_w h + \gamma_s \left[d + (1 - \pi/4)R \right] + \gamma_p \pi R/2, \tag{10.12}$$

where h is the depth of the water in the river.

The average pore pressure below the tunnel is determined by the volume of the space occupied by the tunnel and everything above it, up to the water surface,

$$p = \gamma_w h + \gamma_w \left[d + (1 - \pi/4)R \right] + \gamma_w \pi R/2, \tag{10.13}$$

The average effective stress below the tunnel now is

$$\sigma' = (\gamma_s - \gamma_w) \left[d + (1 - \pi/4)R \right] + (\gamma_p - \gamma_w)\pi R/2. \tag{10.14}$$

The condition that this must be positive, because the particles can not transmit any tensile force, leads again to the criterion (10.11).

Problems

10.1 A block of wood, having a volume of 0.1 m^3, is kept in equilibrium below water in a basin of water by a cord attached to the bottom of the basin. The volumetric weight of the wood is 9 kN/m^3. Calculate the force in the cord.

10.2 The basin is filled with salt water (volumetric weight 10.2 kN/m^3), and fresh water above it. The separation of salt and fresh water coincides with the top of the block of wood. What is now the force in the cord?

10.3 A tunnel of square cross section, $8 \text{ m} \times 8 \text{ m}$, has a weight (above water) of 50 ton per meter length. The tunnel is being floated to its destination. Calculate the draught.

10.4 The tunnel of the previous problem is sunk into a trench that has been dredged in the sand at the bottom of the river, and then covered with sand. The volumetric weight of the sand is 20 kN/m^3. Determine the minimum cover of sand necessary to prevent floatation of the tunnel.

11 Flow net

11.1 Potential and stream function

Two-dimensional groundwater flow through a homogeneous soil can often be described approximately in a relatively simple way by a *flow net*, that is a net of potential lines and stream lines. The principles will be discussed briefly in this chapter.
The groundwater potential, or just simply the potential, Φ is defined as

$$\Phi = kh, \tag{11.1}$$

where k is the permeability coefficient (or hydraulic conductivity), and h is the groundwater head. It is assumed that the hydraulic conductivity k is a constant throughout the field. If this is not the case the concept of a potential can not be used.
Darcy's law, see (9.1), can now be written as

$$\begin{aligned} q_x &= -\frac{\partial \Phi}{\partial x}, \\ q_z &= -\frac{\partial \Phi}{\partial z}, \end{aligned} \tag{11.2}$$

or, using vector notation,

$$\vec{q} = -\nabla\vec{\Phi}. \tag{11.3}$$

In mathematical physics any quantity whose gradient is a vector field (for example forces or velocities), is often denoted as a *potential*. For that reason in groundwater theory Φ is also called the potential. In some publications the groundwater head h itself is sometimes called the potential, but strictly speaking that is not correct, even though the difference is merely the constant k.
The equations (11.2) indicate that no groundwater flow will flow in a direction in which the potential Φ is not changing. This means that in a figure with lines of constant potential (these are denoted as *potential lines*) the flow is everywhere perpendicular to these potential lines, see Figure 11-1.
The flow can also be described in terms of a *stream function*. This can best be introduced by noting that the flow must always satisfy the equation of continuity, see (9.2), i.e.

$$\frac{\partial q_x}{\partial x} + \frac{\partial q_z}{\partial z} = 0. \tag{11.4}$$

This means that a function Ψ must exist such that

$$\begin{aligned} q_x &= -\frac{\partial \Psi}{\partial z}, \\ q_z &= +\frac{\partial \Psi}{\partial x}. \end{aligned} \tag{11.5}$$

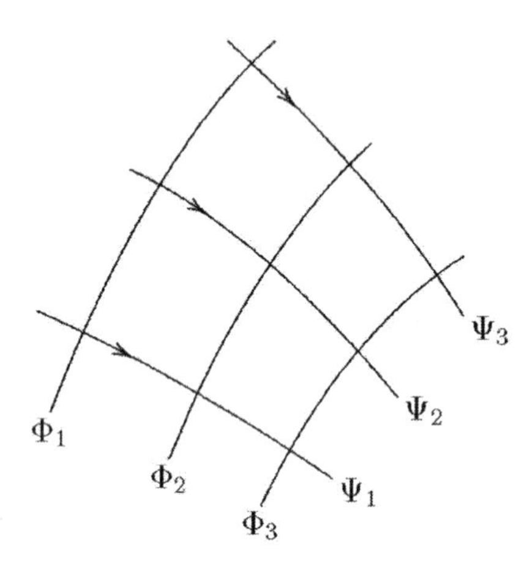

Figure 11-1. Potential lines and stream lines.

By the definition of the components of the specific discharge in this way, as being derived from this function Ψ, the *stream function*, the continuity equation (11.4) is automatically satisfied, as can be verified by substitution of eqs. (11.5) into (11.4).
It follows from (11.5) that the flow is precisely in x-direction if the value of Ψ is constant in x-direction. This can be checked by noting that the condition $q_z = 0$ can only be satisfied if $\partial\Psi/\partial x = 0$. Similarly, the flow is in z-direction only if Ψ is constant in z-direction, because it follows that $q_x = 0$ if $\partial\Psi/\partial x = 0$. This suggests that in general the stream function Ψ is constant in the direction of flow. Along the stream lines in Figure 11-1 the value of Ψ is constant. Formally this property can be proved on the basis of the total differential

$$d\Psi = \frac{\partial \Psi}{\partial x} dx + \frac{\partial \Psi}{\partial z} dz = q_z dx - q_x dz. \tag{11.6}$$

This will be zero if $\partial z/\partial x = q_z/q_x$, and that means that the direction in which $\partial\Psi = 0$ is given by q_z/q_x, which is precisely the direction of flow. It can be concluded that in a mesh of potential lines and stream lines the value of Ψ is constant along the stream lines.
If the x-direction coincides with the direction of flow, the value of q_z is 0. It then follows from (11.2) and (11.5) that in that case Φ is constant in z-direction, and that Ψ is constant in x-direction. Furthermore, in that case one may write, approximately

$$\frac{\Delta\Phi}{\Delta x}=\frac{\Delta\Psi}{\Delta z}. \tag{11.7}$$

It now follows that if the intervals $\Delta\Phi$ and $\Delta\Psi$ are chosen to be equal, then $\Delta x = \Delta z$, i.e. the potential line and the stream line locally form a small square. That is a general property of the system of potential lines and streamlines (the *flow net*): potential lines and stream lines form a system of "squares".
The physical meaning of $\Delta\Phi$ can be derived immediately from its definition, see equation 10.2. If the difference in head between two potential lines, along a stream line, is Δh, then $\Delta\Phi = k\Delta h$. The physical meaning of $\Delta\Psi$ can best be understood by considering a point in which the flow is in x-direction only. In such a point $q = q_x = -\Delta\Psi/\Delta z$, or $\Delta\Psi = -q\Delta z$. In general one may write

$$\Delta\psi = -q\Delta n, \tag{11.8}$$

where n denotes the direction perpendicular to the flow direction, with the relative orientation of n and s being the same as for z and x. If the thickness of the plane of flow is denoted by B, the area of the cross section between two stream lines is ΔnB. It now follows that

$$\Delta Q_{\Delta n} = -B\Delta\Psi. \tag{11.9}$$

The quantity $\Delta Q_{\Delta n}$ appears to be equal to the discharge being transported between two stream lines. It will appear that this will enable to determine the total discharge through a system.

11.2 Flow under a structure

As an example the flow under a structure will be considered, see Figure 11-2. In this case a sluice has been constructed into the soil. It is assumed that the water level on the left side of the sluice is a distance H higher than the water on the right side. At a certain depth the permeable soil rests on an impermeable layer. To restrict the flow under the sluice a sheet pile wall has been installed on the upstream side of the sluice bottom. It reduces the risk of erosion of soil particles below the sluice. This effect is called piping and can dangerously undermine the structure. By placing the sheet pile at the upstream side, also the pore pressures below the sluice are reduced, in order to maximise the friction capacity of the sluice.

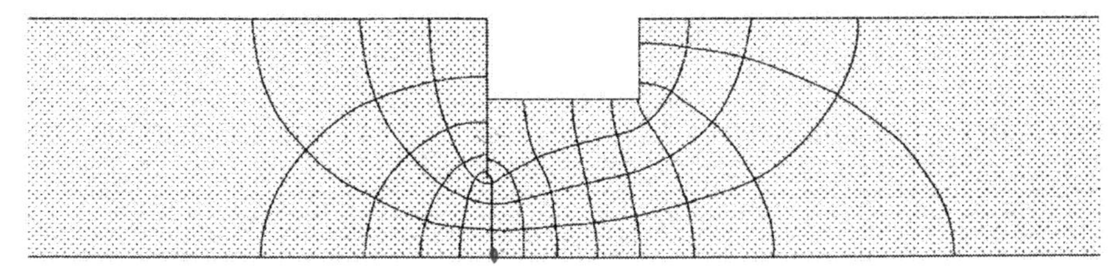

Figure 11-2. Flow net.

The flow net for a case like this can be determined iteratively. The best procedure is by sketching a small number of stream lines, say 2 or 3, following an imaginary water particle from the upstream boundary to the downstream boundary. These stream lines of course must follow the direction of the constraining boundaries at the top and the bottom of the flow field. The knowledge that the stream lines must everywhere be perpendicular to the potential lines can be used by drawing the stream lines perpendicular to the horizontal potential lines to the left and to the right of the sluice. After sketching a tentative set of stream lines, the potential lines can be sketched, taking care that they must be perpendicular to the stream lines. In this stage the distance between the potential lines should be tried to be taken equal to the distance between the stream lines. In the first trial this will not be successful, at least not everywhere, which means that the original set of stream lines must be modified. This then must be done, perhaps using a new sheet of transparent paper superimposed onto the first sketch. The stream lines can then be taken such that a better approximation of squares is obtained.

The entire process must be repeated a few times, until finally a satisfactory system of squares is obtained, see Figure 11-2. Near the corners in the boundaries some special "squares" may be obtained, sometimes having 5 sides. This must be accepted, because the boundary imposes the bend in the boundary. In the case of Figure 11-2 at the right end of the net one half of a square is left. It turns out that there are 12.5 intervals between potential lines, which means that the interval between two potential lines is

$$\Delta\Phi = \frac{kH}{12.5}. \tag{11.10}$$

Because the flow net consists of squares it follows that $\Delta\Psi = \Delta\Phi$, so that

$$\Delta\Psi = \frac{kH}{12.5}. \tag{11.11}$$

Because there appear to be 4 stream bands, the total discharge, according to (11.9), now is

$$Q = \frac{4}{12.5} kHB = 0.32kHB, \tag{11.12}$$

in which B is the width perpendicular to the plane of the figure. The value of the discharge Q must be independent of the number of stream lines that has been chosen, of course. This is indeed the case, as can be verified by repeating the process with 4 interior stream lines rather than 3. It will then be found that the number of potential intervals will be larger, about in the ratio 5 to 4. The ratio of the number of squares in the direction of flow to the number of squares in the direction perpendicular to the flow remains (approximately) constant.

From the completed flow net the groundwater head in every point of the field can be determined. For instance, it can be observed that between the point at the extreme left below the bottom of the sluice and the exit point at the right, about 6 squares can be counted (5 squares and 2 halves). This means that the groundwater head in that point is

$$h = \frac{6}{12.5} H = 0.48 H, \tag{11.13}$$

if the head is measured with respect to the water level on the right side.
The pore water pressure can be derived if the head is known, as well as the elevation, because $h = z + p/\gamma_w$. The evaluation of the water pressure may be of importance for the structural engineer designing the concrete floor, and for the geotechnical engineer who wishes to know the effective stresses, so that the deformations of the soil can be calculated.
From the flow net the force on the particles can also be determined (the seepage force). According to equation (7.15) the seepage force is

$$\begin{aligned} j_x &= -\gamma_w \frac{\partial h}{\partial x}, \\ j_z &= -\gamma_w \frac{\partial h}{\partial z}. \end{aligned} \tag{11.14}$$

In the case illustrated in Figure 11-2 it can be observed that at the right hand exit, next to the structure, in the last (half) square $\Delta h = -H/(2 \times 12.5)$ and $\Delta z = 0.3d$, if d is the depth of the structure into the ground. Then, approximately, $i_z = \partial h/\partial z = -0.133 H/d$, so that $j_z = 0.133 \gamma_w H/d$. This is a positive quantity, indicating that the force acts in upward direction, as might be expected. The particles at the soil surface are also acted upon by gravity, which leads to a volume force of magnitude $-(\gamma_s - \gamma_w)$, negative because it is acting in downward direction. It seems tempting to conclude that there is no danger of erosion of the soil particles if the upward force is smaller than the downward force. This would mean, assuming that $\gamma_s/\gamma_w = 2$, so that $i_{cr} = -(\gamma_s - \gamma_w)/\gamma_w = -1$, that the critical value of H/d would be about 7.5. Only if the value of H/d would be larger than 7.5 erosion of the soil would occur, with the possible loss of stability of the floor foundation at the right hand side.
In reality the danger may be much greater. If the soil is not completely homogeneous, the gradient $\partial h/\partial z$ at the downstream exit may be much larger than the value calculated here. This will be the case if the soil at the downstream side is less permeable than the average. In that case a pressure may build up below the impermeable layer, and the situation may be much more dangerous. On the basis of continuity one might say, very roughly, that the local gradient will vary inversely proportional to the value of the hydraulic conductivity, because $k_1 i_1 = k_2 i_2$. This means that locally the gradient may be much larger than the average value that will be calculated on the basis of a homogeneous average value of the permeability. Locally soil may be eroded, which will then attract more water, and this may lead to further erosion. The phenomenon is called *piping*, because a pipe may be formed,

just below the structure. Piping is especially dangerous if a structure is built directly on the soil surface. If the structure of Figure 11-2 were built on the soil surface, and not into it, the velocities at the downstream side would be even larger (the squares would be very small), with a greater risk of piping.
Prescribing a safe value for the gradient is not so simple. For that reason large safety factors are often used. In the case of vertical outflow, as in Figure 11-2, a safety factor 2, or even larger, is recommended. In cases with horizontal outflow the safety factor must be taken much larger, because in that case there is no gravity to oppose erosion. In many cases piping has been observed, even though the maximum gradient was only about 0.1, assuming homogeneous conditions. Technical solutions are reasonably simple, although they may be costly. A possible solution is that on the upstream side, or near the upstream side, the resistance to flow is enlarged, for instance by putting a blanket of clay on top of the soil, or into it. Another class of solutions is to apply a drainage at the downstream side, for instance by the installation of a gravel pack near the expected outflow boundary. In the case of Figure 11-2 a perfect solution would be to make the sheet pile wall longer, so that it reaches into the impermeable layer. A large dam built upon a permeable soil should be protected by an impermeable core or sheet pile wall, *and* a drain at the downstream side. The large costs of these measures are easily justified when compared to the cost of loosing the dam.

Problems

11.1 Sketch a flow net for the situation shown in Figure 9-7, and calculate the total discharge. Compare the result with the estimate made at the end of the previous chapter.

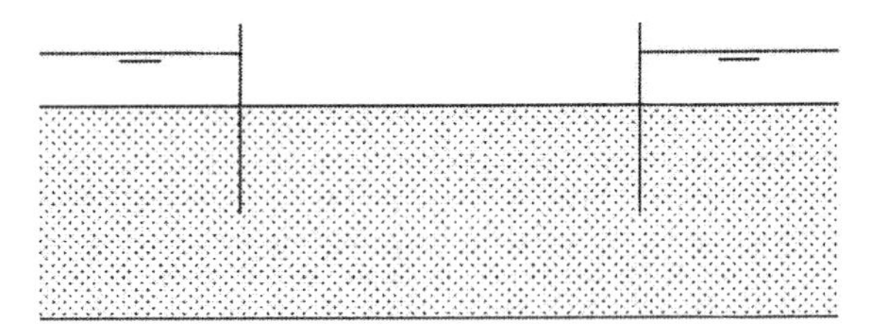

Figure 11-3. Building pit of problem 11-2.

11.2 A building pit in a lake is being constructed, using a sheet pile wall surrounding the building pit. Inside the wall the water level is lowered (by pumping) to the level of the ground surface. Outside the sheet pile wall the water level is 5 m higher. It has been installed to a depth of 10 m below ground surface. The thickness of the soil layer is 20 m. Sketch a flow net, and determine the maximum gradient inside the sheet pile wall.

11.3 Suppose that in a case as considered in the previous problem the soil consists of 1 m clay on top of a thick layer of homogeneous sand. In that case the capacity of the pumps will be much smaller, which is very favourable. Are there any risks involved?

12 Flow towards well

For the theoretical analysis of groundwater flow several computational methods are available, analytical or numerical. Studying groundwater flow is of great importance for soil mechanics problems, because the influence of the groundwater on the behaviour of a soil structure is very large. Many dramatic accidents have been caused by higher pore water pressures than expected. For this reason the study of groundwater flow requires special attention, much more than given in the few chapters of this book. In this chapter one more example will be presented: the flow caused by wells. Direct applications include the drainage of a building pit, or the production of drinking water by a system of wells.

The solutions to be given here apply to a homogeneous sand layer, confined between two impermeable clay layers, see Figure 12-1. This is denoted as a confined aquifer, assuming that the pressure in the groundwater is sufficiently large to ensure complete saturation in the sand layer.

In this case the groundwater flows in a horizontal plane. In this plane the cartesian coordinate axes are denoted as x and y. The groundwater flow is described by Darcy's law in the horizontal plane,

$$q_x = -k\frac{\partial h}{\partial x},$$
$$q_y = -k\frac{\partial h}{\partial z}, \qquad (12.1)$$

and the continuity equation for an element in the horizontal plane,

$$\frac{\partial q_x}{\partial x} + \frac{\partial q_y}{\partial y} = 0. \qquad (12.2)$$

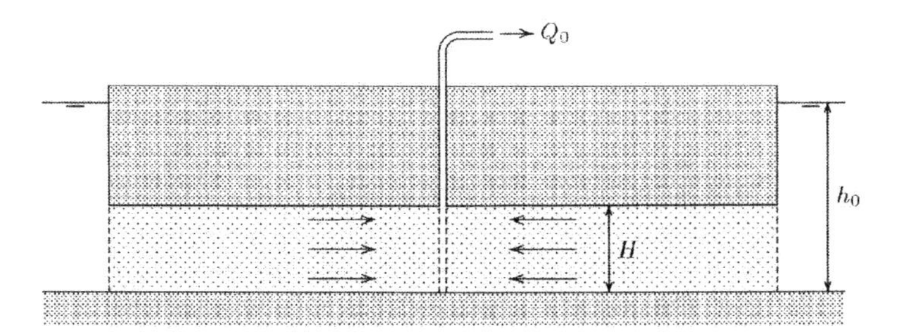

Figure 12-1. Single well in aquifer.

It now follows, if it is assumed that the hydraulic conductivity k is constant, that the partial differential equation governing the flow is

$$\frac{\partial^2 h}{\partial x^2}+\frac{\partial^2 h}{\partial y^2}=0. \tag{12.3}$$

This is again Laplace's equation, but this time in a horizontal plane.
The problem to be considered concerns the flow in a circular region, having a radius R, to a well in the centre of the circle. This is an important basic problem of groundwater mechanics. The boundary conditions are that at the outer boundary (for $r=R$) the groundwater head is fixed: $h=h_0$, and that at the inner boundary, the centre of the circle, a discharge Q_0 is being extracted from the soil.
It is postulated that the solution of this problem is

$$h=h_0+\frac{Q_0}{2\pi kH}\ln\left(\frac{r}{R}\right), \tag{12.4}$$

where Q_0 is the discharge of the well, k the hydraulic conductivity of the soil, H the thickness of the layer, h_0 the value of the given head at the outer boundary ($r=R$), and r is a polar coordinate,

$$r=\sqrt{x^2+y^2}. \tag{12.5}$$

That the expression (12.4) indeed satisfies the differential equation (12.3) can be verified by substitution of this solution into the differential equation. The solution also satisfies the boundary condition at the outer boundary, because for $r=R$ the value of the logarithm is 0 ($\ln(1)=0$). The boundary condition at the inner boundary can be verified by first differentiating the solution (12.4) with respect to r. This gives

$$\frac{dh}{dr}=\frac{Q_0}{2\pi kHr}. \tag{12.6}$$

This means that the specific discharge in r-direction is, using Darcy's law,

$$q_r=-k\frac{dh}{dr}=-\frac{Q_0}{2\pi Hr}. \tag{12.7}$$

The total amount of water flowing through a cylinder of radius r and height H is obtained by multiplication of the specific discharge q_r by the area $2\pi rH$ of such a cylinder,

$$Q=2\pi rHq_r=-2\pi kHr\frac{dh}{dr}=-Q_0 \tag{12.8}$$

This quantity appears to be constant, independent of r, which is in agreement with the continuity principle. It appears that through every cylinder, whatever the radius, an amount of water $-Q_0$ is flowing in the positive r-direction. That means that an amount of water $+Q_0$ is flowing towards the centre of the circle. That is precisely the

required boundary condition, and it can be conclude that the solution satisfies all conditions, and therefore must be correct.
The flow rate very close to the centre is very large, because there the discharge Q_0 must flow through a very small surface area. At the outer boundary the available is very large, so that there the flow rate will be very small, and therefore the gradient will also be small. This makes it plausible that the precise form of the outer boundary is not so important. The solution (12.4) can also be used, at least as a first approximation, for a well in a region that is not precisely circular, for instance a square. Such a square can then be approximated by a circle, taking care that the total circumference is equal to the circumference of the square.
It may be noted that everywhere in the aquifer $r < R$. Then the logarithm in eq. (12.4) is negative, and therefore $h < h_0$, as could be expected. This confirms that by pumping the groundwater head will indeed be lowered.
It is important to note that the differential equation (12.3) is linear, which means that solutions can be added. This is the *superposition principle.* Using this principle solutions can be obtained for a system of many wells, for instance for a drainage system. All wells should be operating near the centre of a large area, the outer boundary of which is schematised to a circle of radius *R*. For a system of *n* wells the solution is

$$h = h_0 + \sum_{j=1}^{n} \frac{Q_j}{2\pi kH} \ln(\frac{r_j}{R}). \tag{12.9}$$

Here Q_j is the discharge of well *j*, and r_j is the distance to that well. The influence of all wells has simply been added to obtain the solution. The discharge Q_j may be positive if the well extracts water, or negative, for a recharging well. At the outer boundary of the system all the values r_j are approximately equal to *R*, the radius of the area, provided that the wells are all located in the vicinity of the centre of that area. Then all logarithms are 0, and the solution satisfies the condition that $h = h_0$ at the outer boundary, at least approximately.
In Figure 12-2 the potential lines and the stream lines have been drawn for the case of a system of a single well and a single recharge well in an infinite field, assuming that the discharges of the well and the recharge well are equal. In mathematical physics these singularities are often denoted as a *sink* and a *source.*

The example of Figure 12-1 shows a confined aquifer. In case of no clay layers on top and only sand, the height for groundwater transport is everywhere the same as the hydraulic head ($H = h$), changing expression (12.4) into

$$h^2 = h_0^2 + \frac{Q_0}{\pi k} \ln\left(\frac{r}{R}\right). \tag{12.10}$$

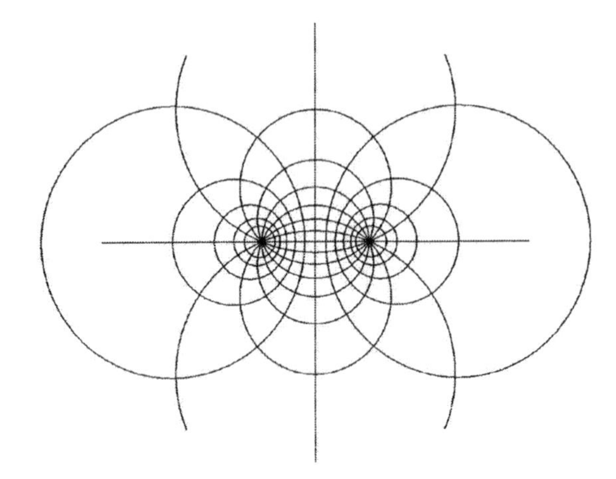

Figure 12-2. Sink and source.

Problems

12.1 For a system of air conditioning water is extracted from a confined layer of 10 m thickness, having a hydraulic conductivity of $1\ \text{m/d}$. The discharge is 50 m^3/d. At a distance of 100 m the water is being injected into the same layer by a recharge well. What is the influence on the groundwater head in the point just between the two wells?

12.2 A well in a (confined) circular area of radius 1000 m appears to lead to a lowering of the groundwater table (a drawdown) of 1 m at a distance of 10 m from the well. What is the drawdown at a distance of 100 m?

12.3 Make a sketch of the solution (12.4) for values of r/R from 0.001 to 1. The value 0.001 applies to the value $r = r_w$, where r_w is the radius of the tube through which the water is being produced. Assume that $h_0 = 20\ \text{m}$, $H = 10\ \text{m}$, and $Q_0/2\pi kH = 1\ \text{m}$. What is the limiting value of the head when the radius of the tube is very small, $r_w \to 0$?

12.4 If $R \to \infty$ the solution (12.4) can not be used because $\ln(0) = -\infty$. Does this mean that in a very large island (Australia) no groundwater can be produced?

Part IV
Stiffness and settlement

13 Stress-strain relations

As stated in previous chapters, the deformations of soils are determined by the *effective stresses*, which are a measure for the contact forces transmitted between the particles. The soil deformations are a consequence of the local displacements at the level of individual particles. In this chapter some of the main aspects of these deformations will be discussed, and this will lead to qualitative properties of the relations between stress and strain. In later chapters these relations will be formulated in a quantitative sense.

13.1 Compression and distorsion

In the contact point of two particles a normal force and a shear force can be transmitted, see Figure 13-1. The normal force can only be a compressive force. Tension can not be transmitted, unless the soil particles are glued together. Such soils do exist (e.g. calcareous soils near the coast of Brazil or Australia), but they are not considered here. The magnitude of the shear force that can be transmitted depends upon the magnitude of the normal force. It can be expected that if the ratio of shear force and normal force exceeds a certain value (the *friction coefficient* of the material of the particles), the particles will start to slide and roll over each other, which will lead to relatively large deformations. The deformations of the particles caused by their compression can be disregarded compared to these sliding deformations. The particles might as well be considered as incompressible.

Figure 13-1. Particle contact.

This can be further clarified by comparing the usual deformations of soils with the possible elastic deformations of the individual particles. Consider a layer of soil of a normal thickness, say 20 m, that is being loaded by a surcharge of 5 m dry sand. The additional stresses caused by the weight of the sand is about 100 kN/m^2, of 0.1 MPa. Deformations of the order of magnitude of 0.1 % or even 1 % are not uncommon for soils. For a layer of 20 m thickness a deformation of 0.1 % means a settlement of 2 cm, and that is quite normal. Many soil bodies show such settlements, or even much more, for instance when a new embankment has been built. Settlements of 20 cm may well be observed, corresponding to a strain of 1 %. If one writes, as a first approximation $\sigma = E\varepsilon$, a stress of 0.1 Mpa and a strain of 0.1 % suggests a deformation modulus $E \approx 100$ MPa. For a strain of 1 % this would be $E \approx 10$ MPa. The modulus of elasticity of the particle material can be found in an encyclopaedia or handbook. This gives about 20 GPa, about one tenth of the modulus of elasticity of steel, and about the same order of magnitude as concrete. That value is a factor 200 or 2000 as large as the value of the soil body as a whole. It can be concluded that the deformations of soils are not caused by deformations of the individual particles, but rather by a rearrangement of the system of particles, with the particles rolling and sliding with respect to each other.

On the basis of this principle many aspects of the behaviour of soil can be explained. It can, for instance, be expected that there will be a large difference between the behaviour in compression and the behaviour in shear. Compression is a deformation of an element in which the volume is changing, but the shape remains the same. In pure compression the deformation in all directions is equal, see Figure 13-2. It can be expected that such compression will occur if a soil element is loaded isotropically, i.e. by a uniform normal stress in all directions, and no shear stresses. In Figure 13-2 the load has been indicated on the original element, in the left part of the figure.

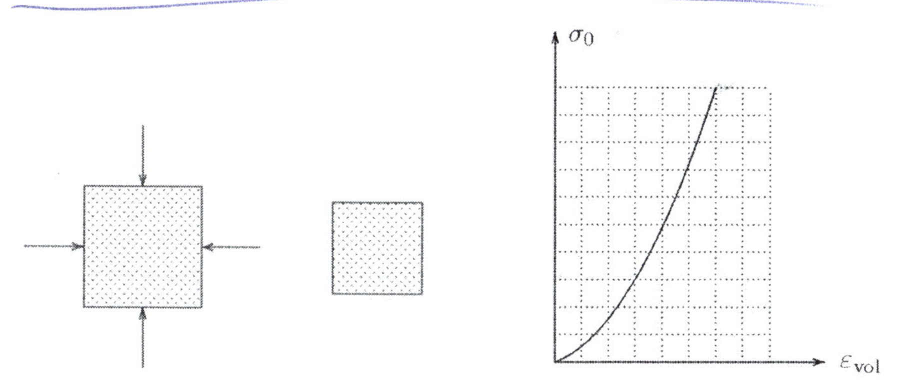

Figure 13-2. Compression.

Figure 13-3. Stiffness in compression.

With such a type of loading, there will be little cause for a change of direction of the forces in the particle contacts. Because of the irregular character of the grain skeleton there may be local shear forces, but these need not to increase to carry increasing compressive forces. If all forces, normal forces and shear forces, increase proportionally, an ever larger compressive external pressure can be transmitted. If the particles were completely incompressible there would be no deformation in that case. In reality the particles do have a small compressibility, and the forces transmitted by the particle contacts are not distributed homogeneously. For these reasons there may be some local sliding and rolling even in pure compression. But it is to be expected that the soil will react much stiffer in compression than in shear, when shear stresses are applied. When external shear stresses are applied to a soil mass, the local shear forces must increase on the average, and this will lead to considerable deformations. In tests it appears that soils are indeed relatively stiff under pure compression, at least when compared to the stiffness in shear. When compared to materials such as steel, soils are highly deformable, even in pure compression.

It can also be expected that in a continuing process of compression the particles will come closer together, increasing the number of contacts, and enlarging the areas of contact. This suggests that a soil will become gradually stiffer when compressed. Compression means that the porosity decreases, and it can be expected that a soil with a smaller porosity will be stiffer than the same assembly of particles in a structure with a larger porosity. It can be concluded that in compression a relation between stress and strain can be expected as shown in Figure 13-3. The quantity σ_0 is the normal stress, acting in all three directions. This is often denoted as the *isotropic stress*. The quantity ε_{vol} is the *volume strain*, the relative change of volume (the change of the volume divided by the original volume).

$$\varepsilon_{vol} = \frac{\Delta V}{V}. \tag{13.1}$$

It may be concluded that the stiffness of soils will increase with continuing compression, or with increasing all round stress. Because in the field the stresses usually increase with depth, this means that in nature it can be expected that the stiffness of soils increases with depth. All these effects are indeed observed in nature, and in the laboratory.

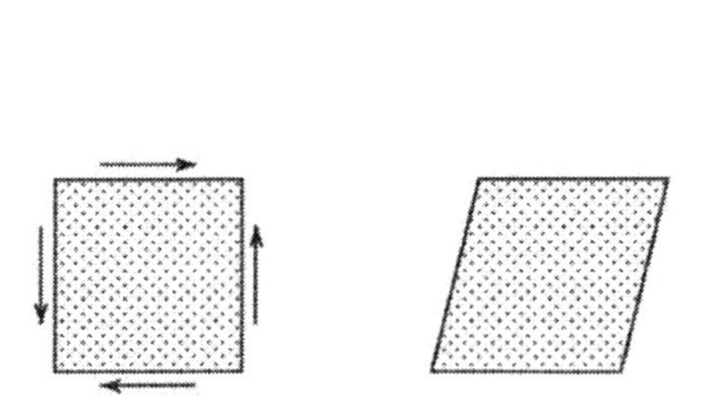

Figure 13-4. Distorsion.

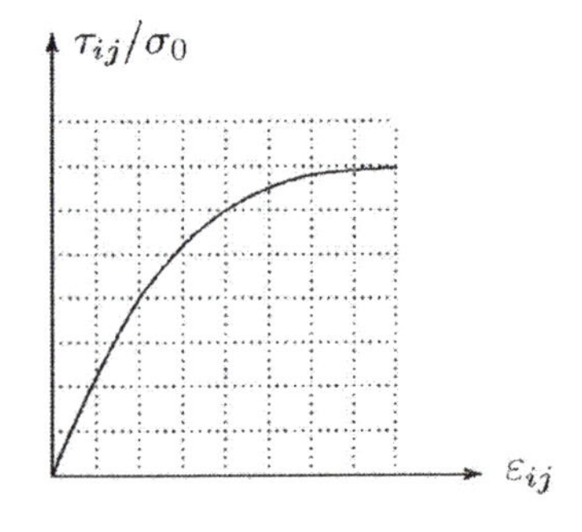

Figure 13-5. Stiffness in distorsion.

Quite a different type of loading is pure distorsion, or pure shear: a change of shape at constant volume, see Figure 13-4. When a soil is loaded by increasing shear stresses it can be expected that in the contact points between the particles the shear forces will increase, whereas the normal forces may remain the same, on the average. This leads to a tendency for sliding in the contact points, and thus there will be considerable deformations. It is even possible that the sliding in one contact point leads to a larger shear force in a neighboring contact point, and this may slide in its

turn. All this means that there is more cause for deformation than in compression. There may even be a limit to the shear force that can be transmitted, because in each contact point the ratio of shear force to normal force can not be larger than the friction angle of the particle material.
During distorsion of a soil a relation between stresses and strains as shown in Figure 13-5 can be expected. In this figure the quantity on the vertical axis is a shear stress, indicated as τ_{ij}, divided by the isotropic stress σ_0. The idea is that the friction character of the basic mechanism of sliding in the contact points will lead to a maximum for the ratio of shearing force to normal force, and that as a consequence for the limiting state of shear stress the determining quantity will be the ratio of average shear stress to the isotropic stress. Tests on dry sand confirm that large deformations, and possible failure, at higher isotropic stresses indeed require proportionally higher shear stresses. By plotting the relative shear stress (i.e. τ_{ij} divided by the isotropic stress σ_0) against the shear deformation, the results of various tests, at different average stress levels, can be represented by a single curve. It should be noted that this is a first approximation only, but it is much better than simply plotting the shear stress against the shear deformation. In daily life the proportionality of maximum shear stress to isotropic can be verified by trying to deform a package of coffee, sealed under vacuum, and to compare that with the same package when the seal has been broken.

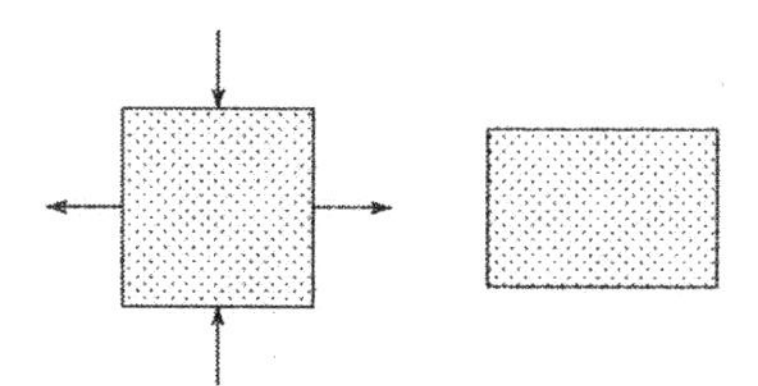

Figure 13-6. Distorsion.

It must be noted that Figure 13-4 represents only one possible form of distorsion. A similar deformation can, of course, also occur in the two other planes of a three-dimensional soil sample. Moreover, the definition of distorsion as change of shape at constant volume means that a deformation in which the width of a sample increases and the height decreases, is also a form of distorsion, see Figure 13-5, because in this case the volume is also constant. That there is no fundamental difference with the shear deformation of Figure 13-4 can be seen by connecting the centres of the four sides in Figure 13-5, before and after the deformation. It will appear that again a square is deformed into a diamond, just as in Figure 13-4, but rotated over an angle of 45°.

In the relations between stresses and strains, as described above, it is of great importance to distinguish compression and distorsion. The behaviour in these two modes of deformation is completely different. The deformations in distorsion (or shear) are usually much larger than the deformations in compression. Also, in compression the material becomes gradually stiffer, whereas in shear it becomes gradually softer.

13.2 Unloading and reloading

Because the deformations of soils are mostly due to changes in the particle assembly, by sliding and rolling of particles, it can be expected that after unloading a soil will not return to its original state. Sliding of particles with respect to each other is an irreversible process, in which mechanical energy is dissipated, into heat. It is to be expected that after a full cycle of loading and unloading of a soil a permanent deformation is observed. Tests indeed confirm this.
When reloading a soil there is probably less occasion for further sliding of the particles, so that the soil will be much stiffer in reloading than it was in the first loading (*virgin loading*). The behaviour in unloading and reloading, below the maximum load sustained before, often seems practically elastic, see Figure 13-7, although there usually is some additional plastic deformation after each cycle. In the figure this is illustrated for shear loading.

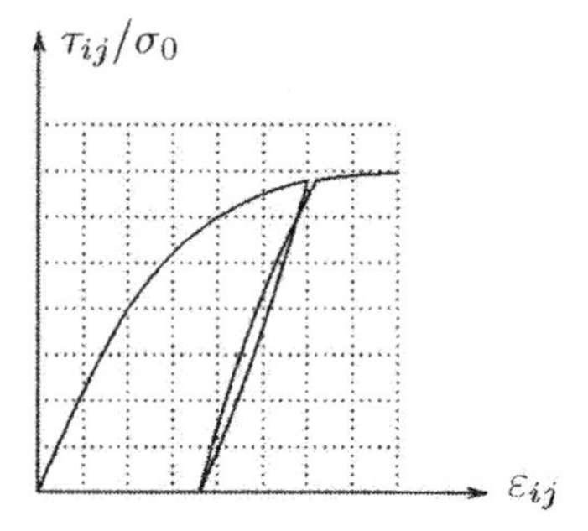

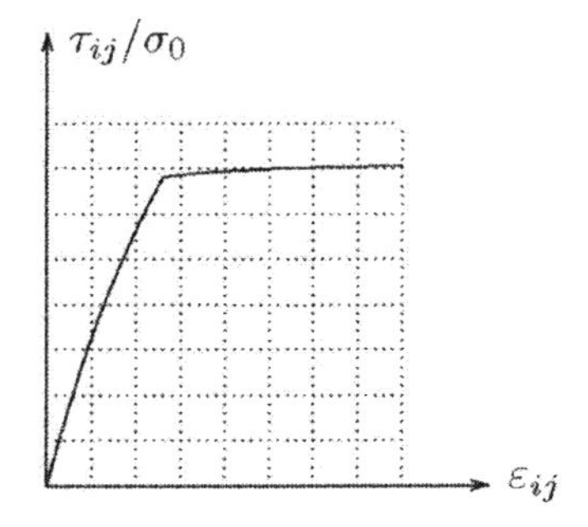

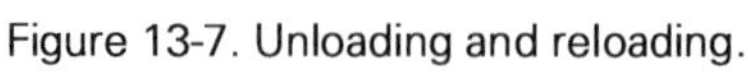
Figure 13-7. Unloading and reloading.

Figure 13-8. Preload.

If the final load is higher than the maximum load experienced before, a relation such as indicated in Figure 13-7 may be observed, with the discontinuity in the curve indicating the level of the previous maximum load, the *preload*. The soil is said to be *overconsolidated* as long as the stresses remain below this preconsolidation load. This type of behaviour is often observed in soils that have been covered in earlier times (an ice age) by a thick layer of ice. The soil is reasonably stiff, but beyond the preconsolidation load the behaviour will become much softer. The soil is now normally consolidated.
A good example of irreversible deformations of soils from engineering practice is the deformation of guard rails along highways. After a collision the guard rail will have been deformed, and has absorbed the kinetic energy of the vehicle. The energy is dissipated by the rotation of the foundation pile through the soil. After removal of the damaged vehicle the rail will not rotate back to its original position, but it can easily be restored by pulling it back. That is the principle of the structure: kinetic energy is dissipated into heat, by the plastic deformation of the soil. That seems much better

than to transfer the kinetic energy of the vehicle into damage of the vehicle and its passengers. The dissipated energy can be observed in the figure as the area enclosed by the branches of loading and unloading, respectively.

It is interesting to note that after unloading and subsequent reloading, the deformations again are much larger if the stresses are increased beyond the previous maximum stress, see Figure 13-8. This is of great practical importance when a soil layer that in earlier times has been loaded and unloaded, is loaded again. If the final load is higher than the maximum load experienced before, a relation such as indicated in Figure 13-8 may be observed, with the discontinuity in the curve indicating the level of the previous maximum load, the *preload*. The soil is said to be *overconsolidated*. As long as the stresses remain below the preconsolidation load the soil is reasonably stiff, but beyond the preconsolidation load the behaviour will be much softer. This type of behaviour is often observed in soils that have been covered in earlier times (an ice age) by a thick layer of ice.

13.3 Dilatancy

One of the most characteristic phenomena in granular soils is *dilatancy*, first reported by Reynolds around 1885. Dilatancy is the volume increase that may occur during shear. In most engineering materials (such as metals) a volume change is produced by an all round (isotropic) stress, and shear deformations are produced by shear stresses, and these two types of response are independent. The mechanical behaviour of soils is more complicated.

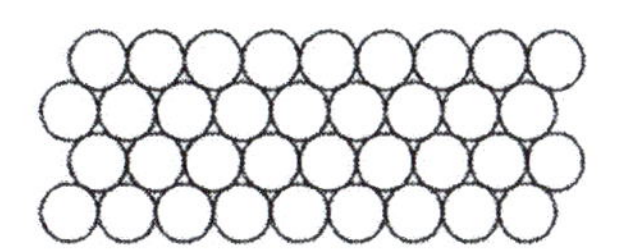

Figure 13-9. Densely packed sand.

This can most conveniently be illustrated by considering a densely packed sand, see Figure 13-9. Each particle is well packed in the space formed by its neighbours. When such a soil is made to shear, by shear stresses, the only possible mode of deformation is when the particles slide and roll over each other, thereby creating some moving space between them. Such a dense material is denoted as *dilatant*.

Dilatancy may have some unexpected results, especially when the soil is saturated with water. A densely packed sand loaded by shear stresses can only sustain these shear stresses by a shear deformation. Through dilatancy this can only occur if it is accompanied by a volume increase, i.e. by an increase of the porosity. In a saturated soil this means that water must be attracted to fill the additional pore space. This phenomenon can be observed on the beach, when walking on the sand in the area flooded by the waves. The soil surrounding the foot may be dried by the suction of the soil next to and below the foot, which must carry the load, see Figure 13-10. For sand at greater depth, for instance the sand below the foundation of an offshore platform, the water needed to fill the pore space can not be attracted in a short time, and this means that an under pressure in the water is being produced. After a certain time this will disappear, when sufficient amounts of water have been supplied. For short values of time the soil is almost incompressible, because it takes time for the water to be supplied, and the shear deformation will lead to a decrease of the pore water pressure. This will be accompanied by an increase of the effective stress, as the total stress remains approximately constant, because the total load must be carried. The soil appears to be very stiff and strong, at least for short values of time. That may be interpreted as a positive effect, but it should be noted that the effect disappears at later times, when the water has flowed into the pores.

Figure 13-10. Dilatancy on the beach.

The phenomenon that in densely packed saturated sand the effective stresses tend to increase during shear is of great importance for the dredging process. When cutting densely packed strata of sand under water an under pressure is generated in the pore water, and this will lead to increasing effective stresses. This increases the resistance of the sand to cutting. A cutting dredger may have great difficulty in removing the sand. The effect can be avoided when the velocity of the cutting process is very small, but then the production is also small. Large production velocities will require large cutting forces.
The reverse effect can occur in case of very loosely packed sand, see Figure 13-11. When an assembly of particles in a very loose packing is being loaded by shear stresses, there will be a tendency for volume decrease. This is called *contractancy*. The assembly may collapse, as a kind of card house structure. Again the effect is most dramatic when the soil is saturated with water. The volume decrease means that there is less space available for the pore water. This has to flow out of the soil, but that takes some time, and in the case of very rapid loading the tendency for volume decrease will lead to an increasing pore pressure in the water. The effective stresses will decrease, and the soil will become weaker and softer. It can even happen that the effective stresses are reduced to zero, so that the soil looses all of its coherence. This is called *liquefaction* of the soil. The soil then behaves as a heavy fluid (*quick sand*), having a volumetric weight about twice as large as water. A person will sink into the liquefied soil, to the waist.

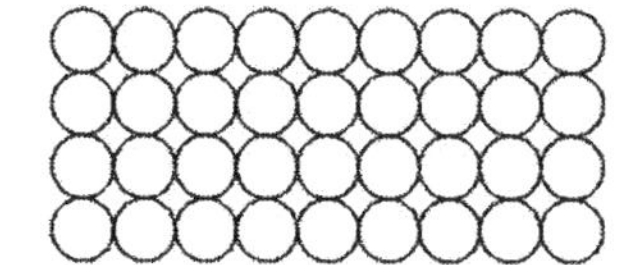

Figure 13-11. Loosely packed sand.

Figure 13-12. Mytilus.

The phenomenon of increasing pore pressures, caused by contractancy of loose soils, can have serious consequences for the stability of the foundation of structures. For example, the sand in the estuaries in the South West of the Netherlands is loosely packed because of the ever continuing process of erosion by tidal currents and deposition of the sand at the turning of the tide. For the construction of the storm surge barrier in the Eastern Scheldt the soil has been densified by vibration before the structure could safely be built upon it. For this purpose a special vessel has been constructed, the Mytilus, see Figure 13-12, containing a series of vibrating needles. Other examples are the soils in certain areas in Japan, for instance the soil in the artificial Port Island in the bay near Kobe. During the earthquake of 1995 the loosely packed sand liquefied, causing great damage to the quay walls and to many buildings. In the area where the soil had previously been densified the damage was much less. For the Chek Lap Kok airport of Hong Kong, an artificial sand island has been constructed in the sea, and to prevent damage by earthquakes the soil has been densified by vibration, at large cost.

It can be concluded that the density of granular soils can be of great importance for the mechanical behaviour, especially when saturated with water, and especially for short term effects. Densely packed sand will have a tendency to expand (dilatancy), and loosely packed sand will have a tendency to contract (contractancy). At continuing deformations both dense and loose sand will tend towards a state of average density, sometimes denoted as the *critical density*. This is not a uniquely defined value of the density, however, as it also depends upon the isotropic stress. At high stresses the critical density is somewhat smaller than at small stress. The branch of soil mechanics studying these relations is *critical state soil mechanics*.

It may be interesting to mention that during *cyclic loads* soils usually tend to contract after each cycle, whatever the original density is. It seems that in a full cycle of loading a few particles may find a more dense packing than before, resulting in a continuing volume decrease. The effect becomes smaller and smaller if the number of cycles increases, but it seems to continue practically forever. It can be compared to the situation in a full train, where there seems to be no limit to the number of passengers that can be transported. By some more pressing a full train can always accommodate another passenger. The cyclic effect is of great importance for the foundation of offshore structures, which may be loaded by a large number of wave loads. During a severe storm each wave may generate a small densification, or a small increase of the pore pressure, if the permeability of the soil is small. After a great many of these wave loads the build up of pore pressures may be so large that the stability of the structure is endangered.

Problems

13.1 A soil sample is loaded in a laboratory test, by an isotropic stress. If the stress is increased from 100 kPa to 200 kPa, the volume decrease is 0.1 %. Suppose that the stress is further increased to 300 kPa. Will the additional volume decrease then be smaller than, larger than, or equal to 0.1 %?

13.2 A part of a guard rail along a highway has been tested by pulling sideways. A force of 10 kN leads to a lateral displacement of 1 cm. What will be the additional displacement is the force is increased to 20 kN, more or less than 1 cm?

13.3 A plastic bottle contains saturated sand, and water reaching into the neck of the bottle, above the sand. When squeezing the bottle, the water level appears to go down. Explain this phenomenon. Is this sand suitable for the foundation of a bridge pier?

13.4 In a laboratory quick sand is being produced in a large cylindrical tank, by pumping water into it from below, while the excess of water flows over the top of the tank, back in to the reservoir. How deep will a student sink into the fluidised mixture of sand and water?

14 Tangent-moduli

The difference in soil behaviour in compression and in shear suggests to separate the stresses and deformations into two parts, one describing compression, and another describing shear. This will be presented in this chapter. Dilatancy will be disregarded, at least initially.

14.1 Strain and stress

The components of the *displacement vector* will be denoted by u_x, u_y and u_z. If these displacements are not constant throughout the field there will be *deformations*, or *strains*. In Figure 14-1 the strains in the x, y-plane are shown.

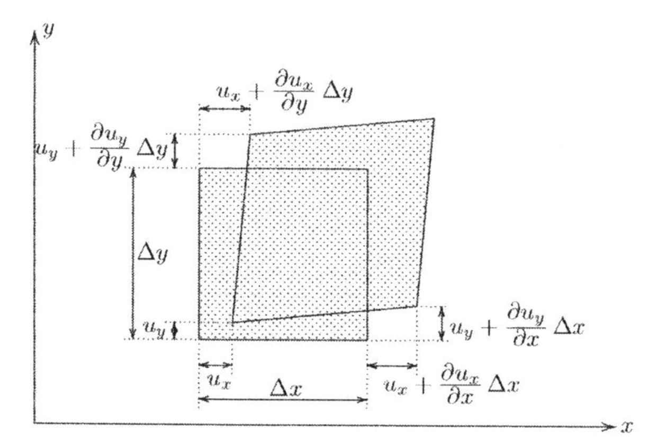

Figure 14-1. Strains.

The change of length of an element of original length Δx, divided by that original length, is the horizontal strain ε_{xx}. This strain can be expressed into the displacement difference, see Figure 14-1, by

$$\varepsilon_{xx} = -\partial u_x / \partial x.$$

The minus sign has been introduced because, opposed to continuum mechanics, settlements and volume decrease have positive signs in soil mechanics. The change of length of an element of original length Δy, divided by that original length, is the vertical strain ε_{yy}. Its definition in terms of the displacement is, see Figure 14-1,

$$\varepsilon_{yy} = -\partial u_y / \partial y.$$

Because u_x can increase in y-direction, and u_y in x-direction, the right angle in the lower left corner of the element may become somewhat smaller. One half of this decrease is denoted as the *shear strain* ε_{xy},

$$\varepsilon_{xy} = -\frac{1}{2}(\partial u_x/\partial y + \partial u_y/\partial x).$$

Similar strains may occur in the other planes, of course, with similar definitions.
In the general three-dimensional case the definitions of the strain components are

$$\begin{aligned} \varepsilon_{xx} &= -\frac{\partial u_x}{\partial x}, \quad \varepsilon_{xy} = -\frac{1}{2}(\frac{\partial u_x}{\partial y} + \frac{\partial u_y}{\partial x}), \\ \varepsilon_{yy} &= -\frac{\partial u_y}{\partial y}, \quad \varepsilon_{yz} = -\frac{1}{2}(\frac{\partial u_y}{\partial z} + \frac{\partial u_z}{\partial y}), \\ \varepsilon_{zz} &= -\frac{\partial u_z}{\partial z}, \quad \varepsilon_{zx} = -\frac{1}{2}(\frac{\partial u_z}{\partial x} + \frac{\partial u_x}{\partial z}). \end{aligned} \tag{14.1}$$

All derivatives, $\partial u_x/\partial x$, $\partial u_x/\partial y$, etc., are assumed to be small compared to 1. Then the strains are also small compared to 1. Even in soils, in which considerable deformations may occur, this is usually valid, at least as a first approximation.
The volume of an elementary small block may increase if its length increases, or it width increases, or its height increases. The total volume strain is the sum of the strains in the three coordinate directions,

$$\varepsilon_{vol} = \frac{\Delta V}{V} = \varepsilon_{xx} + \varepsilon_{yy} + \varepsilon_{zz}. \tag{14.2}$$

This volume strain describes the compression of the material. The remaining part of the strain tensor describe the distorsion. For this purpose the deviator strains are defined as

$$\begin{aligned} e_{xx} &= \varepsilon_{xx} - \frac{1}{3}\varepsilon_{vol}, \quad & e_{xy} &= \varepsilon_{xy}, \\ e_{yy} &= \varepsilon_{yy} - \frac{1}{3}\varepsilon_{vol}, \quad & e_{yz} &= \varepsilon_{yz}, \\ e_{zz} &= \varepsilon_{zz} - \frac{1}{3}\varepsilon_{vol}, \quad & e_{zx} &= \varepsilon_{zx}. \end{aligned} \tag{14.3}$$

These deviator strains do not contain any volume change, because $e_{xx} + e_{yy} + e_{zz} = 0$.
In a similar way *deviator stresses* can be defined,

$$\begin{aligned} \tau_{xx} &= \sigma_{xx} - \sigma_0, \quad & \tau_{xy} &= \sigma_{xy}, \\ \tau_{yy} &= \sigma_{yy} - \sigma_0, \quad & \tau_{yz} &= \sigma_{yz}, \\ \tau_{zz} &= \sigma_{zz} - \sigma_0, \quad & \tau_{zx} &= \sigma_{zx}. \end{aligned} \tag{14.4}$$

Here σ_0 is the *isotropic stress*,

$$\sigma_0 = \frac{1}{3}\left(\sigma_{xx} + \sigma_{yy} + \sigma_{zz}\right). \tag{14.5}$$

The isotropic stress σ_0 is the average normal stress. In an isotropic material volume changes are determined primarily by changes of the isotropic stress. This means that the volume strain ε_{vol} is a function of the isotropic stress σ_0 only.
Even though this may seem almost trivial, for soils it is in general not true, as it excludes dilatancy and contractancy. It is nevertheless assumed here, as a first approximation.

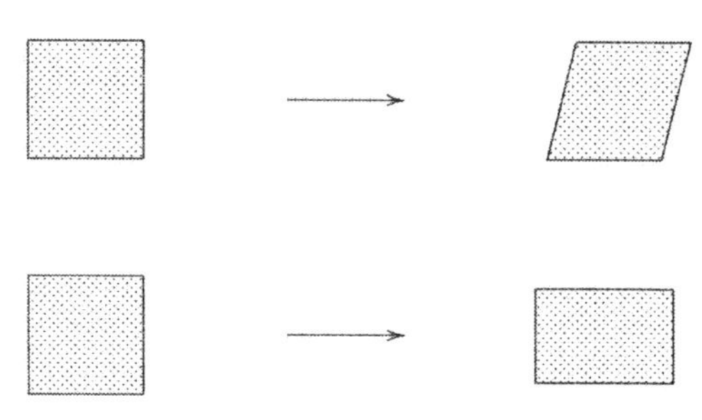

Figure 14-2. Distorsion.

The remaining part of the stress tensor, after subtraction of the isotropic stress, see (14.4), consists of the deviator stresses. These are responsible for the distorsion, i.e. changes in shape, at constant volume.
There are many forms of distorsion: shear strains in the three directions, but also a positive normal strain in one direction and a negative normal strain in a second direction, such that the volume remains constant. Some of these possibilities are shown in Figure 14-2. In the other three planes similar forms of distorsion may occur.

14.2 Linear elastic material

The simplest possible relation between stresses and strains in a deformable continuum is the linear elastic relation for an isotropic material. This can be described by two positive constants, the *compression modulus* K and the *shear modulus* G. The compression modulus K gives the relation between the volume strain and the isotropic stress,

$$\sigma_0 = K\varepsilon_{vol}. \tag{14.6}$$

The shear modulus G (perhaps distorsion modulus would be a better word) gives the relation between the deviator strains and the deviator stresses,

$$\tau_{ij} = 2Ge_{ij}. \tag{14.7}$$

Here i and j can be all combinations of x, y or z, so that, for instance, $\tau_{xx} = 2Ge_{xx}$ and $\tau_{xy} = 2Ge_{xy}$. The factor 2 appears in the equations for historical reasons.

In applied mechanics the relation between stresses and strains of an isotropic linear elastic material is usually described by *Young's modulus E*, and *Poisson's ratio* ν. The usual form of the equations for the normal strains then is

$$
\begin{aligned}
\varepsilon_{xx} &= \frac{1}{E}\left[\sigma_{xx} - \nu\left(\sigma_{yy} + \sigma_{zz}\right)\right], \\
\varepsilon_{yy} &= \frac{1}{E}\left[\sigma_{yy} - \nu\left(\sigma_{zz} + \sigma_{xx}\right)\right], \\
\varepsilon_{zz} &= \frac{1}{E}\left[\sigma_{zz} - \nu\left(\sigma_{xx} + \sigma_{yy}\right)\right].
\end{aligned}
\tag{14.8}
$$

It can easily be verified that the equations (14.8) are equivalent to (14.6) and (14.7) with

$$
K = \frac{E}{3(1-2\nu)}, \tag{14.9}
$$

$$
G = \frac{E}{2(1+\nu)}. \tag{14.10}
$$

For the description of compression and distorsion, which are so basically different in soil mechanics, the parameters K and G are more suitable than E and ν. In continuum mechanics they are sometimes preferred as well, for instance because it can be argued, on thermodynamical grounds, that they both must be positive, $K > 0$ and $G > 0$.

14.3 A non-linear material

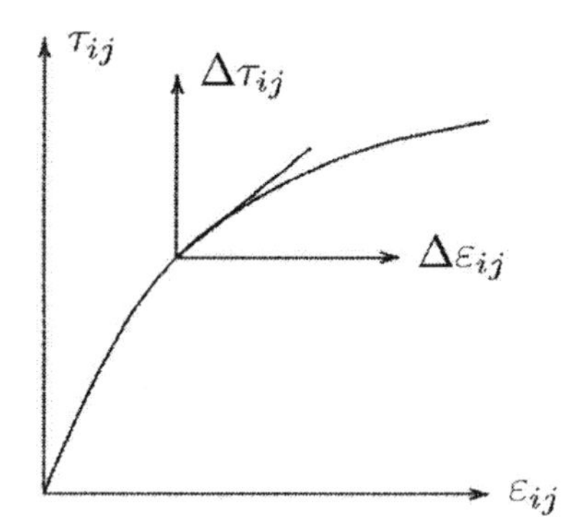

Figure 14-3. Tangent modulus.

In the previous chapter it has been argued that soils are non-linear and non-elastic. Furthermore, soils are often not isotropic, because during the formation of soil deposits it may be expected that there will be a difference between the direction of deposition (the vertical direction) and the horizontal directions. As a simplification this anisotropy will be disregarded here, and the irreversible deformations due to a difference in loading and unloading are also disregarded. The behaviour in compression and distorsion will be considered separately, but they will no longer be described by constant parameters. As a first improvement on the linear elastic model the modulus will be assumed to be dependent upon the stresses. A non-linear relation between stresses and strains is shown schematically in Figure 14-3. For a small change in stress the tangent to the curve might be used. This means that one could write, for the incremental volume change,

$$\Delta\sigma_0 = K\Delta\varepsilon_{vol}, \tag{14.11}$$

Similarly, for the incremental shear strain one could write

$$\Delta\tau_{ij} = 2G\Delta e_{ij}. \tag{14.12}$$

The parameters K and G in these equations are not constants, but they depend upon the initial stress, as expressed by the location on the curve in Figure 14-3. These type of constants are denoted as *tangent moduli*, to indicate that they actually represent the tangent to a non-linear curve. They depend upon the initial stress, and perhaps also on some other physical quantities, such as time, or temperature. As mentioned in the previous chapter, it can be expected that the value of K increases with an increasing value of the isotropic stress, see Figure 13-3. Many researchers have found, from laboratory tests, that the stiffness of soils increases approximately linear with the initial stress, although others seem to have found that the increase is not so strong, approximately proportional to the square root of the initial stress. If it is assumed that the stiffness in compression indeed increases linearly with the initial stress, it follows that the stiffness in a homogeneous soil deposit will increase about linearly with depth. This has also been confirmed by tests in the field, at least approximately.

For distorsion it can be expected that the shear modulus G will decrease if the shear stress increases. It may even tend towards zero when the shear stress reaches its maximum possible value, see Figure 13-5.

It should be emphasized that a linearisation with two tangent moduli K and G, dependent upon the initial stresses, can only be valid in case of small stress increments. That is not an impossible restriction, as in many cases the initial stresses in a soil are already relatively large, because of the weight of the material. It should also be mentioned that many effects have been disregarded, such as anisotropy, irreversible (plastic) deformations, creep and dilatancy. An elastic analysis using K and G, or E and ν, at best is a first approximate approach. It may be quite valuable, however, as it may indicate the trend of the development of stresses. In the last decades of the 20th century more advanced non-linear methods of analysis have been developed, for instance using finite element modelling, that offer more realistic computations.

Problems

14.1 A colleague in a foreign country reports that the Young's modulus of a certain layer has been back-calculated from the deformations of a stress increase due to a surcharge, from 20 kPa to 40 kPa. This modulus is given as $E = 2000$ kPa . A new surcharge is being planned, from 40 kPa to 60 kPa, and your colleague wants your advice on the value of E to be used then. What is your suggestion?

14.2 A soil sample is being tested in the laboratory by cyclic shear stresses. In each cycle there are relatively large shear strains. What do you expect for the volume change in the 100th cycle? And what would that mean for the value of Poisson's ratio ν ?

15 One-dimensional compression

In the previous chapters the deformation of soils has been separated into pure compression and pure shear. Pure compression is a change of volume in the absence of any change of shape, whereas pure shear is a change of shape, at constant volume. Ideally laboratory tests should be of constant shape or constant volume type, but that is not so simple. An ideal compression test would require isotropic loading of a sample, that should be free to deform in all directions. Although tests on spherical samples are indeed possible, it is more common to perform a compression test in which no horizontal deformation is allowed, by enclosing the sample in a rigid steel ring, and then deform the sample in vertical direction. In such a test the deformation consists mainly of a change of volume, although some change of shape also occurs. The main mode of deformation is compression, however.

15.1 Confined compression test

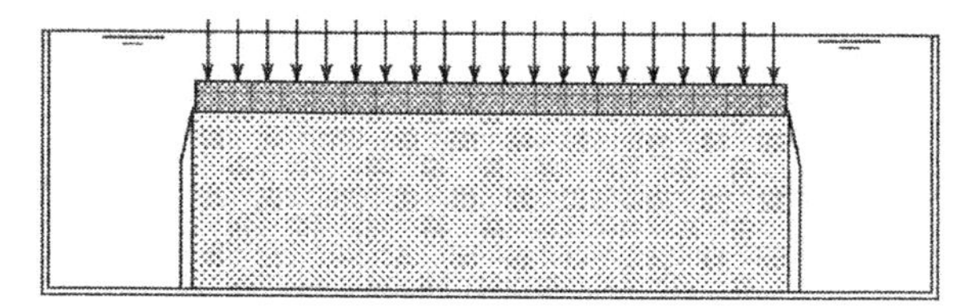

Figure 15-1. Confined compression test.

In the confined compression test, or oedometer test a cylindrical soil sample is enclosed in a very stiff steel ring, and loaded through a porous plate at the top, see Figure 15-1. The equipment is usually placed in a somewhat larger container, filled with water. Pore water may be drained from the sample through porous stones at the bottom and the top of the sample. The load is usually applied by a dead weight pressing on the top of the sample. This load can be increased in steps, by adding weights. The ring usually has a sharp edge at its top, which enables to cut the sample from a larger soil body.
In this case there can be no horizontal deformations, by the confining ring,

$$\varepsilon_{xx} = \varepsilon_{yy} = 0. \tag{15.1}$$

This means that the only non-zero strain is a vertical strain. The volume strain will be equal to that strain,

$$\varepsilon_{vol} = \varepsilon = \varepsilon_{zz}. \tag{15.2}$$

For convenience this strain will be denoted as ε. The load of the sample is a vertical stress σ_{zz}, which will be denoted as σ,

$$\sigma = \sigma_{zz}. \tag{15.3}$$

When performing the test, it is observed, as expected, that the increase of vertical stress caused by a loading from say 10 kPa to 20 kPa leads to a larger deformation than a loading from 20 kPa to 30 kPa. The sample becomes gradually stiffer, when the load increases. Often it is observed that an increase from 20 kPa to 40 kPa leads to the same incremental deformation as an increase from 10 kPa to 20 kPa. And increasing the load from 40 kPa to 80 kPa gives the same additional deformation. Each doubling of the load has about the same effect.

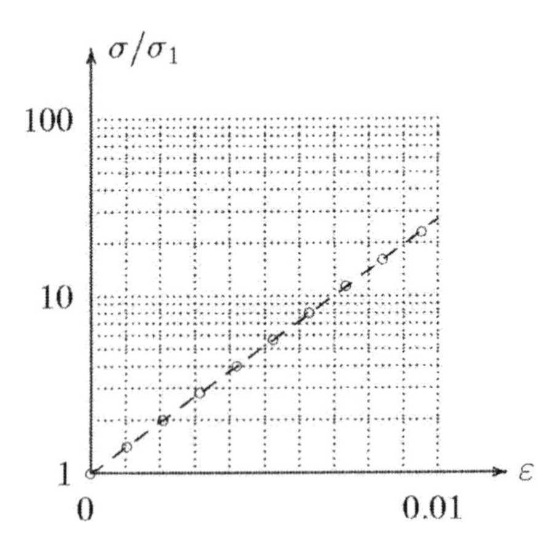

Figure 15-2. Results.

This suggests to plot the data on a semi-logarithmic scale, see Figure 15-2. In this figure $\log(\sigma/\sigma_1)$ has been plotted against ε, where σ_1 denotes the initial stress. The test results appear to form a straight line, approximately, on this scale. The logarithmic relation between vertical stress and strain has been found first by Terzaghi, around 1930.
It means that the test results can be described reasonably well by the formula

$$\varepsilon = \frac{1}{C}\ln(\frac{\sigma}{\sigma_1}). \tag{15.4}$$

Using this formula each doubling of the load, i.e. loadings following the series 1,2,4,8,16,. . . , gives the same strain. The relation (15.4) is often denoted as Terzaghi's logarithmic formula. Its approximate validity has been verified by many laboratory tests.

In engineering practice the formula is sometimes slightly modified by using the common logarithm (of base 10), rather than the natural logarithm (of base e), perhaps because of the easy availability of semi-logarithmic paper on the basis of the common logarithm. The formula then is

$$\varepsilon = \frac{1}{C_{10}} \log(\frac{\sigma}{\sigma_1}). \tag{15.5}$$

Because $\log(x) = \ln(x)/2.3$ the relation between the constants is

$$C_{10} = \frac{C}{2.3}, \tag{15.6}$$

or

$$C = 2.3 \times C_{10}. \tag{15.7}$$

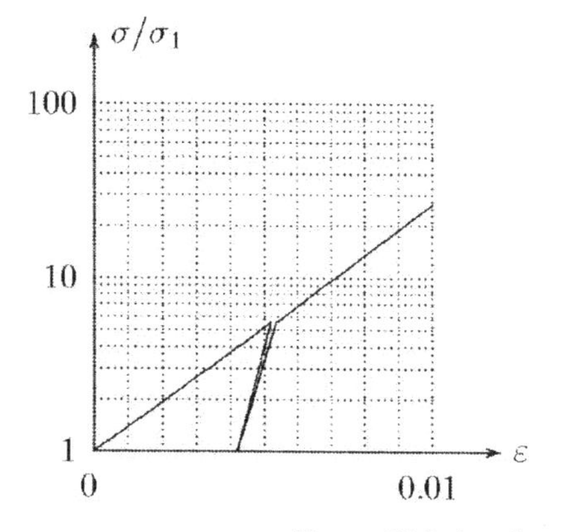

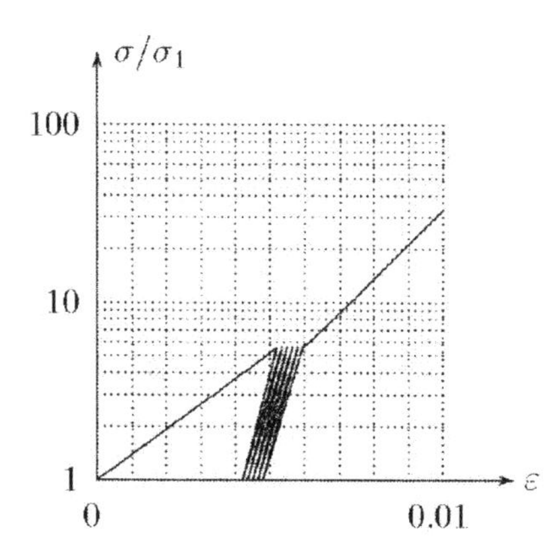

Figure 15-3. Loading, unloading and cyclic loading.

Type of soil	C	C_{10}
sand	50-100	20-200
silt	25-125	10-50
clay	10-100	4-40
peat	2-25	1-10

Table 15-1: Compression constants.

The *compression constants* C and C_{10} are dimensionless parameters. Some average values are shown in Table 15-1. The large variation in the compressibility suggests that the table has only limited value. The compression test is a simple test, however, and the constants can easily be determined for a particular soil, in the laboratory. The circumstance that there are two forms of the formula, with a factor 2.3 between the values of the constants, means that great care must be taken that the same logarithm is being used by the laboratory and the consultant or the design engineer.
The values in Table 15-1 refer to virgin loading, i.e. cases in which the load on the soil is larger than the previous maximum load. If the soil is first loaded, then

unloaded, and next is loaded again, the results, when plotted on a logarithmic scale for the stresses, are as shown in Figure 15-3. Just as in loading, a straight line is obtained during the unloading branch of the test, but the stiffness is much larger, by a factor of about 10. When a soil is unloaded or reloaded below its preconsolidation load the stress strain relation can best be described by a logarithmic formula similar to the ones presented above, but using a coefficient A rather than C, where the values of A are about a factor 10 larger than the values given in Table 15-1. Such large values can also be used in cyclic loading. A typical response curve for cyclic loading is shown in the right part of Figure 15-3. After each full cycle there will be a small permanent deformation, because of the stiff behaviour. When loading the soil beyond the *pre-consolidation pressure* the response is again much softer.
In some countries, such as the Scandinavian countries and the USA, the results of a confined compression test are described in a slightly different form, using the void ratio e to express the deformation, rather than the strain ε. The formula used is

$$e_1 - e = C_c \log(\frac{\sigma}{\sigma_1}), \tag{15.8}$$

where e_1 represents the void ratio at the initial stress σ_1. In this representation the test results also lead to a straight line, when using a logarithmic scale for the stresses. The formula indicates that the void ratio decreases when the stress increases, which corresponds to a compression of the soil. The coefficient C_c is denoted as the *compression index*. The coefficient is the opposite of a stiffness constant. A highly compressible soil will have a large value of C_c. As seen before the behaviour is much stiffer for unloading and reloading below the pre-consolidation pressure. The compression index is then much smaller (by about a factor 10). Three typical branches of the response are shown in Figure 15-4. The relationship shown in the figure is often denoted as an $e-\log(p)$ diagram, where the notation p has been used to indicate the effective stress.

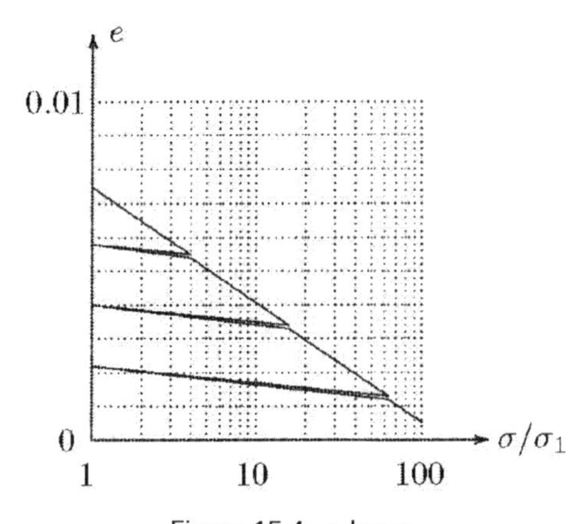

Figure 15-4. e–log p.

To demonstrate that eq. (15.8) is in agreement with the formula (15.5), given before, it may be noted that the strain ε has been defined as $\varepsilon = -\Delta V/V$, where V is the volume of the soil. This can be expressed as $V = (1+e)V_p$, where e is the void ratio, and V_p is the volume of the particles. Because the particle volume is constant (the particles are practically incompressible) it follows that $\Delta V = \Delta e V_k$, so that

$$\varepsilon = -\frac{\Delta e}{1+e}. \qquad (15.9)$$

Equation (15.8) therefore can also be written as

$$\varepsilon = \frac{C_c}{1+e}\log(\frac{\sigma}{\sigma_1}), \qquad (15.10)$$

Comparison with eq. (15.5) shows that the relation between C_c and C_{10} is

$$\frac{1}{C_{10}} = \frac{C_c}{1+e}. \qquad (15.11)$$

It is of course unfortunate that different coefficients are being used to describe the same phenomenon. This can only be explained by the historical developments in different parts of the world. It is especially inconvenient that in both formulas the constant is denoted by the character C, but in one form it appears in the numerator, and in the other one in the denominator. A large value for C_{10} corresponds to a small value for C_c. It can be expected that the compression index C_c will prevail in the future, as this has been standardized by ISO, the International Organization for Standardization.

It may also be noted that in a well known model for elasto-plastic analysis of deformations of soils, the Cam clay model, developed at Cambridge University, the compression of soils is described in yet another somewhat different form,

$$e_1 - e = \lambda \ln(\frac{\sigma}{\sigma_1}) \quad \text{or} \quad \varepsilon = \lambda^* \ln\frac{\sigma}{\sigma_1} \quad \text{with} \quad \lambda^* = \frac{\lambda}{1+e} \qquad (15.12)$$

The difference with eq. (15.8) is that a natural logarithm is used rather than the common logarithm (the difference being a factor 2.3), and that the deformation is expressed by the strain ε rather than the void ratio e. The difference between these two quantities is a factor $1+e$.

The logarithmic relations given in this chapter should not be considered as fundamental physical laws. Many non-linear phenomena in physics produce a straight line when plotted on semi-logarithmic paper, or if that does not work, on double logarithmic paper. This may lead to very useful formulas, but they need not have much fundamental meaning. The error may well be about 1 % to 5 %. It should be noted that the approximation in Terzaghi's logarithmic compression formula is of a different nature than the approximation in Newton's laws. These last are basic physical laws (even though Einstein has introduced a small correction). The logarithmic compression formula is not much more than a convenient approximation of test results.

15.2 Elastic analysis

In a confined compression test on a sample of an isotropic linear elastic material, the lateral stresses are, using (14.8), and noting that $\varepsilon_{xx} = \varepsilon_{yy} = 0$,

$$\sigma_{xx} = \sigma_{yy} = K\sigma_{zz} \quad \text{with} \quad K = \frac{\nu}{1-\nu}. \tag{15.13}$$

From the last equation of the system (14.8) it now follows that

$$\varepsilon_{zz} = \frac{1}{E_{oed}}\sigma_{zz} \quad \text{with} \quad E_{oed} = \frac{(1-\nu)}{(1+\nu)(1-2\nu)}E. \tag{15.14}$$

When expressed into the constants K and G this can also be written as

$$\sigma_{zz} = (K + \frac{4}{3}G)\varepsilon_{zz}. \tag{15.15}$$

The elastic coefficient for one-dimensional confined compression appears to be $K + \frac{4}{3}G$. This is sometimes denoted as E_{oed} or D, the *constrained modulus*,

$$E_{oed} = K + \frac{4}{3}G = \frac{(1-\nu)}{(1+\nu)(1-2\nu)}E = 3K(\frac{1-\nu}{1+\nu}). \tag{15.16}$$

When $\nu = 0$ it follows that $E_{oed} = E$; if $\nu > 0 : E_{oed} > E$. In the extreme case that $\nu = \frac{1}{2}$ the value of $E_{oed} \to \infty$. Such a material is indeed incompressible.
Similar to the considerations in the previous chapter on tangent moduli the logarithmic relationship (15.4) may be approximated for small stress increments. The relation can be linearised by differentiation. This gives

$$\frac{d\varepsilon}{d\sigma} = \frac{1}{C\sigma}. \tag{15.17}$$

So that

$$\Delta\sigma = C\sigma\Delta\varepsilon. \tag{15.18}$$

Comparing eqs. (15.15) and (15.18) it follows that for small incremental stresses and strains one write, approximately,

$$E_{oed} = K + \frac{4}{3}G = C\sigma. \tag{15.19}$$

This means that the stiffness increases linearly with the stress, and that is in agreement with many test results (and with earlier remarks). The formula (15.19) is of considerable value to estimate the elastic modulus of a soil. Many computational methods use the concepts and equations of elasticity theory, even when it is acknowledged that soil is not a linear elastic material. On the basis of eq. (15.19) it is possible to estimate an elastic "constant". For sand it is often estimated that

$$\Delta\sigma = E_{oed}\Delta\varepsilon \quad \text{with} \quad E_{oed} \approx 250\sigma'. \tag{15.20}$$

For a layer of sand at 10 m depth, for instance, it can be estimated that the effective stress will be about 100 kPa (assuming that groundwater table is very high). This means that the elastic modulus E_{oed} is about 25000 kPa = 25 MPa. This is a useful first estimate of the elastic modulus for virgin loading. As stated before, the soil will be about a factor 10 stiffer for cyclic loading. This means that for problems of wave propagation the elastic modulus to be used may be about 250 MPa. It should be noted that these are only first estimates. The true values may be larger or smaller by a factor 2. And nothing can beat measuring the stiffness in a laboratory test or a field test, of course.

Problems

15.1 In a confined compression test a soil sample of 2 cm thickness has been preloaded by a stress of 100 kPa. An additional load of 20 kPa leads to a vertical displacement of 0.030 mm. Determine the value of the compression constant C_{10}.

15.2 If the test of the previous problem is continued with a next loading step of 20 kPa, what will then be the displacement in that step? What should be the additional load to again cause a displacement of 0.030 mm?

15.3 A clay layer of 4 m thickness is located below a sand layer of 10 m thickness. The volumetric weights are all 20 kN/m^3, and the groundwater table coincides with the soil surface. The compression constant of the clay is $C_{10} = 20$. Predict the settlement of the soil by compression of the clay layer due to an additional load of 40 kPa.

15.4 A sand layer is located below a road construction of weight 20 kPa. The sand has been densified by vibration before the road was built. Estimate the order of magnitude of the elastic modulus of the soil that can be used for the analysis of traffic vibrations in the soil.

15.5 The book Soil Mechanics by Lambe & Whitman (Wiley, 1968) gives the value $C_c = 0.47$ for a certain clay. The void ratio is about 0.95. Estimate C_{10}, and verify whether this value is in agreement with Table 15-1.

16 Consolidation

In the previous chapters it has been assumed that the deformation of a soil is uniquely determined by the stress. This means that a time dependent response has been excluded. In reality the behaviour is strongly dependent on time, however, especially for clay soils. This can be creep, but in a saturated soil the deformations can also be retarded by the time that it takes for the water to flow out of the soil. In compression of a soil the porosity decreases, and as a result there is less space available for the pore water. This pore water may be expelled from the soil, but in clays this may take a certain time, due to the small permeability. The process is called consolidation. Its basic equations are considered in this chapter. The analysis will be restricted to one-dimensional deformation, assuming that the soil does not deform in lateral direction. It is also assumed that the water can only flow in vertical direction. This will be the case during an oedometer test, or in the field, in case of a surcharge load over a large area, see Figure 16-1.

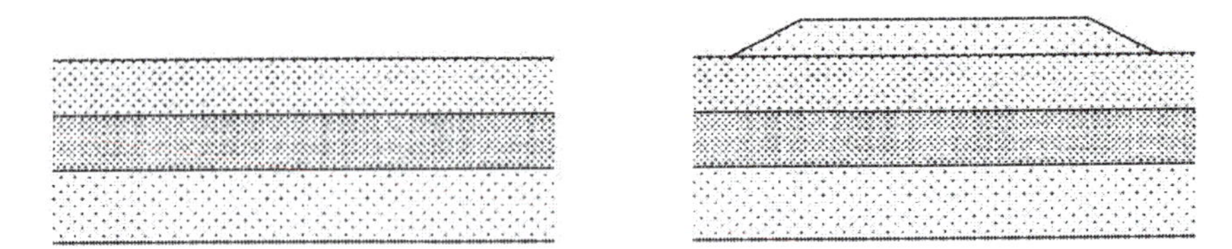

Figure 16-1.Uniform load.

16.1 Differential equation

To simplify the analysis it will be assumed that the change in stress is small compared to the initial stress. In that case the stress-strain relation may be linearised, using an elastic coefficient $E_{oed} = K + \frac{4}{3}G$, see (15.19). The precise value of that coefficient depends upon the initial stress. The relation between the increment of effective stress $\Delta\sigma'$ and the increment of strain $\Delta\varepsilon$ can now be written as

$$\Delta\sigma' = E_{oed}\Delta\varepsilon \quad \text{with} \quad E_{oed} = (K + \frac{4}{3}G). \tag{16.1}$$

In the remainder of this chapter the notation Δ will be omitted. Thus the increment of the effective stress will be denoted simply by σ', and the increment of the strain by ε,

$$\sigma' = E_{oed}\varepsilon. \tag{16.2}$$

Using stresses and strains with respect to some initial state is very common in soil mechanics. For the strains there is actually no other possibility. Strains can only be measured with respect to some initial state, and in this initial state the soil is not stress free. Gravity is always acting, and the stresses due to gravity have been developed gradually during geological history. The logical procedure is to regard the state of stress including the influence of the weight of the soil layers as a given initial state, and to regard all effects of engineering activity with respect to that initial state. It should be noted that to obtain the true stresses in the field the initial stresses should be added to the incremental stresses.

In the analysis of consolidation it is also customary to write equation (16.2) in its inverse form,

$$\varepsilon = m_v \sigma' \quad \text{with} \quad m_v = \frac{1}{E_{oed}}, \tag{16.3}$$

where m_v is denoted as the *compressibility coefficient*. If the incremental vertical total stress is denoted by σ, and the incremental pore pressure by p, then Terzaghi's principle of effective stress is

$$\sigma' = \sigma - p. \tag{16.4}$$

It follows from (16.3) that

$$\varepsilon = m_v(\sigma - p). \tag{16.5}$$

The total stress σ is often known, as a function of time. Its value is determined by the load. Let it be assumed that initially $\sigma = 0$, indicating no additional load. During the application of the load the total stress σ is supposed to be increased by a given amount, in a very short time interval, after which the total stress remains constant. The pore pressure may vary during that period. To describe its generation and dissipation the continuity of the water must be considered.

Consider an elementary volume V in the soil, see Figure 16-2. The volume of water is $V_w = nV$, where n is the porosity. The remaining volume, $V_p = (1-n)V$ is the total volume of the particles. As usual, the particles are considered as incompressible. This means that the
volume V can change only if the porosity changes. This is possible only if the water in the pores is compressed, or if water flows out of the element.

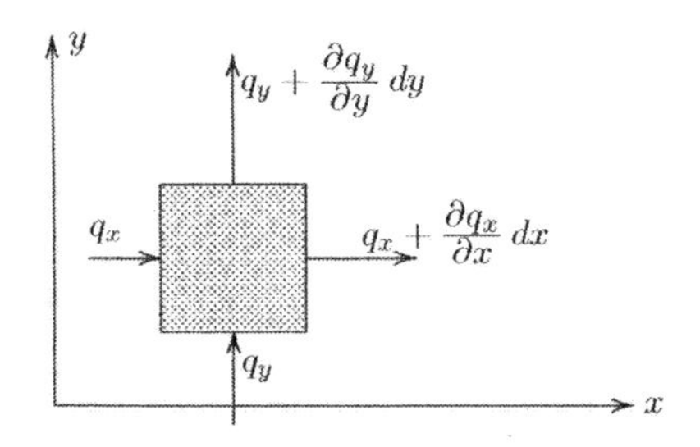

Figure 16-2.Outflow.

The first possibility, a volume change by compression of the pore water, can be caused by a change of the pore pressure p. It can be expected that the change of volume is proportional to the change of the pressure, and to the original volume, i.e.

$$\Delta V_1 = \beta V_w \Delta p = \beta n V \Delta p, \tag{16.6}$$

Where β represents the ***compressibility*** of the water. For pure water handbooks of physics give , which is very small. Water is practically incompressible. However, when the water contains some small bubbles of gas (air or natural gas), the value of β may be much larger, approximately

$$\beta = S\beta_0 + \frac{(1-S)}{p_0}, \tag{16.7}$$

Where β_0 is the compressibility of pure water, S is the degree of saturation, and p_o is the absolute pressure in the water, considered with respect to vacuum (this means that under atmospheric conditions $p_0 = 100 \text{ kPa}$). If $S = 0.99$ and the pressure is $p_0 = 100 \text{ kPa}$, then $\beta = 10^{-7} \text{ m}^2/\text{N}$.

That is still a small value, but about 200 times larger than the compressibility of pure water. The apparent compressibility of the water is now caused by the compression of the small air bubbles. The formula (16.7) can be derived on the basis of Boyle's gas law. Taking into account the compressibility of the fluid, even though the effect is small, makes the analysis more generally applicable.

The second possibility of a volume change, as a result of a net outflow of water, is described by the divergence of the specific discharge, see Figure 16-2. There is a net loss of water when the outflow from the element is larger than the inflow into it. In a small time Δt the volume change is

$$\Delta V_2 = (\nabla \cdot \vec{q}) V \Delta t = (\frac{\partial q_x}{\partial x} + \frac{\partial q_y}{\partial y} + \frac{\partial q_z}{\partial z}) V \Delta t. \tag{16.8}$$

The positive value of $\nabla \cdot \vec{q}$ indicates that there is a net outflow, which means that the volume will decrease. The volume increase ΔV_2 then is negative.

The total volume change in a small time Δt now is

$$\Delta \varepsilon_{vol} = \frac{\Delta V}{V} = \frac{\Delta V_1 + \Delta V_2}{V} = n\beta \Delta p + (\frac{\partial q_x}{\partial x} + \frac{\partial q_y}{\partial y} + \frac{\partial q_z}{\partial z}) \Delta t. \tag{16.9}$$

After division by Δt, and passing into the limit $\Delta t \to 0$, the resulting equation is

$$\frac{\partial \varepsilon_{vol}}{\partial t} = n\beta \frac{\partial p}{\partial t} + (\frac{\partial q_x}{\partial x} + \frac{\partial q_y}{\partial y} + \frac{\partial q_z}{\partial z}). \tag{16.10}$$

This is an important basic equation of the theory of consolidation, the *storage equation.* It expresses that a volume change ($\partial e/\partial t$) can be caused by either a pressure change (the factor n indicating how much water is present, and the factor β indicating its compressibility), or by a net outflow of water from the pores.

In the one-dimensional case of vertical flow only, the storage equation reduces to

$$\frac{\partial \varepsilon_{vol}}{\partial t} = n\beta \frac{\partial p}{\partial t} + \frac{\partial q_z}{\partial z}. \tag{16.11}$$

The value of the specific discharge q_z depends upon the pressure gradient, through Darcy's law,

$$q_z = -\frac{k}{\gamma_w}\frac{\partial p}{\partial z}. \tag{16.12}$$

It should be noted that it is not necessary to take into account a term for the pressure gradient due to gravity, because p indicates the increment with respect to the initial state, in which gravity is taken into account.
It follows from (16.11) and (16.12), assuming that the hydraulic conductivity k is constant,

$$\frac{\partial \varepsilon_{vol}}{\partial t} = n\beta \frac{\partial p}{\partial t} - \frac{k}{\gamma_w}\frac{\partial^2 p}{\partial z^2}. \tag{16.13}$$

This equation contains two variables, the volume strain ε_{vol} and the fluid pressure p. Another equation is needed for a full description of the problem. This second equation is provided by the relation of the deformation of the soil to the stresses.

In the one-dimensional case considered here the lateral strains are zero, so that the volume strain ε_{vol} is equal to the vertical strain ε,

$$\varepsilon_{vol} = \varepsilon. \tag{16.14}$$

It now follows from eq. (16.5), (16.13) and (16.14), if it is assumed that the compressibility m_v is constant in time,

$$\frac{\partial p}{\partial t} = \frac{m_v}{m_v + n\beta}\frac{\partial \sigma}{\partial t} + c_v \frac{\partial^2 p}{\partial z^2}, \tag{16.15}$$

where c_v is the consolidation coefficient,

$$c_v = \frac{k}{\gamma_w(m_v + n\beta)}. \tag{16.16}$$

Equation (16.15) is the basic differential equation for the one-dimensional consolidation process. From this equation the pore pressure p must be determined.
The simplest type of loading occurs when the total stress σ is constant during the entire process. This will be the case if the load does not change after its initial application. Then

$$\frac{\partial p}{\partial t} = c_v \frac{\partial^2 p}{\partial z^2}, \tag{16.17}$$

In mathematical physics an equation of this type is denoted as a ***diffusion equation.*** The same equation describes the process of heating or cooling of a strip of metal. The variable then is the temperature.
It may be noted that the differential equation does not become simpler when the water is assumed to be incompressible ($\beta = 0$). Only the coefficient c_v is affected. The compressibility of the water does not complicate the mathematics.

16.2 Boundary conditions and initial condition

To complete the formulation of the problem, the boundary conditions and initial conditions must be added to the differential equation (16.17). In the case of an oedometer test, see Figure 16-3, the sample is usually drained at the top, using a thin sheet of filter paper and a steel porous plate, or a porous stone. In the container in which the sample and its surrounding ring are placed, the water level is kept constant. This means that at the top of the sample the excess pore pressure is zero,

$$z = h \quad : \qquad p = 0. \tag{16.18}$$

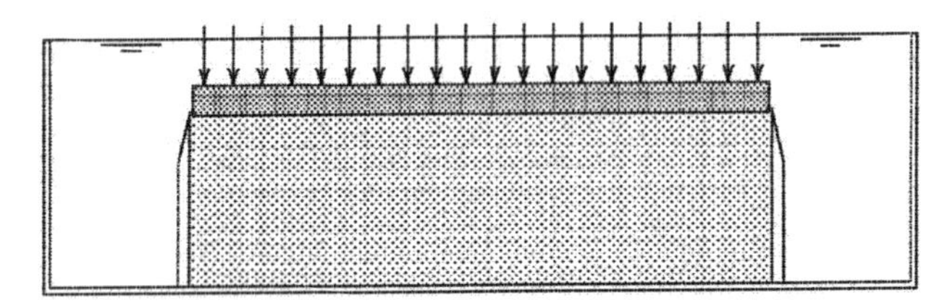

Figure 16-3.Oedometer test.

The soil sample may also be drained at its bottom, but alternatively, it may be supported by an impermeable plate. In that case the boundary condition at the bottom of the sample is

$$z = 0 \quad : \qquad \frac{\partial p}{\partial z} = 0, \tag{16.19}$$

indicating no outflow at the bottom of the sample. These two boundary conditions are physically sufficient. In general a second order differential equation requires two boundary conditions.
The initial condition is determined by the way of loading. A common testing procedure is that a load is applied in a very short time (by placing a weight on the loading plate). After this loading the load is kept constant. At the time of loading an immediate increase of the pore pressure is generated, that can be determined in the following way. The storage equation (16.10) is integrated over a short time interval Δt, giving

$$\varepsilon_{vol} = n\beta p + \int_0^{\Delta t} \frac{\partial q_z}{\partial z}\, dt. \tag{16.20}$$

The integral represents the amount of water that has flowed out of the soil in the time interval Δt. If $\Delta t \rightarrow 0$ this must be zero, so that

$$t=0 \quad : \qquad \varepsilon_{vol} = n\beta p. \tag{16.21}$$

On the other hand, it follows from (16.5), taking into account that in this case $\varepsilon_{vol} = \varepsilon$,

$$\varepsilon_{vol} = m_v(\sigma - p). \tag{16.22}$$

From equations (16.20) and (16.21) it now follows that

$$t=0 \quad : \qquad p = \frac{\sigma}{1+n\beta/m_v}. \tag{16.23}$$

This is the initial condition. It means that at the time of loading, $t=0$, the pore water pressure p is given.
If the water is considered as completely incompressible (that is a reasonable assumption when the soil is completely saturated with water) Eq. (16.22) reduces to

$$t=0, \beta=0 \quad : \qquad p=\sigma. \tag{16.24}$$

In that case the initial pore pressure equals the given load. That can be understood by noting that in case of an incompressible pore fluid there can be no immediate volume change. This means that there can be no vertical strain, as the volume change equals the vertical strain in this case of a sample that is laterally confined by the stiff steel ring. Hence there can be no vertical strain at the moment of loading, and therefore the effective stress can not increase at that instant. In this case, of lateral confinement and incompressible water, the entire load is initially carried by the water in the pores. It should be noted that throughout this chapter the deformation and the flow are one-dimensional. In a more general three-dimensional case, there may be lateral deformations, and an immediate deformation is very well possible, although the volume must remain constant if the fluid is incompressible. There can then be an immediate change of the effective stresses. The water will then carry only part of the load. The three-dimensional theory of consolidation is an interesting topic for further study.

17 Analytical solution

In this chapter an analytical solution of the one-dimensional consolidation problem is given. In soil mechanics this solution was first given by Terzaghi, in 1923. In mathematics the solution had been known since the beginning of the 19th century. Fourier developed the solution to determine the heating and cooling of a metal strip, which is governed by the same differential equation.

17.1 The problem

The mathematical problem of one-dimensional consolidation has been established in the previous chapter. The differential equation is

$$\frac{\partial p}{\partial t} = c_v \frac{\partial^2 p}{\partial z^2}, \tag{17.1}$$

with the initial condition

$$t = 0 \quad : \qquad p = p_0 = \frac{q}{1 + n\beta/m_v}, \tag{17.2}$$

in which q the load applied at time $t = 0$. It is assumed that the load remains constant for $t > 0$.

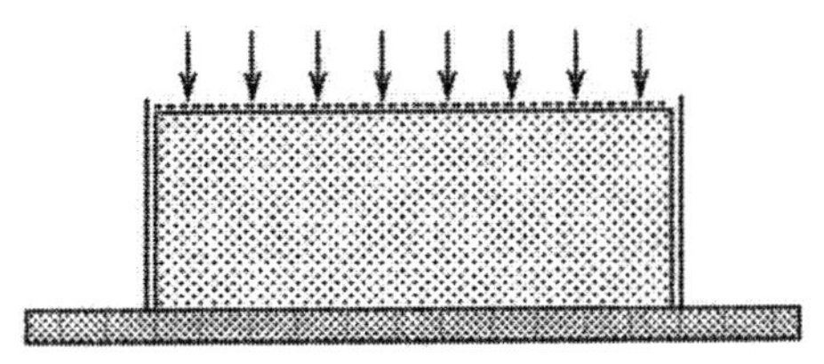

Figure 17-1.Consolidation.

The boundary conditions are, for the case of a sample of height h, drained at its top and impermeable at the bottom,

$$z = 0 \quad : \qquad \frac{\partial p}{\partial z} = 0, \tag{17.3}$$

$$z = h \quad : \qquad p = 0. \tag{17.4}$$

These equations describe the consolidation of a soil sample in an oedometer test, or a confined compression test, with a constant load, and drained only at the top of the sample. The equations also apply to a sample of thickness $2h$, drained both at its top and bottom ends. The top half of such a sample drains to the upper boundary, and the lower half drains to the lower boundary. The centre line acts as an impermeable boundary. The same problem occurs in case of a layer of clay between two very permeable layers, when the soil is loaded, in a very short time and over a very large area, by a constant load. If the area is very large it can be assumed that there will be no lateral deformations, and vertical flow only. The load can be a surcharge by an additional sand layer, applied in a very short time.

17.2 The solution

The problem defined by the equations (17.1)-(17.4) can be solved, for instance, by separation of variables, or, even better, by the Laplace transform method. This last method will be used here, without giving the details.
The Laplace transform $\overline{p}$ of the pressure p is defined as

$$\overline{p} = \int_0^\infty \exp(-st)\,dt. \tag{17.5}$$

The basic principle of the Laplace transform method is that the differential equation (17.1) is multiplied by $\exp(-st)dt$, and then integrated from $t=0$ to $t=\infty$. This gives, using partial integration and the initial condition (17.2),

$$s\overline{p} - p_0 = c_v \frac{d^2\overline{p}}{dz^2}. \tag{17.6}$$

The partial differential equation (17.1) has now been transformed into an ordinary differential equation. Its solution is

$$\overline{p} = \frac{p_0}{s} + A\exp(z\sqrt{s/c_v}) + B\exp(-z\sqrt{s/c_v}). \tag{17.7}$$

Here A and B are integration constants, that do not depend upon z, but may depend upon the transform parameter s. These constants may be determined from the boundary conditions (17.3) and (17.4),

$$A = -\frac{p_0}{2s\cosh(h\sqrt{s/c_v})}, \tag{17.8}$$

$$B = -\frac{p_0}{2s\cosh(h\sqrt{s/c_v})}. \tag{17.9}$$

The transform of the pore pressure now is

$$\frac{\overline{p}}{p_0} = \frac{1}{s} - \frac{\cosh(z\sqrt{s/c_v})}{s\cosh(h\sqrt{s/c_v})}. \tag{17.10}$$

The remaining problem now is the inverse transformation of the expression (17.10). This is a mathematical problem, that requires some experience with the Laplace transform method, including the inversion theorem. Without giving any details, it is postulated here that the final result is

$$\frac{\bar{p}}{p_0} = \frac{4}{\pi}\sum_{j=1}^{\infty}\frac{(-1)^{j-1}}{2j-1}\cos[(2j-1)\frac{\pi}{2}\frac{z}{h}]\exp[-(2j-1)^2\frac{\pi^2}{4}\frac{c_v t}{h^2}]. \tag{17.11}$$

This is the analytical solution of the problem, see Figure 17-2. At a first glance the solution (17.11) may not seem to give much insight, but after some closer inspection many properties of the solution can be obtained from it. It is for instance easy to see that for $z=h$ the pressure $p=0$, which shows that the solution satisfies the boundary condition (16.4). The cosine of each term of the series (17.11) is zero if $z=h$, because $\cos(\pi/2)=0$, $\cos(3\pi/2)=0$, $\cos(5\pi/2)=0$, etc. It can also be verified easily that the solution (17.11) satisfies the differential equation (17.1), because each individual term satisfies that equation. That the boundary condition (17.3) is satisfied can most easily be checked by noting that after differentiation with respect to z each term will contain a factor $\sin(...z)$, and these are all zero if $z=0$. To check the initial condition is not so easy, because for $t=0$ the series converges rather slowly. The verification can best be performed by writing a simple computer program, and then calculating the values for $t=0$.

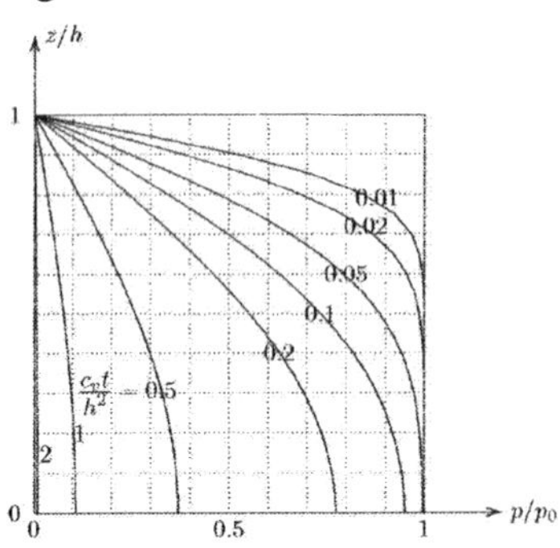

Figure 17-2. Analytical solution.

A good impression of the solution can be obtained by investigating its behaviour for large values of time. Because the exponential functions contain a factor $(2j-1)^2$, i.e. factors 1,9,16,..., all later terms can be disregarded if the first term is small. This means that for large values of time the series can be approximated by its first term,

$$\frac{c_v t}{h^2} \gg 0.1 \quad : \quad \frac{\bar{p}}{p_0} \approx \frac{4}{\pi}\cos(\frac{\pi}{2}\frac{z}{h})\exp(-\frac{\pi^2}{4}\frac{c_v t}{h^2}). \tag{17.12}$$

After a sufficiently long time only one term of the series remains, which is a cosine function in z-direction. Its values tend to zero if $t \to \infty$. The approximation (17.12) can be used if time t is not too small. In practice, it can already be used if $c_v t/h^2 > 0.2$.

The pore pressures are shown in Figure 17-2 as a function of z/h and the dimensionless time parameter $c_v t/h^2$. The values for this figure have been calculated by a simple computer program, in BASIC, see Program 17-1. The program gives the values of the pore water pressure as a function of depth, for a certain value of time. In the program the terms of the infinite series are taken into account until the argument of the exponential function reaches the value 20. This is based upon the notion that al terms containing a factor $\exp(-20)$, or smaller, can be disregarded.

```
100 CLS:PRINT "One-dimensional Consolidation"
110 PRINT "Analytical solution":PRINT
120 INPUT "Thickness of layer ..............  ";H
130 INPUT "Consolidation coefficient .......  ";C
140 INPUT "Number of subdivisions ..........  ";N
150 INPUT "Value of time ...................  ";T
160 PRINT:TT=C*T/(H*H):PI=4*ATN(1):A=4/PI:PP=PI*PI/4
170 FOR K=0 TO N:Z=K/N:P=0:C=-1:J=0
180 J=J+1:C=-C:JJ=2*J-1:JT=JJ*JJ*PP*TT
190 P=P+(A*C/JJ)*COS(JJ*PI*Z/2)*EXP(-JT)
200 IF JT<20 THEN GOTO 180
210 PRINT " z/h = ";Z;" - p/po = ";P
220 NEXT K:END
```

Program 17-1: Analytical solution for one-dimensional consolidation.

17.3 The deformation

Once that the pore pressures are known, the deformations can easily be calculated. The vertical strain is given by

$$\varepsilon = m_v(\sigma - p). \tag{17.13}$$

This means that the total deformation of the sample is

$$\Delta h = \int_0^h \varepsilon\, dz = m_v hq - m_v \int_0^h p\, dz. \tag{17.14}$$

The first term on the right hand side is the final deformation, which will be reached when all pore pressures have been reduced to zero. That value will be denoted by Δh_∞,

$$\Delta h_\infty = m_v hq. \tag{17.15}$$

Immediately after the application of the load the pore pressure $p = p_0$, see Eq. (17.2) . The deformation then is, with (17.14),

$$\Delta h_0 = m_v h q \frac{n\beta/m_v}{1+n\beta/m_v}. \tag{17.16}$$

If the water is incompressible ($\beta = 0$), this is zero, as expected. The expressions (17.15) and (17.16) are negative if $q > 0$, which indicates that the sample will become shorter when loaded.

To describe the deformation as a function of time, a useful quantity is the *degree of consolidation*, defined as

$$U = \frac{\Delta h - \Delta h_0}{\Delta h_\infty - \Delta h_0}. \tag{17.17}$$

This is a dimensionless quantity, varying between 0 (for $t = 0$) and 1 (for $t \to \infty$). The degree of consolidation indicates how far the consolidation process has been progressed.

With Eq. (17.14), (17.15) and (17.16) one obtains

$$U = \frac{1}{h}\int_0^h \frac{p_0 - p}{p_0} dz. \tag{17.18}$$

And with (17.11) this gives

$$U = 1 - \frac{8}{\pi^2}\sum_{j=1}^{\infty}\frac{1}{(2j-1)^2}\exp[-(2j-1)^2\frac{\pi^2}{4}\frac{c_v t}{h^2}]. \tag{17.19}$$

For $t \to \infty$ this is indeed equal to 1. The value $U = 0$ for $t = 0$ can be verified from the series

$$\sum_{j=1}^{\infty}\frac{1}{(2j-1)^2} = 1 + \frac{1}{3^2} + \frac{1}{5^2} + \frac{1}{7^2} + \frac{1}{9^2} + \ldots = \frac{\pi^2}{8}. \tag{17.20}$$

The degree of consolidation, which is a function of the dimensionless time parameter $c_v t/h^2$ only, is shown in Figure 17-3. The data have been calculated by the Program 17-2. The program also gives an approximate value (U'), see the next section. Theoretically speaking the consolidation process takes infinitely long to be completed. For engineering practice, however, it is sufficient if the first (and largest) term in Eq. (17.19), the infinite series, is about 0.01. Then 99% of the final deformation has been reached. It can be seen that this is the case if $c_v t/h^2 = 1.784$, or roughly speaking $c_v t/h^2 = 2$. This means that

$$t_{99\%} = \frac{2h^2}{c_v} = \frac{2h^2(m_v + n\beta)\gamma_w}{k}. \tag{17.21}$$

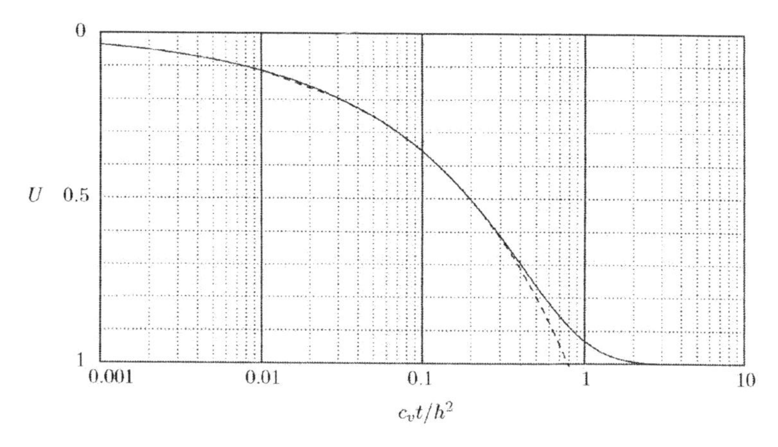

Figure 17-3. Degree of consolidation.

```
100 CLS:PRINT "One-dimensional Consolidation"
110 PRINT "Consolidation ratio":PRINT
120 INPUT "Thickness of layer .............. ";H
130 INPUT "Consolidation coefficient ....... ";C
140 INPUT "Value of time ................... ";T
160 PRINT:TT=C*T/(H*H):PI=4*ATN(1):PP=PI*PI/4
170 A=8/(PI*PI):J=0:U=1
180 J=J+1:JJ=2*J-1:JT=JJ*JJ*PP*TT
190 U=U-A*EXP(-JT)/(JJ*JJ):IF JT<20 THEN GOTO 180
200 PRINT " c*t/(h*h) = ";TT;" U = ";U;
210 PRINT " U' = ";2*SQR(T/PI):END
```

Program 17-2: Degree of consolidation.

This very useful formula is a summary of the process of consolidation. Because the coefficient of consolidation c_v is the quotient of the permeability k and the compressibility m_v, it can be seen from Eq. (17.21) that the consolidation process takes longer if the permeability is smaller, or if the compressibility is larger. This is understandable if one realizes that the consolidation process consists of compression of the soil, retarded by the outflow of water. If the permeability is smaller the outflow is slower, and the consolidation therefore takes longer. And if the compressibility is large much water must be expelled, and that takes a long time.

For engineering practice it is also very important that the time t appears in the formula (17.19) in the combination $c_v t/h^2$. This means that the process will take 4 times as long if the layer is a factor 2 thicker. It also means that if in a laboratory test on a sample of 2 cm thickness, the consolidation process has been found to be finished after 1 hour (this can best be measured by measuring the pore pressures, and then waiting until it is practically zero), the consolidation of that soil in the field for a layer of 2 m thickness, will take 10,000 times as long, that is more than 1 year.

Another important consequence of the fact that the consolidation process is governed by the factor $c_v t/h^2$ is that the duration of the consolidation process can be shortened considerably by reducing the drainage length h. As an example one may consider the consolidation process of a clay layer of 10 m thickness. Suppose that the permeability k is about 10^{-9} m/s. Let it furthermore be expected that the final deformation of the clay layer by a load of 50 kPa (the weight of 3 m dry sand) is 20 cm. This means that the value of the compressibility m_v is, with Eq. (16.3): $m_v = 0.0004\ \text{m}^2/\text{kN}$. The coefficient of consolidation then is, with (16.16), $c_v = 0.25\times 10^{-6}\ \text{m}^2/\text{s}$. The consolidation time is, with (17.21), $t_{99\,\%} = 2\times 10^8$ s. That is about 6 years, which means that it will take many years before the deformation reaches its final value of 20 cm. To speed up the consolidation process a large number of vertical drains may be installed, in the form of plastic filter material. If these drains are installed in a pattern with mutual distances of about 1.60 m, the drainage length becomes about a factor 6 smaller (0.80 m rather than 5 m). If it is assumed that the horizontal permeability is equal to the vertical permeability, the duration of the consolidation process will be a factor 36 shorter, that is about 2 months. For a new road, or a new town extension this means that the settlements are concentrated in a much shorter time span.

17.4 Approximate formulas

If the time parameter $c_v t/h^2$ is very small, many terms are needed in the analytical solutions to obtain accurate results. That may not be a great disadvantage if the computations are performed by a computer program, but it does not give much insight into the solution. A more convenient approximation can be obtained using a theorem from Laplace transform theory saying that an approximation for small values of t can be obtained by assuming the value of s in the transformed solution as very large. Again, the details are omitted here. The result for the degree of consolidation is found to be

$$U \approx \frac{2}{\sqrt{\pi}}\sqrt{\frac{c_v t}{h^2}} \quad \text{if} \quad U < 0.5. \tag{17.22}$$

It appears that in the beginning of the consolidation process its advance increases with the square root of time. This property can be used with some advantage later. The computer Program 17-2 computes the approximate value as well, and in Figure 17-3 the approximate values are represented by a dotted line. The approximation appears to be very good, until values of about 70 %.

The approximate formula (17.22) also enables to estimate how short the loading time of a load must be to be considered as instantaneous. It can be seen that only 1 % of the consolidation process has been competed if $c_v t/h^2 = 10^{-4}\pi/4$, or about $t = t_{1\,\%}$, with

$$t_{1\%} = 10^{-4} \frac{h^2}{c_v}. \tag{17.23}$$

A load that is applied faster than this value of time can be considered as an instantaneous load.
For the second half of the consolidation process another good approximate formula exists,

$$U \approx 1 - \frac{8}{\sqrt{\pi^2}} \exp\left(-\frac{\pi^2}{4} \frac{c_v t}{h^2}\right) \quad \text{if} \quad U > 0.5. \tag{17.24}$$

Problems

17.1 A clay sample of 2 cm thickness is being tested in an oedometer. The sample is drained on both sides. The coefficient of consolidation is $c_v = 10^{-7}$ m²/s. At a certain moment of time the sample is loaded. Calculate the time for the pore water pressure in the centre of the sample to be reduced to 50 % of its initial value.

17.2 What would be the answer to the previous problem if the sample were drained on one side only?

17.3 In a test on a clay sample of 2 cm thickness it has been measured that after 15 minutes the pore pressures are practically zero. What will be the duration of the consolidation process for a layer of the same clay, of 5 m thickness?

17.4 In a laboratory test on a clay sample it has been omitted to measure the deformation immediately after the application of the load. The measurement after 1 minute was a deformation of 0.06 mm, and after 4 minutes a deformation of 0.08 mm. Estimate the initial deformation.

17.5 Determine the relative error in the approximation (17.12) for $c_v t/h^2 = 0.2$, by calculating the second term in the series, for $z = 0$.

17.6 The computer programs in this chapter can not be used if $t = 0$, because then the loop will continue forever. The series solutions do converge, however. Formulate a better criterion for terminating the series, and install this improvement in the programs.

17.7 Extend the computer programs of this chapter with facilities for graphical output, or output on a printer.

18 Numerical solution

The dissipation of the pore water pressures during the consolidation process can be calculated very simply by a numerical solution procedure, using the finite difference method. This is presented in this chapter. The technique is kept as simple as possible.

18.1 Finite differences

The differential equation for one-dimensional consolidation is equation **(16.17)**,

$$\frac{\partial p}{\partial t} = c_v \frac{\partial^2 p}{\partial z^2}. \tag{18.1}$$

The time derivative can be approximated by

$$\frac{\partial p}{\partial t} \approx \frac{p_i(t+\Delta t) - p_i(t)}{\Delta t}, \tag{18.2}$$

where the index i indicates that the values refer to the pressures in the point $z = z_i$. Equation (18.2) can be considered as the definition of the partial derivative $\partial p/\partial t$, except that the limit $t \to 0$ has been omitted.
Finite differences will also be used in the z-direction. For this purpose the thickness h of the sample is subdivided into n small elements of thickness Δz,

$$\Delta z = \frac{h}{n}. \tag{18.3}$$

The second derivative with respect to z can be approximated by

$$\frac{\partial^2 p}{\partial z^2} \approx \frac{p_{i+1}(t) - 2p_i(t) + p_{i-1}(t)}{(\Delta z)^2}. \tag{18.4}$$

This relation is illustrated in Figure 18-1. The formula can most simply be found by noting that the second derivative is the derivative of the first derivative. This means that the second derivative is the difference of the slope in the upper part of the figure and the slope in the lower part of the figure, divided by the distance Δz. It can also be verified from the figure that for a straight line the expression (18.4) indeed gives a value zero, because then the value in the centre is just the average of the values at the two values above it and below it.
Substitution of Eq. (18.2) and (18.4) into (18.1) gives

$$p_i(t+\Delta t) = p_i(t) + \alpha\{p_{i+1}(t) - 2p_i(t) + p_{i-1}(t)\}, \tag{18.5}$$

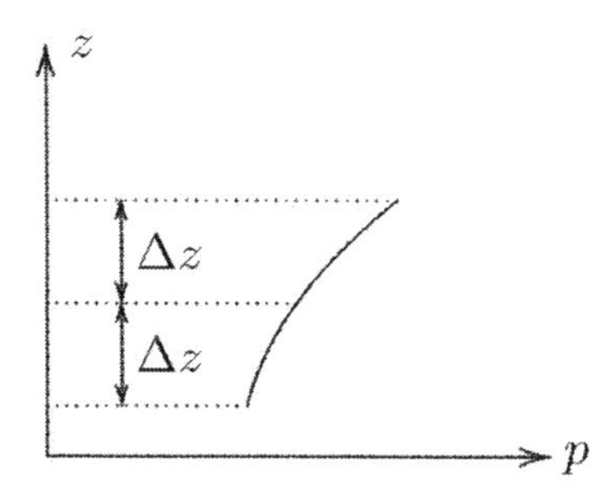

Figure 18-1. Second derivative.

where

$$\alpha = c_v \frac{\Delta t}{(\Delta z)^2}. \tag{18.6}$$

The expression Eq. (18.5) is an explicit formula for the new value of the pore pressure in the point *i*, if the old values (at time *t*) in that point and in the two points just above and just below it are known.
The boundary conditions must also be represented in a numerical way. For the boundary condition at the upper boundary, where the pressure *p* must be zero, see Eq. (16.18), this is very simple,

$$p_n = 0. \tag{18.7}$$

The boundary condition at the bottom of the sample is that for $z = 0$ the derivative $\partial p/\partial z = 0$, see Eq. (16.18). That can best be approximated by continuing the numerical subdivision by one more interval below $z = 0$, so that in a point at a distance Δz below the lower boundary a value of the pore pressure is defined, say p_{-1}. By requiring that $p_{-1} = p_1$, whatever the value of p_0 is, the condition $\partial p/\partial z = 0$ is satisfied at the symmetry axis $z = 0$. This means that the numerical equivalent of the boundary condition at $z = 0$ is

$$p_{-1} = p_1. \tag{18.8}$$

The general algorithm (18.5) for the point $i = 0$ can now be written as

$$p_0(t+\Delta t) = p_0(t) + \alpha\{2p_1(t) - 2p_0(t)\}. \tag{18.9}$$

The two boundary conditions Eqs. (18.7) and (18.9), which are valid at all values of time, complete the algorithm (18.5), together with the initial conditions

$$t = 0 \quad : \qquad p_i = p_0, \quad i = 0,1,2,...n-1, \qquad p_n = 0. \tag{18.10}$$

At the initial time $t = 0$ all values are known : all values of the pressure are p_0, except the one at the top, where the pressure is zero. The new values, after a time step Δt, can be calculated using the algorithm (18.5). This can be applied for all values of i in the interval $0,1,2,...,n-1$. At the top, for $i = n$, the value of the pressure remains zero.

The numerical process has been executed, for a layer of 1 m thickness, subdivided into 10 layers, in Table 18-1. The table gives the values of p/p_0 after the first 4 time steps, for the case that $\alpha = 0.25$.

x	$t=0$	$t=\Delta t$	$t=2\Delta t$	$t=3\Delta t$	$t=4\Delta t$
1.0	0.000	0.000	0.000	0.000	0.000
0.9	1.000	0.750	0.625	0.547	0.492
0.8	1.000	1.000	0.973	0.875	0.820
0.7	1.000	1.000	1.000	0.984	0.961
0.6	1.000	1.000	1.000	1.000	0.996
0.5	1.000	1.000	1.000	1.000	1.000
0.4	1.000	1.000	1.000	1.000	1.000
0.3	1.000	1.000	1.000	1.000	1.000
0.2	1.000	1.000	1.000	1.000	1.000
0.1	1.000	1.000	1.000	1.000	1.000
0.0	1.000	1.000	1.000	1.000	1.000
-0.1	1.000	1.000	1.000	1.000	1.000

Table 18-1: Numerical solution , $\alpha = 0.25$.

The process appears to progress rather slowly, which suggests to let the calculations be performed by a computer program, for instance a spreadsheet program, or a special program.
Because the process is so slow (after 4 time steps some of the values are still equal to their initial values 1.000) it may seem that the process can be made to run faster by taking a larger value of the dimensionless parameter α, say $a = 1$. That is very risky, however, as will be seen later.
A simple computer program, in BASIC, is shown in Program 18-1. In this program the general algorithm is represented in line 200, and the boundary condition at the upper boundary is taken into account by simply never changing the value of P(N) from its initial zero value. The boundary condition at the lower boundary is taken into account by assuming that for $i = -1$ there is an image point below the boundary where $p_{-1} = p_1$, to create symmetry. The algorithm for point $i = 0$ then is modified to the statement given in line 200. The program also calculates the degree of consolidation, using Eq. (17.18) and a simple numerical integration rule.
The numerical results are compared with the analytical results in Figure 18-2. The values of the dimensionless time $c_v t/h^2$ for which the pore pressures are shown, are the same as those used in Figure 17-2.
The numerical data have been calculated by subdividing the height h in 20 equal parts, $\Delta z = h/20$. The value of α has been chosen as $\alpha = 0.2$. This means that $\Delta t = 0.0005 h^2/c_v$. It turns out that in that case about $2/0.0005 = 10000$ time steps are needed to complete the entire consolidation process, until the pore pressures have been reduced to practically zero (for $c_v t/h^2 = 2$), but even this many time steps are executed very quickly on a computer. The numerical data appear to agree very well

with the analytical results, see Figure 18-2. The same is true for the numerical values of the degree of consolidation that are calculated by Program 18-1. The accuracy of the numerical solution, and its simplicity, may serve to explain the popularity of the numerical method.

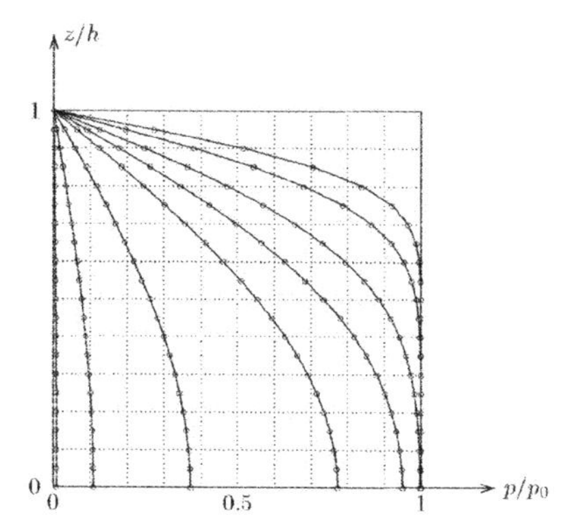

Figure 18-2. Comparison of numerical and analytical solution.

```
100 CLS:PRINT "One-dimensional Consolidation"
110 PRINT "Numerical solution":PRINT
120 INPUT "Thickness of layer ............. ";H
130 INPUT "Consolidation coefficient ....... ";C
140 INPUT "Number of subdivisions .......... ";N
150 T=0:DZ=H/N:DT=0.25*DZ*DZ/C:DIM P(N),PA(N)
160 PRINT "Suggestion for time step ........ ";DT
170 INPUT "Time step ...................... ";DT
180 A=C*DT/(DZ*DZ):FOR I=0 TO N:P(I)=1:NEXT I:P(N)=0
190 U=1:T=T+DT:FOR I=1 TO N-1:PA(I)=P(I)+A*(P(I+1)-2*P(I)+P(I-1))
200 NEXT I:PA(0)=P(0)+A*(P(1)-2*P(0)+P(1))
210 CLS:FOR I=0 TO N:PRINT " z = ";I*DZ;" p = ";PA(I):P(I)=PA(I)
220 U=U-P(I)/N:NEXT I:PRINT:PRINT " t = ";T;" U = ";U:GOTO 190
```

Program 18-1: Numerical solution for one-dimensional consolidation.

18.2 Numerical stability

In Program 18-1 the value of the factor α in the algorithm (18.5) is being assumed to be 0.25. Using this value the program calculates a suggestion for the time step Δt, and then the user of the program may enter a value for the time step. The user may follow that suggestion, but this is not absolutely necessary, of course. The suggestion is being given because the process is numerically unstable if the value of α is too large. This can easily be verified by running the program and then entering a larger value for the time step, for instance by a factor 4 larger than the suggested value. It

then appears that the values jump from positive to negative values, and that these values become very large. These results seem to be inaccurate.
The instability can be investigated by calculating the development of a small error by the numerical process. For this purpose it may be assumed that near the end of the consolidation process, when all pore pressures should be zero, some errors remain, with $p_i(t)=\varepsilon$ and $p_{i+1}(t)=p_{i-1}(t)=-\varepsilon$. The algorithm (18.5) then gives $p_i(t+\Delta t)=(1-4\alpha)\varepsilon$. The error will decrease if the new value is smaller than the old one, in absolute value. This will be the case if

$$|1-4\alpha|<1. \tag{18.11}$$

This means that

$$0<\alpha<\frac{1}{2}. \tag{18.12}$$

Of course, all distributions of errors should gradually be reduced to zero, and it is not certain that the requirement (18.12) is sufficient for stability. However, more fundamental investigations show that the criterion (18.12) is sufficient to guarantee that for all possible distributions of errors, they will eventually be reduced to zero.
The criterion (18.12) means that the algorithm used in this chapter is stable only if the time step is positive (that seems to be self-evident), and not too large,

$$\Delta t<\frac{1}{2}\frac{(\Delta z)^2}{c_v}. \tag{18.13}$$

To satisfy this criterion the value of the factor α in the Program 18-1 has been defined as 0.25. It is a simple matter to modify the program, and take a somewhat larger value, α larger than $\frac{1}{2}$. It will then appear immediately that the process is unstable. The pore pressures will become larger and larger, alternating between negative and positive values. If the time step is chosen such that the criterion Eq. (18.13) is satisfied, the numerical process is always stable, as can be verified by running the program with different values of the time step. The numerical results are always very accurate as well, provided that he stability criterion Eq. (18.13) is satisfied.
As may be evident from this chapter and the previous one, the numerical solution method is simpler than the analytical solution, and perhaps much easier to use. It may be added that the numerical solution method can easier be generalized than the analytical method. It is, for instance, rather simple to develop a numerical solution for the consolidation of a layered soil, with different values for the permeability and the compressibility in the various layers. The analytical solution for such a layered system can also be constructed, at least in principle, but this is a reasonably complex mathematical exercise.
The numerical solution presented in this chapter appears to be stable only if a certain stability criterion is satisfied. It may be mentioned that there exist other numerical procedures that are unconditionally stable. By using a different type of finite

differences, such as a backward finite difference or a central finite difference for the time derivative, a stable process is obtained. The numerical procedures then are somewhat more complicated, however. Another effective method is to use a formulation by finite elements. This also makes it very simple to include variable soil properties. There is sufficient reason for a further study of consolidation theory, or of numerical methods.

Problems

18.1 The consolidation process of a clay layer of 4 meter thickness is solved by a numerical procedure. The consolidation coefficient is $c_v = 10^{-6}$ m²/s. The layer is subdivided into 20 small layers. What is the maximum allowable magnitude of the time step?

18.2 To make a more accurate calculation of the previous problem the subdivision in layers can be made finer, say in 40 layers. What effect does that have on the time step?

18.3 What is the effect of taking twice as many layers on the total duration of the numerical calculations, if these are continued until the pore pressures have been reduced to 1 % of their initial value?

18.4 Execute the problem mentioned above, using a computer program, and using various values of the parameter α, say $\alpha = 0.25$ and $\alpha = 1.00$.

19 Consolidation coefficient

If the theory of consolidation, presented in the previous chapters, were a perfect description of the physical behaviour of soils, it should be rather simple to determine the value of the coefficient of consolidation cv from the data obtained in a consolidation test. For instance, one could measure the time at which 50 % of the final deformation has taken place. From the theory it follows that this is reached when $c_v t/h^2 = 0.197$, because then the value of $U = 0.5$, see formula (17.19). As the values of time t and the sample thickness h are known, it is then possible to determine the value of c_v. Unfortunately, there are some practical and some theoretical difficulties. The procedure would require an accurate determination of the initial deformation and of the ultimate deformation, and that is not so simple as it may seem. The initial deformation of the sample, Δh_0, is the deformation at the moment of application of the load, and at the moment of loading the indicator of the deformation will suddenly start to move, with a sudden jump followed by a continuous increase. It is difficult to decide what the value at the exact moment of loading is, as the moment is gone when the indicator starts to move. Also, it usually appears that no final constant value of the deformation, Δh_∞ is reached, as the deformation seems to continue, even when the pore water pressures have been dissipated completely. For these reasons somewhat modified procedures have been developed to define the *initial deformation* and the *final deformation*. In this chapter the two most common procedures are presented.

19.1 Log(t)-method

A first method to overcome the difficulties of determining the initial value and the final value of the deformation has been proposed by Casagrande. In this method the deformation of the sample, as measured as a function of time in a consolidation test, is plotted against the logarithm of time, see Figure 19-1. It usually appears that there is no horizontal asymptote of the curve, as the classical theory predicts, but for very large values of time a straight line is obtained, see also the next chapter. It is now postulated, somewhat arbitrarily, that the intersection point of the straight line asymptote for very large values of time, with the straight line that can be drawn tangent to the measurement curve at the inflection point (that is the steepest possible tangent), is considered to determine the final deformation of the primary consolidation process. The continuing deformation beyond that deformation is denoted as secondary consolidation, representing deformation at practically zero pore pressures. This procedure is indicated in Figure 19-1, leading to the value Δh_∞ for the final deformation.

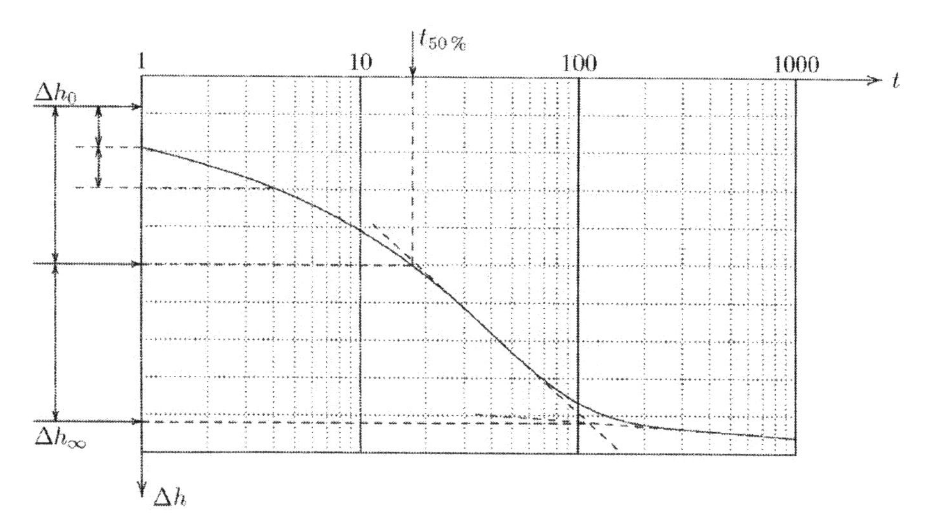

Figure 19-1. Log(t)-method.

In order to define the initial settlement of the loaded sample use is made of the knowledge, see chapter 17, that in the beginning of the consolidation process the degree of consolidation increases proportional to $\sqrt{t}$. This means that between $t = 0$ and $t = t_1$ the deformation will be equal to the deformation between $t = t_1$ and $t = 4t_1$. If the deformation is measured after 1 minute and after 4 minutes, it can be assumed that between $t = 0$ and $t = 1$ minute the deformation would have been the same as the deformation between $t = 1$ minute and $t = 4$ minutes. This procedure has been indicated in Figure 19-1, leading to the value Δh_0 for the initial deformation.

From the values of the initial deformation Δh_0 and the final deformation Δh_∞, it is simple to determine the moment at which the degree of consolidation is just between these two values, which would mean that $U = 0.5$. This is also indicated in Figure 19-1, giving a value for $t_{50\,\%}$. The value of the coefficient of consolidation then follows from $c_v t_{50\%} / h^2 = 0.197$, or

$$c_v = 0.197 \frac{h^2}{t_{50\%}}. \tag{19.1}$$

It should be noted that the quantity h in this expression represents the thickness of the sample, for the case of a sample drained on one side only. The consolidation process would be the same in a sample of thickness $2h$ and drainage to both sides. The original solution of Terzaghi considers that case, and the solution of the consolidation problem is given in that form in many textbooks. Because of the symmetry of that problem there is no difference with the problem and the solution considered here.

19.2 $\sqrt{t}$ -method

A second method to determine the value of the coefficient of consolidation is to use only the results of a consolidation test for small values of time, and to use the fact that in the beginning of the process its progress is proportional to the square root of time. In this method the measurement data are plotted against $\sqrt{t}$, see Figure 19-2. The basic formula is, see (17.22),

$$\Delta h - \Delta h_0 = (\Delta h_\infty - \Delta h_0)\frac{2}{\sqrt{\pi}}\sqrt{\frac{c_v t}{h^2}}. \tag{19.2}$$

In principle the value of the coefficient of consolidation c_v could be determined from the slope of the straight line in the figure, but this again requires the value of the initial deformation and the final deformation, as these appear in the formula (19.2). The value of the initial deformation Δh_0 can be determined from the intersection point of the straight tangent to the curve with the axis $\sqrt{t} = 0$. The final deformation Δh_∞, however, can not be obtained directly from the data. In order to circumvent this difficulty Taylor has suggested to use the result following from the theoretical curve and its approximation that for $U = 0.90$, i.e. for 90 % of the consolidation, the value of pt according to the exact solution is 15% larger that the value given by the approximate formula (19.2). The exact formula (17.11) gives that $U = 0.90$ if $c_v t/h^2 = 0.8481$, and the approximate formula (19.2) gives that $U = 0.90$ for $c_v t/h^2 = 0.6362$. The ratio of these two values is 1.333, which is the square of 1.154.

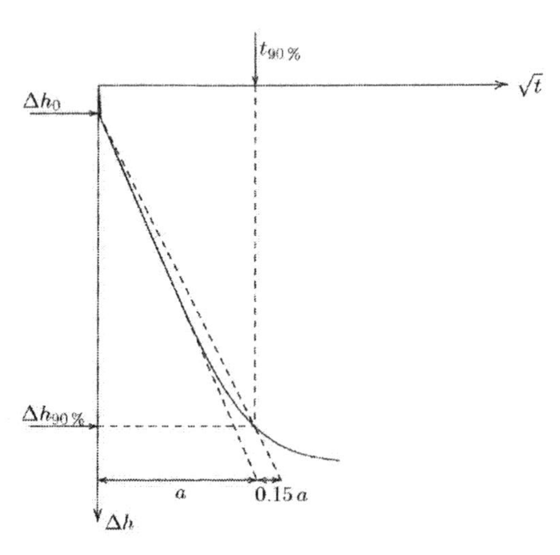

Figure 19-2. t-method.

This means that if in Figure 19-2 a straight line is plotted at a slope that is 15% smaller than the tangent to the measurement data for small values of time, this line should intersect the measured curve in the point for which $U = 0.90$. The corresponding value of the time parameter $c_v t/h^2$ is 0.848, and therefore the consolidation coefficient can be determined as

$$c_v = 0.848 \frac{h^2}{t_{90\%}}. \tag{19.3}$$

If the theory of consolidation were an exact description of the real behaviour of soils, the two methods described above should lead to precisely the same value for the coefficient of consolidation c_v. Usually this appears to be not the case, however, with errors of the order of magnitude up to 10 or 20 %. This indicates that the measurement data may be imprecise, especially when the deformations are very small, or that the theory is less than perfect. Perhaps the weakest point in the theory is the assumption of a linear relation between stress and strain.

19.3 Determination of m_v and k

In both of the two methods, the $\log(t)$-method and the $\sqrt{t}$-method, the procedure includes a value for the final consolidation settlement of the sample, even though it is realized that the deformations may continue beyond that value. In the $\log(t)$-method this final value forms part of the analysis, in the $\sqrt{t}$-method the final value of the deformation can be determined by adding 10 % to the difference of the level of 90% consolidation and the initial deformation,

$$\Delta h_\infty = \Delta h_0 + \frac{10}{9}(\Delta h_{90\%} - \Delta h_0). \tag{19.4}$$

In general the final deformation is

$$\Delta h_\infty = h m_v q, \tag{19.5}$$

so that the value of the compressibility mv follows from

$$m_v = \frac{\Delta h_\infty}{hq}. \tag{19.6}$$

Because the coefficient of consolidation c_v has been determined before, it follows that the permeability k can be determined as

$$k = \gamma_w m_v c_v. \tag{19.7}$$

The determination of the permeability k and the compressibility mv may be theoretically unique, but due to theoretical approximations and inaccuracies in the measurement data the accuracy in the actual values may not be very large.

Problems

19.1 A consolidation test, on a sample of 2 cm thickness, with drainage on both sides, has resulted in the following deformations, under a load of 10 kPa.

t(s)	Δh(mm)
10	0.070
20	0.082
30	0.089
40	0.094
60	0.105
120	0.127
240	0.157
600	0.201
1200	0.230
1800	0.240
3600	0.258
7200	0.271

Determine the coefficient of consolidation, using the $\log(t)$-method, and using the $\sqrt{t}$-method.

19.2 Determine the value of the final deformation Δh_∞ of the consolidation process (ignoring creep), and then determine the values of the compressibility and the permeability, using the two methods.

19.3 Using equation (17.19) determine the value of the degree of consolidation for various values of the dimensionless time parameter $c_v t/h^2$. Assume that c_v = 10^{-6} m^2/s, $h = 2$ m, $\Delta h_0 = 0.005$ m and $\Delta h_\infty = 0.05$ m. Make a graphical representation of the deformation, using a logarithmic time scale, and then verify whether the procedure described for the $\log(t)$-method leads to the correct value of c_v.

19.4 Make a graphical representation of the deformations in the previous example using a scale of $\sqrt{t}$ for time. Verify whether the procedure described for the $\sqrt{t}$-method leads to the correct value of c_v.

20 Secular effect (creep)

As mentioned in the previous chapter, in a one-dimensional compression test on clay, under a constant load, the deformation usually appears to continue practically forever, even if the pore pressures have long been reduced to zero, see Figure 19-1. Similar types of behaviour are found in other materials, such as plastics, concrete, etcetera. The phenomenon is usually denoted as creep. For many materials this behaviour can be modelled reasonably well by the theories of visco-elasticity or visco-plasticity. In such models the creep is represented by a viscous element, in which part of the stress is related to the rate of deformation of the material. Although the behaviour of soils may contain such a viscous component, the creep behaviour of soils is usually modelled by a special type of model, that has been based upon the observations in laboratory testing and in field observations.

20.1 Keverling Buisman

In 1936 Keverling Buisman, of the Delft University of Technology, found that the deformations of clay in a consolidation test did not approach a constant final value, but that the deformations continued very long. On a semi-logarithmic scale the deformations can be approximated very well by a straight line, see Figure 20-1.
This suggests that the relation between strain and stress increment, after very long values of time, can be written as

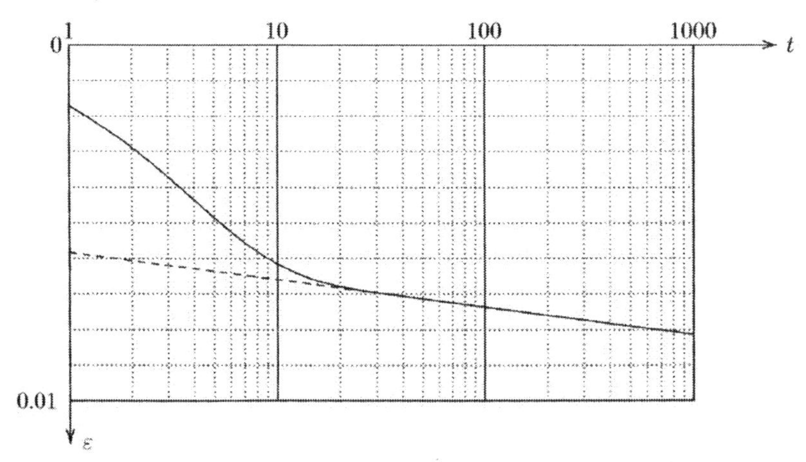

Figure 20-1. Secular effect.

$$\varepsilon = \varepsilon_p + \varepsilon_s \log(\frac{t}{t_0}). \tag{20.1}$$

Here ε_p is the *primary strain*, and ε_s is the *secular strain*, or the *secondary strain*. The quantity t_0 is a reference time, usually chosen to be 1 day. Keverling Buisman denoted the continuing deformations after the dissipation of the pore pressures as the secular effect, with reference to the Latin word seculum (for century). In most international literature it is denoted as secondary consolidation, the primary consolidation being Terzaghi's pore pressure dissipation process.

The primary strain ε_p is the deformation due to the consolidation of the soil. This is being retarded by the outflow of groundwater from the soil, as described in Terzaghi's theory of consolidation. Afterwards the deformation continues, and this additional deformation can be described, in a first approximation, by a semi-logarithmic relation, see Figure 20-1, using the secular strain parameter ε_s. The phenomenon can be modelled at the microscopic level by the outflow of water from micro pores to a system of larger pores, or by a slow creeping deformation of clay elements (plates) under the influence of elementary forces at the microscopic level.

From a theoretical point of view the formula (20.1) is somewhat peculiar, because for $t \to \infty$ the strain would become infinitely large. It seems as if one can calculate the time span after which the thickness of the sample will have been reduced to zero, when the deformation becomes as large as the original thickness of the sample. For $t < t_1$ the behaviour of the formula is also peculiar, because then the strain would be negative. Attempts have been made to adjust the formula for very large values of time, but in engineering practice the original formula, in its simple form (20.1) is perfectly usable, as long as it is assumed that $t \geq t_1$ and that the values of time in practice will be limited to say a few thousands (or perhaps millions) of years.

The magnitude of the parameters ε_p and ε_s can be determined from the data of a compression test at two different values of time, for instance at time $t = t_1$ (= 1 day) and $t = 10t_1$ (= 10 days). In the case illustrated in Figure 20-1 this gives $\varepsilon_p = 0.0058$ and $\varepsilon_p + \varepsilon_s = 0.0066$ (the values are positive because the sample becomes thinner), so that $\varepsilon_s = 0.0008$. If the results are extrapolated to a value of $t = 100$ years the strain will be, after 100 years, $\varepsilon = 0.0094$. And after 1000 years the strain is $\varepsilon = 0.0102$. Predictions over longer periods of time are unusual in civil engineering practice. The time span of a structure is usually considered to be several hundreds of years.

In many countries the secondary strain is often denoted by C_α, the *secondary compression index*. In The Netherlands research results by Keverling Buisman, Koppejan and Den Haan have lead to the introduction of slightly different parameters, to be presented below.

Both the primary strain ε_p and the secondary strain ε_s can, of course, depend upon the magnitude of the applied load. For this reason Keverling Buisman wrote his formula in the form

$$\varepsilon = \sigma'[\alpha_p + \alpha_s \log(\frac{t}{t_1})], \tag{20.2}$$

in which σ' represents the load increment. This may suggest that the relation between stress and strain is linear, which in general is not the case. The coefficients α_p and α_s therefore depend upon the stress, and on the stress history.

The dependence of the stiffness has been considered earlier in the discussion on Terzaghi's logarithmic compression formula, see Chapter 15. It can be considered that the deformation considered in that chapter (for sand soils) is a special case of the more general case considered here, in the absence of creep, i.e. with $\varepsilon_s = 0$. It then appears that the primary strain ε_p is proportional to the logarithm of the stress, with proportionality constants that are different for virgin loading and for unloading and reloading. Koppejan suggested to combine the formulas of Terzaghi and Keverling Buisman to

$$\varepsilon = [\frac{1}{C_p} + \frac{1}{C_s}\log(\frac{t}{t_1})]\ln(\frac{\sigma}{\sigma_1}). \tag{20.3}$$

The coefficients C_p and C_s should be understood to have quite different values for virgin loading (above the pre-consolidation pressure p_{pcp}) and for unloading and reloading (below the pre-consolidation pressure). If the degree of consolidation is added, the formula of Koppejan, which is often used in the Netherlands, is found

$$\varepsilon = U[\frac{1}{C_p} + \frac{1}{C_s}\log(\frac{t}{t_1})]\ln(\frac{\sigma}{\sigma_1}). \tag{20.4}$$

It is agreed to define $t_1 = 1$ day.
It can be noticed that equation (20.4) contains the natural logarithm as well as the common logarithm (of base 10).
In many countries the deformation is often expressed into the void ratio e. A familiar form of the compression formula is Bjerrum's relation

$$e_1 - e = C_r \log(\frac{\sigma_{\text{pcp}}}{\sigma_1}) + C_c \log(\frac{\sigma}{\sigma_{\text{pcp}}}) + C_\alpha \log(\frac{t}{t_1}), \tag{20.5}$$

in which e_1 is the void ratio at the initial stress σ_1, for $t = t_1$. C_r is the parameter for reversible (elastic) behaviour. C_c is the parameter for irreversible (plastic) behaviour. C_α is the parameter which accounts for the secular effect. Using these parameters the pre-consolidation pressure σ_{pcp} can be calculated, if the time span since the last loading is known. If t = 100 years ago the effective stress is raised to the present σ_1, the pre-consolidation pressure follows from Bjerrum's relation, using $e = e_1$

$$C_\alpha \log(\frac{t}{t_1}) = (C_c - C_r)\log(\frac{\sigma_{\text{pcp}}}{\sigma_1}), \tag{20.6}$$

The pre-consolidation pressure is the stress above which both elastic and plastic deformations occur and below which only elastic deformations take place. If the present effective stress σ_1 is increased to σ, Bjerrum's relation describes the settlement as a function of time. Just like Koppejan's formula it is possible to add the degree of consolidation U to Bjerrum's relation.
In Chapter 15 the relation between the change of the void ratio e and the strain ε has been shown to be

$$\varepsilon = \frac{\Delta e}{1+e}, \tag{20.7}$$

see equation (15.9). Using this relation the various expressions given in this chapter can be shown to be equivalent, and the various coefficients can be expressed into each other.
It is, of course, regrettable that slightly different formulas and different constants are being used for the same phenomenon, especially as there is general agreement on the basic form of relationships, with a logarithm of time. This is mainly a consequence of national traditions and experiences. In engineering practice some care must be taken that it is sometimes necessary to translate local experience with certain constants into a formula using different constants. The conversion is simple, however.
One of the main applications in engineering practice is the prediction of the settlement of a layered soil due to an applied load. The standard procedure is to collect a sample of each of the soil layers, to apply the initial load to each of the samples, and then to load each sample by an additional load corresponding to the load in the field. In this way the stress dependence of the stiffness is taken into account by subjecting each sample to the same stress increment in the laboratory and in the field. In general the settlement appears to increase with the logarithm of time after application of the load, in agreement with the formula (20.1). The deformation in the field can then be predicted using this formula. The contribution of each layer to the total settlement is obtained by multiplying the strain of the layer by its thickness. The total settlement is obtained by adding the deformations of all layers.
The prediction of the deformations can be complicated because the stiffness of the soil depends on the stress history. In an area with a complex stress history (for instance a terrain that has been used for different purposes in history, or a field that has been subject to high preloading in an earlier geologic period) this means that the behaviour of the soil may be quite different below an unknown earlier stress level and above that stress level. Extrapolation of laboratory results may be inaccurate if the stress history is unknown. For this purpose it is advisable to always simulate the actual stress level and its proposed increase in the field in the laboratory tests. In that case the laboratory tests will be a good representation of the behaviour in the field. As the logarithmic time behaviour is generally observed, the duration of the tests need not be very long. Extrapolation in time is usually sufficiently accurate.

It should be mentioned that all the considerations in this chapter refer only to one-dimensional compression. This means that they apply only if in the field there are no horizontal deformations. In case of a local load it can be expected that there will be lateral deformation as well as vertical deformation. In such cases consolidation and creep should be considered as three-dimensional phenomena. These are considerably more complicated than the one-dimensional case considered here.

Problems

20.1 A terrain consists of 1 meter dry sand ($\gamma = 17\ \text{kN/m}^3$), 4 meter saturated sand ($\gamma = 20\ \text{kN/m}^3$), 2 meter clay ($\gamma = 18\ \text{kN/m}^3$), 5 meter sand ($\gamma = 20\ \text{kN/m}^3$), 4 meter clay ($\gamma = 19\ \text{kN/m}^3$), and finally a thick sand layer. The terrain is loaded by an additional layer of 2 meter dry sand ($\gamma = 17\ \text{kN/m}^3$). The deformations of the clay layers will be analysed by performing oedometer tests on samples from each clay layer. What should be the initial load on each of the two samples, and what should be the additional load?

20.2 In the tests mentioned in the previous problem the test results are that after one day a strain of 2 % is observed, and after 10 days a strain of 3 %, for both clay layers. If it is assumed that the deformation of the sand layers can be neglected, predict the total settlement of the terrain after 1 year, 10 years and 100 years.

20.3 In a certain town it is required that in a period of 20 years after the sale of a terrain the deformation may not be more than 20 cm. For a terrain that has been prepared by the application of a sand layer on a soft soil layer of 7.6 m thickness, it has been found from tests on the soft soil that the deformation after one day is 1.1 %, and after 10 days 2.4 %. How long should the town wait after the application of the sand layer before the terrain can be sold?

Part V
Strength and tests

21 Shear strength

As mentioned before, one of the main characteristics of soils is that the shear deformations increase progressively when the shear stresses increase, and that for sufficiently large shear stresses the soil may eventually fail. In nature, or in engineering practice, dams, dikes, or embankments for railroads or highways may fail by part of the soil mass sliding over the soil below it. As an example, Figure 21-1 shows the failure of a peat dike in The Netherlands. It appears that the strength of the soil was not sufficient to carry the horizontal water pressure against the dike. Other important effects of failure may be the change of the groundwater level in a dam or the weight of the soil of a (steep) slope causing a landslide. In many cases a very small cause, such as a small local excavation or rainfall, may be the cause of a large landslide.
In this chapter the states of stresses causing such failures of the soil are described. In later chapters the laboratory tests to determine the shear strength of soils will be presented.

Figure 21-1. Dike slide, Wilnis 2003

21.1 Coulomb

It seems reasonable to assume that a sliding failure of a soil will occur if on a certain plane the shear stress is too large, compared to the normal stress. On other planes the shear stress is sufficiently small compared to the normal stress to prevent sliding failure. It may be illustrative to compare the analogous situation of a rigid block on a slope, see Figure 21-2. Equilibrium of forces shows that the shear force in the plane of the slope is $T = W \sin\alpha$ and that the normal force acting on the slope is $N = W\cos\alpha$, where W is the weight of the block. The ratio of shear force to normal force is $T/N = \tan\alpha$. As long as this is smaller than a certain critical value, the friction coefficient f, the block will remain in equilibrium. However, if the slope angle α becomes so large that $\tan\alpha = f$, the block will slide down the slope. On steeper slopes the block can never be in equilibrium.

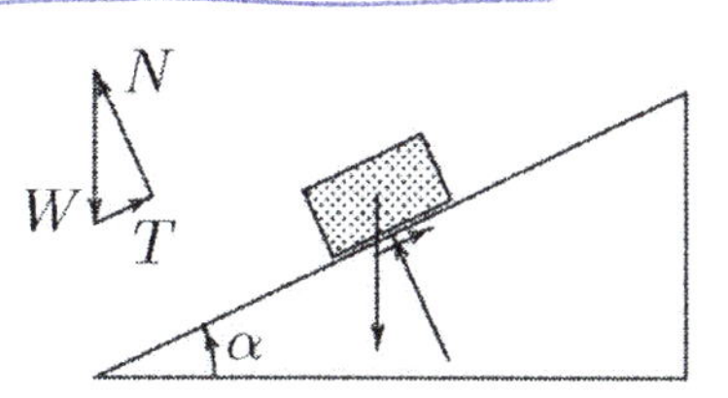

Figure 21-2.Block on slope.

The analogy with a sliding block lead Coulomb to the proposal that the critical shear stress τ_f in a soil body is

$$\tau_f = c + \sigma' \tan\phi. \tag{21.1}$$

Here σ' is the normal (effective) stress on the plane considered. The quantity c is the *cohesion*, and ϕ is the *angle of internal friction* or *the friction angle*. An elementary interpretation is that if the shear stress on a certain plane is smaller than the critical value τ_f, then the deformations will be limited, but if the shear stresses on any single plane reaches the critical value, then the shear deformations are unlimited, indicating shear failure. The cohesion c indicates that even when the normal stress is zero, a certain shear stress is necessary to produce shear failure. In the case of two rough surfaces sliding over each other (e.g. two blocks of wood), this may be due to small irregularities in the surface. In the case of two very smooth surfaces molecular attractions may play a role.

For soils the formula (21.1) should be expressed in terms of effective stresses, as the stresses acting from one soil particle on another determine the eventual sliding. For this reason the soil properties are often denoted as c' and ϕ', in order to stress that these quantities refer to effective stresses.

21.2 Mohr's circle

From the theory of stresses (see Appendix A) it is known that the stresses acting in a certain point on different planes can be related by analytical formulas, based upon the equilibrium equations. In these formulas the basic variable is the angle of rotation of the plane with respect to the principal directions. These principal directions are the directions in which the shear stress is zero, and in which the normal stresses are maximal or minimal. It is assumed here that the maximum principal stress, σ_1, is acting in vertical direction, and hence that the smallest principal stress, σ_3 acts in horizontal direction. The intermediate principal stress (acting in a direction normal to the plane of the figure) is denoted by σ_2. It is possible that $\sigma_2 = \sigma_1$ or $\sigma_2 = \sigma_3$, otherwise $\sigma_3 < \sigma_2 < \sigma_1$. The stresses on two planes having their normal vectors in the x-direction and the y-direction, which make an angle α with the directions of the major and the minor principal stresses, can be expressed into the major and the minor principal stresses by means of the equations of equilibrium, see Figure 21-3.

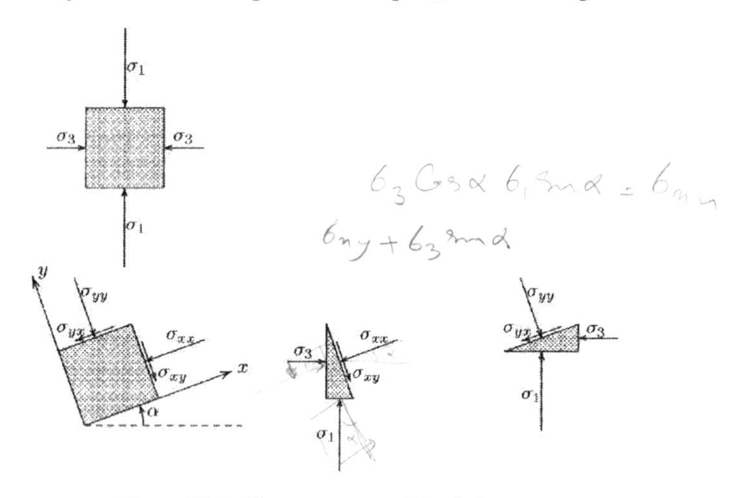

Figure 21-3. Stresses on a rotated plane.

The stress components σ_{xx} and σ_{xy}, acting on a plane with its normal in the x-direction, can be found from the equations of equilibrium of a small elementary triangle, formed by a plane perpendicular to the x-direction and a vertical and a horizontal plane, see the small triangle in the centre of Figure 21-3. The small wedge drawn is a part of the rotated element shown in the lower left part of the figure. If the area of the oblique surface is A, the area of the vertical surface is $A\cos\alpha$, and the area of the horizontal plane is $A\sin\alpha$. Equilibrium of forces in the x-direction now gives

$$\sigma_{xx} = \sigma_1 \sin^2\alpha + \sigma_3 \cos^2\alpha. \tag{21.2}$$

Equilibrium of the forces acting upon the small wedge in the y-direction gives

$$\sigma_{xy} = \sigma_1 \sin\alpha\cos\alpha - \sigma_3 \sin\alpha\cos\alpha. \tag{21.3}$$

The stress components σ_{yy} and σ_{yx}, acting upon a plane having its normal in the y-direction, can be found by considering equilibrium of a small triangular wedge, formed by a plane perpendicular to the y-direction and a vertical and a horizontal

plane, see the small triangle in the lower right part of Figure 21-3.
Equilibrium in y-direction gives

$$\sigma_{yy} = \sigma_1 \cos^2\alpha + \sigma_3 \sin^2\alpha. \quad (21.4)$$

Equilibrium in x-direction gives

$$\sigma_{yx} = \sigma_1 \sin\alpha\cos\alpha - \sigma_3 \sin\alpha\cos\alpha. \quad (21.5)$$

Comparison of (21.5) and (21.3) shows that $\sigma_{xy} = \sigma_{yx}$, which is in agreement with equilibrium of moments of the element in the lower left part of Figure 21-3.
It should be noted that the transformation formulas for rotation of a plane all contain two factors $\sin\alpha$ or $\cos\alpha$. This is a characteristic property of quantities such as stresses and strains, which are second order *tensors*. Unlike a vector (sometimes denoted as a first order tensor), which can be described by a magnitude and a single direction, a (second order) tensor refers to two directions: in this case the direction of the plane on which the stresses are acting, and the direction of the stress vector on that plane. In the equations of equilibrium this is seen in the appearance of a factor $\cos\alpha$ or $\sin\alpha$ because of taking the component of a force in x- or y-direction, but another such factor appears because of the size of the area on which the stress component is acting.
Using the trigonometric formulas

$$\sin 2\alpha = 2\sin\alpha\cos\alpha, \quad (21.6)$$

$$\cos 2\alpha = \cos^2\alpha - \sin^2\alpha = 2\cos^2\alpha - 1 = 1 - 2\sin^2\alpha, \quad (21.7)$$

the transformation formulas can be expressed in 2α,

$$\sigma_{xx} = \frac{1}{2}(\sigma_1 + \sigma_3) - \frac{1}{2}(\sigma_1 - \sigma_3)\cos 2\alpha, \quad (21.8)$$

$$\sigma_{yy} = \frac{1}{2}(\sigma_1 + \sigma_3) + \frac{1}{2}(\sigma_1 - \sigma_3)\cos 2\alpha, \quad (21.9)$$

$$\sigma_{xy} = \sigma_{yx} = \frac{1}{2}(\sigma_1 - \sigma_3)\sin 2\alpha. \quad (21.10)$$

The stress components on planes with different orientations can be represented graphically using Mohr's circle, see Figure 21-4. A simple form of Mohr's diagram occurs if it the positive normal stresses σ_{xx} and σ_{yy} are plotted towards the right on the horizontal axis, that a positive shear stress σ_{xy} is plotted vertically downward, and that a positive shear stress σ_{yx} is plotted vertically upward.
The circle is constructed by first indicating distances corresponding to σ_1 and σ_3 on the horizontal axis. These two points define a circle, with its centre on the horizontal axis, at a distance $\frac{1}{2}(\sigma_1 + \sigma_3)$ from the origin. The radius of the circle is $\frac{1}{2}(\sigma_1 - \sigma_3)$. These happen to be the two values appearing in the formulas (21.8) –

(21.10). If in the centre of the circle an angle of magnitude 2α is measured, it follows that the point A on the circle has the coordinates σ_{xx} and σ_{xy}. The point B, on the opposite side on the circle, has coordinates σ_{yy} en σ_{yx}. It should be noted that this is true only if on the vertical axis σ_{xy} is considered positive in downward direction, and σ_{yx} is considered positive in upward direction. The formulas (21.8) – (21.10) now all are represented by the graphical construction.

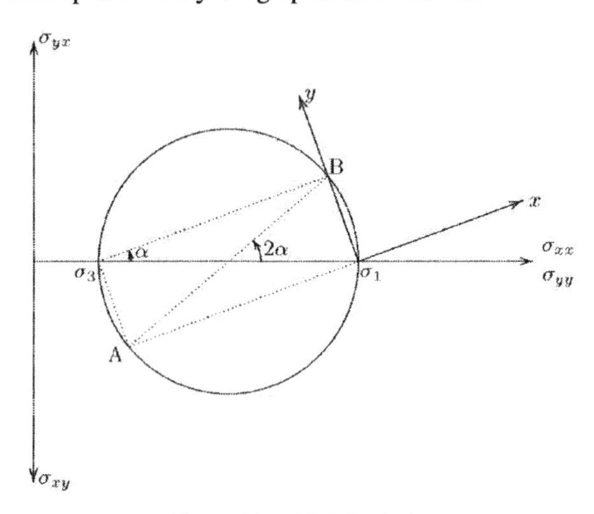

Figure 21-4. Mohr's circle.

Because an inscribed angle on a certain arc is just one half of the central angle, it follows that point B can also be found by drawing a line at an angle α from the leftmost point of the circle, and intersecting that line with the circle. In the same way the point A can be found by drawing a line from the same point perpendicular to the previous line.

The point A, which defines the stress components on a plane with its normal in the x-direction, can also be found by drawing a line from the rightmost point of the circle in the direction of the x-axis. Similarly, the point B, which defines the stress components on a plane with its normal in the y-direction, can be found by drawing a line from that point in the direction of the y-axis, see Figure 21-3. The rightmost point of the circle is therefore sometimes denoted as the pole of the circle. Drawing lines in the directions of two perpendicular axes x and y will lead to two opposite intersection points on the circle, which define the values of the stress components in these two directions. If the axes rotate, i.e. when α increases, these intersection points travel along the circle.

For $\alpha = 0$ the x-axis coincides with the direction of σ_3, and the y-axis coincides with the direction of σ_1. The point A then is located in the leftmost point of the circle, and the point B in the rightmost point. If the angle α now increases from 0 to $\frac{\pi}{2}$ the two stress points A and B travel along the circle, in a half circle. When $\alpha = \frac{\pi}{2}$ point A arrives in the rightmost point and point B arrives in the leftmost point. Then the x-axis points vertically upward, and the y-axis points horizontally towards the left. If α varies from 0 to π the stress points travel along the entire circle.

21.3 Mohr-Coulomb

A point of Mohr's circle defines the normal stress and the shear stress on a certain plane. The stresses on all planes together form the circle, because when the plane rotates the stress points traverse the circle. It appears that the ratio of shear stress to normal stress varies along the circle, i.e. this ratio is different for different planes. It is possible that for certain planes the failure criterion (21.1) is satisfied. In Figure 21-5 this failure criterion has also been indicated, in the form of two straight lines, making an angle ϕ with the horizontal axis. Their intersections with the vertical axis is at distances c. In order to underline that failure of a soil is determined by the effective stresses, the stresses in this figure have been indicated as σ'. There are two planes, defined by the points C and D in Figure 21-5, in which the stress state is critical. On all other planes the shear stress remains below the critical value. Thus it can be conjectured that failure will start to occur whenever Mohr's circle just touches the Coulomb envelope. This is called the *Mohr-Coulomb* failure criterion. If the stress circle is completely within the envelope no failure will occur, because on all planes the shear stress remains well below the critical value, as given by equation (21.1). Circles partly outside the envelope are impossible, as the shear stress on some planes would be larger than the critical value.

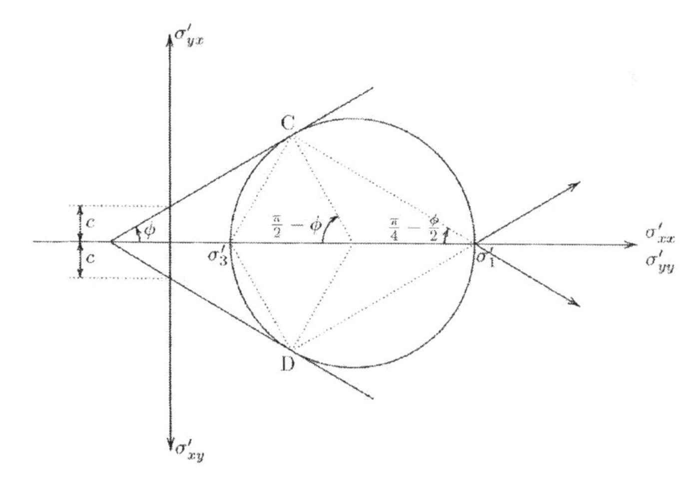

Figure 21-5. Mohr-Coulomb failure criterion.

When the circle just touches the envelope there are two planes, making angles $\frac{\pi}{4} - \frac{\phi}{2}$ with the direction of the major principal stress, on which the stresses are critical. Sliding failure may occur on these planes. It can be expected that the soil may slide in the directions of these two critical planes. In the case represented by the figures in this chapter, in which it is assumed that the vertical direction is the direction of the major principal stress, see Figure 21-3, the planes on which the stresses are most critical make an angle $\frac{\pi}{4}-\frac{\phi}{2}$ with the vertical direction. Thus it can be expected that sliding failure will occur in planes that are somewhat steeper than 45°. If for instance $\phi = 30°$, which is a normal value for sands, failure will occur by sliding along planes that make an angle of 30° with the vertical direction.

21.4 The Mohr-Coulomb criterion

The mathematical formulation of the Mohr-Coulomb failure criterion can be found by noting the radius of Mohr's circle is $\frac{1}{2}(\sigma_1' - \sigma_3')$, and that the distance from the origin to the centre is $\frac{1}{2}(\sigma_1' + \sigma_3')$. Failure will occur if

$$\sin\phi = \frac{\frac{1}{2}(\sigma_1' - \sigma_3')}{c\cot\phi + \frac{1}{2}(\sigma_1' + \sigma_3')}. \tag{21.11}$$

This can also be written as

$$\frac{(\sigma_1' - \sigma_3')}{2} - \frac{(\sigma_1' + \sigma_3')}{2}\sin\phi - c\cos\phi = 0. \tag{21.12}$$

Using this equation the value of σ_3' in the failure state can be expressed into σ_1',

$$\sigma_3' = \sigma_1' \frac{1-\sin\phi}{1+\sin\phi} - 2c\frac{\cos\phi}{1+\sin\phi}. \tag{21.13}$$

On the other hand, the value of σ_1' in the failure state can also be expressed into σ_3', of course,

$$\sigma_1' = \sigma_3' \frac{1+\sin\phi}{1-\sin\phi} - 2c\frac{\cos\phi}{1-\sin\phi}. \tag{21.14}$$

These formulas will be used very often in later chapters.

21.5 Remarks

The Mohr-Coulomb criterion is a rather good criterion for the failure state of sands. For such soils the cohesion usually is practically zero, $c = 0$, and the friction angle usually varies from $\phi = 30°$ to $\phi = 45°$, depending upon the angularity and the roundness of the particles. Clay soils usually have some cohesion, and a certain friction angle, but usually somewhat smaller than sands. Great care is needed in the application of the Mohr-Coulomb criterion for very small stresses. For clay one might find that a Mohr's circle would be possible in the extreme left corner of the diagram, with tensile normal stresses. It is usually assumed that this is not possible, and therefore the criterion should be extended by a vertical cut-off at the vertical axis. To express that the cohesion of soils does not necessarily mean that the soil can withstand tensile stresses, the property is sometimes denoted as *apparent cohesion*, indicating that it is merely a first order schematisation.

In metallurgy it is usually found that the shear strength of metals is independent of the normal stress. The failure criterion then is that there is a given maximum shear stress, $\tau_f = c$. The Mohr-Coulomb criterion reduces to the criterion for metals by taking $\phi = 0$.
The Mohr-Coulomb criterion can also be used, at least in a first approximation, for materials such as rock and concrete. In such materials a tension cut-off is not necessary, as they can indeed withstand considerable tensile stresses. In such materials the cohesion may be quite large, at least compared to soils. The contribution of friction is not so dominant as it is in soils. Also it often appears that the friction angle is not constant, but decreases at increasing stress levels.
In some locations, for instance in offshore coastal areas near Brazil and Australia, calcareous soils are found. These are mostly sands, but the particles have been glued together, by the presence of the calcium. Such materials have very high values of the cohesion c, which may easily be destroyed, however, by a certain deformation. This deformation may occur during the construction of a structure, for instance the piles of an offshore platform. During the exploration of the soil this may have been found to be very strong, but after installation much of the strength has been destroyed. An advantage of true frictional materials is that the friction usually is maintained, also after very large deformations. Soils such as sands may not be very strong, but at least they maintain their strength.
For clays the Mohr-Coulomb criterion is reasonably well applicable, provided that proper care is taken of the influence of the pore pressures, which may be a function of time, so that the soil strength is also a function of time. Some clays have the special property that the cohesion increases with time during consolidation. This leads to a higher strength because of overconsolidation. For very soft clays the Mohr-Coulomb criterion may not be applicable, as the soil behaves more like a viscous liquid.

Problems

21.1 In a sample of sand ($c = 0$) a stress state appears to be possible with $\sigma_{xx} = 10$ kPa, $\sigma_{yy} = 20$ kPa and $\sigma_{xy} = 5$ kPa, without any sign of failure. What can you say of the friction angle ϕ?

21.2 A sand, with $c = 0$ and $\phi = 30°$ is on the limit of failure. The minor principal stress is 10 kPa. What is the major principal stress?

21.3 In a soil sample the state of stress is such that the major principal stress is the vertical normal stress, at a value $3p$. The horizontal normal stress is p. Determine the normal stress and the shear stress on a plane making an angle of $45°$ with the horizontal direction.

21.4 Also determine the normal stress and the shear stress on a plane making an angle of $30°$ with the vertical direction, and determine the angle of the resulting force with the normal vector to that plane.

21.5 If you have solved the previous two problems analytically, using the transformation formulas, then do it again, using Mohr's circle and the pole.

22 Triaxial test

The failure of a soil sample under shear could perhaps best be investigated in a laboratory test in which the sample is subjected to pure distorsion, at constant volume. The volume could be kept constant by taking care that the isotropic stress $\sigma_0 = \frac{1}{3}(\sigma_1 + \sigma_2 + \sigma_3)$ remains constant during the test, or, better still, by using a test setup in which the volume change can be measured and controlled very accurately, so that the volume change can be zero. In principle such a test is possible, but it is much simpler to perform a test in which the lateral stress is kept constant, the *triaxial test*, see Figure 22-1. In order to avoid the complications caused by pore pressure generation, it will first be assumed that the soil is dry sand. The influence of pore water pressures will be considered later.

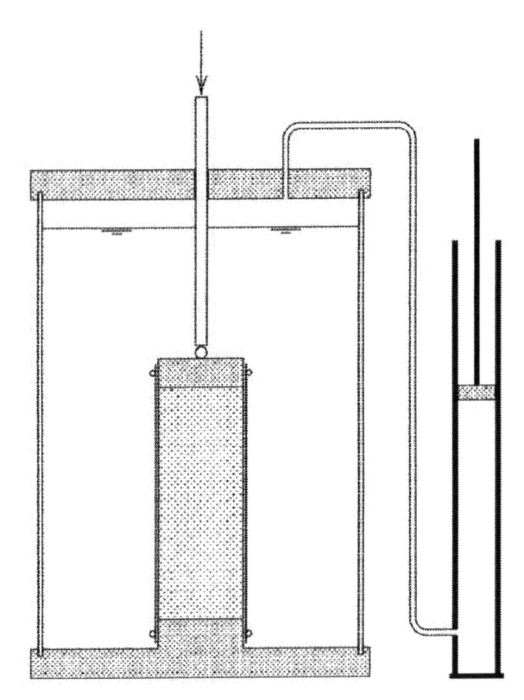

Figure 22-1. Triaxial test.

In the triaxial test a cylindrical soil sample is placed in a glass or plastic cell, with the sample being enclosed in a rubber membrane. The membrane is connected to circular plates at the top and the bottom of the sample, with two o-rings ensuring a water tight connection. The cell is filled with water, with the pressure in the water (*the cell pressure*) being controlled by a pressure unit, usually by a connection to a tank in which the pressure can be controlled. Because the sample is completely surrounded by water, at its cylindrical surface and at the top, a pressure equal to the cell pressure is generated in the sample. The usual, and simplest, test procedure is to keep the cell pressure constant during the test.

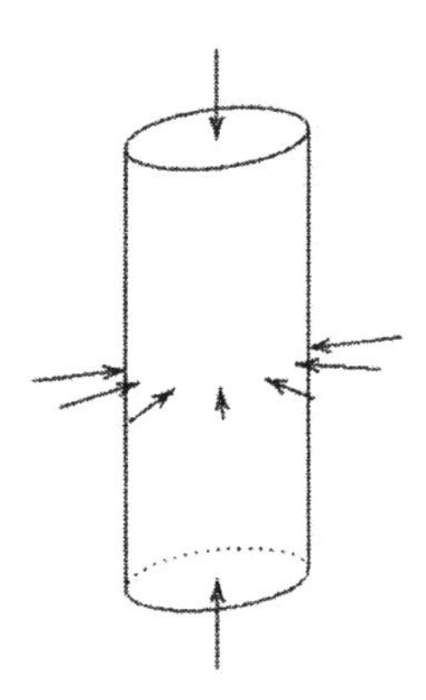

Figure 22-2. Cell pressure.

In addition to the lateral (and vertical) loading by the cell pressure, the sample can also be loaded by a vertical force, by means of a steel rod that passes through the top cap of the cell. The usual procedure is that in the second stage of the test the rod is being pushed down, at a constant rate, by an electric motor. This means that the vertical deformation rate is constant, and that the force on the sample gradually increases. The force can be measured using a strain gauge or a compression ring, and the vertical movement of the top of the sample is measured by a mechanical or an electronic measuring device.
During the test the vertical displacement of the top of the sample increases gradually as a function of time, because the motor drives the steel rod at a very small constant velocity downwards. The vertical force on the sample will also gradually increase, so that the difference of the vertical stress and the horizontal stress gradually increases, but after some time this reaches a maximum, and remains constant afterwards, or shows some small additional increase, or decreases somewhat. The maximum of the vertical force indicates that the sample starts to fail. Usually the test is continued up to a level where it is quite clear that the sample has failed, by recording large deformations, up to 5 % or 10 %. This can often be observed in the shape of the sample too, with the occurrence of some distinct sliding planes. It may also be, however, that the deformation of the sample remains practically uniform, with a considerable shortening and at the same time a lateral extension of the sample. In the interior of the sample many sliding planes may have formed, but these may not be observed at its surface.

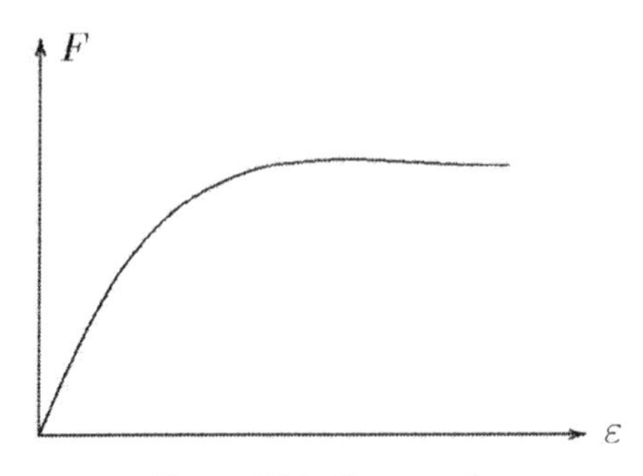

Figure 22-3. Test result.

The test is called the *triaxial test* because stresses are imposed in three directions. This can be accomplished in many different ways, however, and there even exist tests in which the stresses applied in three orthogonal directions onto a cubical soil sample (enclosed in a rubber membrane) can all be different, the *true triaxial test*. This gives many more possibilities, but it is a much more complex apparatus, and the testing procedures are more complex as well.

In the normal triaxial test the sample is of cylindrical shape, and the two horizontal stresses are identical. The usual diameter of the sample is 3.8 cm (or 1.5 inch, as the test was developed in England), but there also exist triaxial cells in which larger size samples can be tested. For tests on gravel a diameter of 3.8 cm seems to be insufficient to achieve a uniform state of stress. For clay and sand it is sufficient to guarantee that in every cross section there is a sufficient number of particles for the stress to be well defined.
If the cell pressure is denoted by σ_c, and the vertical axis is the z-axis, then the lateral stresses in the test are

$$\sigma_{xx} = \sigma_{yy} = \sigma_c, \tag{22.1}$$

and the vertical stress is

$$\sigma_{zz} = \sigma_c + \frac{F}{A}, \tag{22.2}$$

in which F is the vertical force, and A is the cross sectional area of the sample. Because the soil has been supposed to be dry sand, so that there are no pore pressures, these are effective stresses as well as total stresses.
In this case the vertical stress is the major principal stress, and the horizontal stress is the minor principal stress,

$$\sigma_1 = \sigma_c + \frac{F}{A}, \tag{22.3}$$

$$\sigma_3 = \sigma_c. \tag{22.4}$$

It should be noted that the stresses in the sample are assumed to be uniformly distributed. This will be the case only if the sample is of homogeneous composition. Furthermore, it has been assumed that there are no shear stresses on the upper and lower planes of the sample. This requires that the loading plates are very smooth. This can be accomplished by using special materials (e.g. Teflon) or by applying a thin smearing layer.

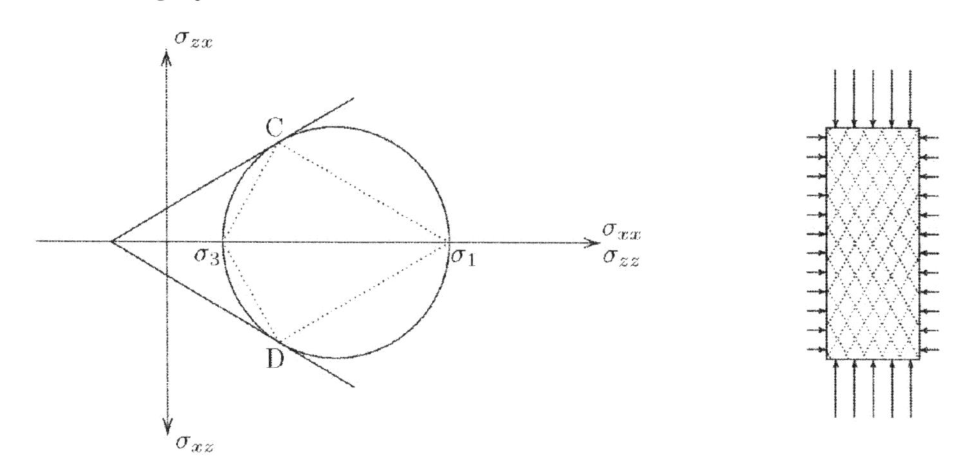

Figure 22-4. Mohr's circle for the triaxial test.

The stresses on planes having an inclined orientation with respect to the vertical axis, can be determined using Mohr's circle, see Figure 22-4. The *pole* for the normal directions coincides with the rightmost point of the circle. On a horizontal plane and on a vertical plane the shear stresses are zero, but on all other planes there are certain shear stresses. If the vertical force F gradually increases during the test, the size of the circle will gradually increase, and if the force is sufficiently large the circle will touch the straight lines indicating the Coulomb criterion, the *Mohr-Coulomb envelope*. In that situation there are two planes on which the combination of shear stress and normal stress is such that the maximum shear stress, according to (21.1) is reached. These are the planes for which the stress points are indicated by C and D in the figure. The direction of the normals to these planes can be found by connecting the points C and D with the pole. The orientation of the planes themselves is perpendicular to these normals. In Figure 22-4 these planes have been indicated in the right half of the figure, by dashed lines.

When several tests are performed on the same material, but at different cell pressures, the various critical circles define the envelope, so that the values of the cohesion c and the friction angle ϕ can be determined. The usual practice is to do two tests, on the same material, at clearly different cell pressures. In each of the tests a value of the major principal stress σ_1 is found, at a certain value of the lateral stress σ_3. The two critical circles can be drawn in a Mohr diagram, and the Mohr-Coulomb envelope can then be determined by drawing straight lines touching these two circles, see Figure 22-5. In this way the values of c and ϕ can be determined.

When doing more than two tests the accuracy of the basic assumption that the envelope is a straight line can be tested. It is often found that for high stresses the value of the friction angle ϕ somewhat decreases.

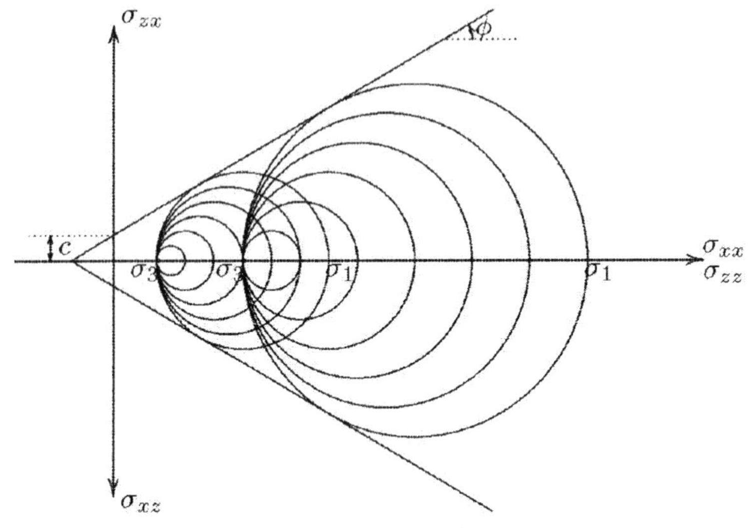

Figure 22-5. Determination c and ϕ from two tests.

For sands the tests usually give that the cohesion c is practically zero, and that the friction angle ϕ varies from about 30° to 45°, depending on the type of sand, and its packing. Sharp sand, i.e. sand with many sharp angles, usually has a much higher friction angle than sand consisting of rounded particles. And densely packed sand has a higher friction angle than loosely packed sand. For clay the cohesion may be of the order of magnitude of 5 kPa to 50 kPa, or even higher, whereas ϕ may vary from 15° to 30°. The friction angle ϕ is a material constant, while the cohesion c is (linearly) related to the pre-consolidation pressure. For the determination of c and ϕ of clay care must be taken that the influence of pore pressures is accounted for, see Chapter 25.

It may be mentioned that the strength of rock can also be determined by triaxial tests. The pressures then are much higher, and the cell wall usually is made of steel rather than glass. In petroleum engineering, where the properties of deep layers of rock are of paramount importance, rock samples are often tested by triaxial tests.
From Mohr's circle, see Figure 22-4, it has been found that the critical planes make angles of $\frac{\pi}{4}-\frac{\phi}{2}$ with the vertical direction. If the failure mechanism would consist only of sliding along one of these planes the test would result in a discontinuity in the deformation pattern in the direction of that plane. This is indeed sometimes found, for rather loose sands, but very often the deformation pattern is disturbed by more or less simultaneous sliding along different planes, by rotations, and by elastic deformations. Even when a clear sliding surface seems to appear, it is not

recommended to try to determine the friction angle by measuring the angle of that surface with the vertical direction, and equating it to $\frac{\pi}{4}-\frac{\phi}{2}$. This often leads to significant errors, as angles between $\frac{\pi}{4}$ and $\frac{\pi}{4}-\frac{\phi}{2}$ may be observed, and repetition of the test may lead to different direction. This can be explained by considering a thin zone in which failure occurs, with sliding along different sliding planes in the interior of that zone. The macroscopic (apparent) sliding angle depends on the relative contribution of each of the two sliding directions. Figure 22-6 shows an example with possible sliding planes at angles of 30° with the vertical direction. The case represented in the figure consists of a combination of a large shearing of the right hand side with respect to the left hand side along one set of planes, and a small shearing of the left hand side with respect to the right hand side along the other set of planes. The result appears to be that an apparent shearing takes place over an angle with the vertical direction that is considerably larger than 30°, that is less steep. If one would consider that angle to be $\frac{\pi}{4}-\frac{\phi}{2}$, the friction angle ϕ would be grossly underestimated.

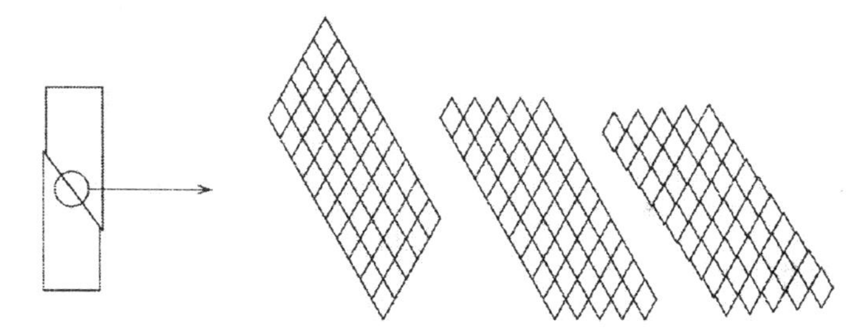

Figure 22-6. Apparent shear plane in triaxial test.

It should be noted that there is absolutely no need to determine the friction angle ϕ from the direction of a possible sliding plane. The merit of the triaxial test is that it provides a relatively simple and accurate method for the determination of the strength parameters c and ϕ from two tests, because in both tests the critical stresses are very accurately measured. The cell pressure and the vertical force can easily be controlled and measured, and therefore the determination of the critical stress states is very accurate. In other tests this may not be the case.

Problems

22.1 On two soils samples, having a diameter of 3.8 cm, triaxial tests are performed, at cell pressures of 10 kPa and 20 kPa, respectively. In the first test failure occurs for an axial force of 22.7 N, and in the second test for an axial force of 44.9 N. Determine c and ϕ of this soil.

22.2 It is given that for a certain sand $c=0$ and $\phi=30°$. A triaxial test is done on this sand, using a cell pressure of 100 kPa. The diameter of the sample is 3.8 cm. What is the axial force at the moment of failure?

22.3 Is it technically possible to perform a test on a sample in a triaxial apparatus such that the vertical stress is smaller than the horizontal stress, which is always equal to the cell pressure?

23 Dutch cell test

An early version of the triaxial test was developed around 1938 by Keverling Buisman, see Figure 23-1. Actually, the triaxial test can be considered to be an improved version of this *Dutch cell test*. The apparatus consists of a container with a cylindrical glass wall (the cell), with a fixed rubber membrane, in which the sample can be installed. The water

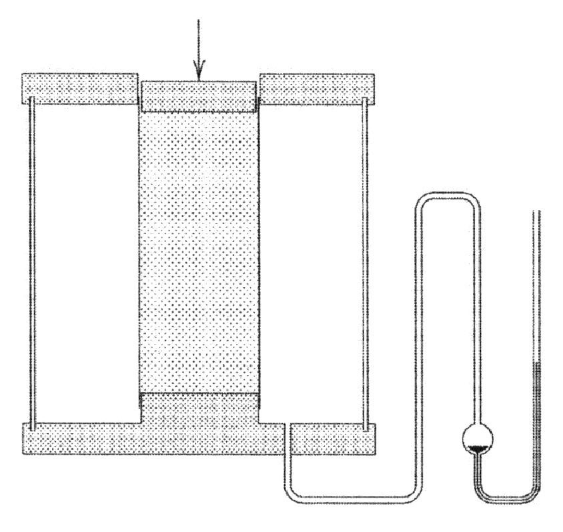

Figure 23-1. Cell test.

pressure in the cell can be controlled. In contrast with the triaxial apparatus, in which the cell pressure also acts on the top of the sample, in the cell test the pressure in the cell only acts on the cylindrical surface of the sample, because of the way of installing the membrane, which is glued to the top and bottom plates of the apparatus. The usual testing procedure consists of applying a vertical load, by means of a gradually increasing dead weight, and to measure the vertical deformation. Note that in the standard triaxial test the rate of vertical deformation is imposed, and the vertical load is measured. The cell test is *stress controlled*, whereas the triaxial test is *strain controlled*. In horizontal direction there is no difference between the two tests: in both the horizontal stress is controlled by the cell pressure. The size of the sample in a cell test usually is 6.5 cm diameter, which is somewhat larger than the size of the samples in a triaxial test (3.8 cm). This is of minor importance, however, and there exist larger cells, especially for tests on course material, such as gravel.

The original purpose of the cell test was to investigate the stresses that would occur in a loaded sample while the strains remained small, without reaching failure. For this purpose the lateral deformations should be kept to a minimum, close to zero. This can be accomplished, approximately, by filling the cell with water, and to prevent volume change of this water, by closing the cell. The cell pressure will then increase if the vertical stress is increased. In this classical form of the cell test the cell pressure and the vertical deformation are measured as a function of the vertical stress. The measured states of stress can be considered as safe, and correspond to small deformations only. A disadvantage of this procedure is that the cell pressure, and thus the state of stress in the sample, depends upon the stiffness of the system of cell, water and rubber membrane. An air bubble in the cell, or a fold in the membrane, may lead to greater lateral deformations and lower horizontal stresses. An advantage is that the sample is never brought to failure, so that the behaviour at various stress levels can be tested on a single sample.

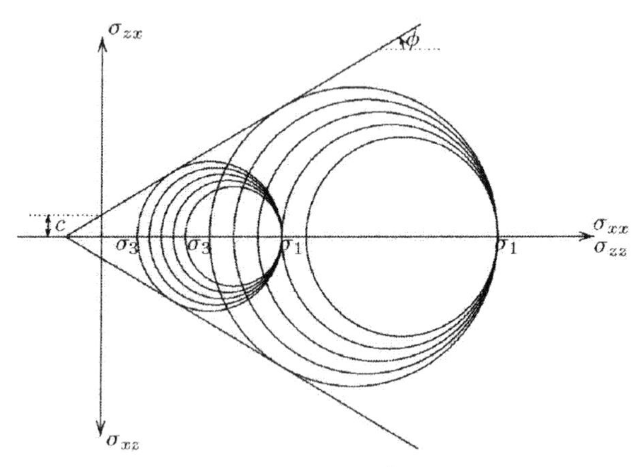

Figure 23-2. Determination c and ϕ from two cell tests.

In the course of time the procedures of the triaxial test and of the cell test have been modified, and the possibilities have been increased, so that an arbitrary combination of stresses can be applied, and all strains can be measured, or that the deformations are imposed and the corresponding stresses are measured. The main difference between the triaxial test and the cell test then is that in the cell test the cell stress does not act on the top of the sample, but only in horizontal direction.

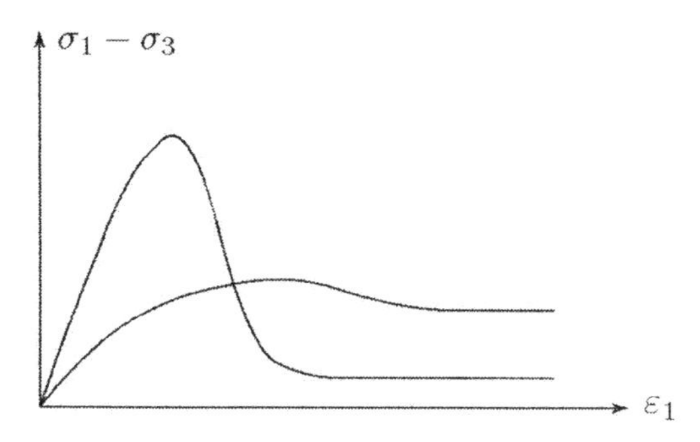

Figure 23-3. Some test results.

Many testing procedures have been developed. For the determination of the shear strength parameters c and ϕ a popular procedure is to do two triaxial tests on two different samples of the same material, at different cell pressures, at a constant vertical deformation rate, as described in Chapter 22. Another procedure is to load a sample in a cell test, keeping the vertical stress constant, and then to carefully bring the sample almost to failure, by letting a small amount of water escape from the cell, in the form of water drops. The cell pressure will decrease, and water can be drained until the cell pressure remains constant. The sample then is on the verge of failure, but the deformations remain small. In a second stage a new test can be done on the same sample, by first increasing the vertical stress, with the drainage tap of the cell closed. The cell pressure will increase, with a value that depends upon the stiffness of the system of cell, water and membrane. This new cell pressure can easily be measured, of course. Next the sample can again be brought to the limit of failure by draining off some water from the cell, in small drops. The Mohr circles for such a testing procedure are shown in Figure 23-2. Because in this type of test actual failure of the sample is just avoided, the values of the shear strength parameters c and ϕ are usually somewhat smaller than the values obtained from triaxial tests.
It may be mentioned that very often laboratory tests are being used to determine the relation between stress and strain for the entire range of strains, from the small deformations in the early stages, up to the large deformations at failure, and perhaps beyond, see Figure 23-3. If the vertical load is applied by imposing the strain (or the strain rate) a possible decrease of the stress after reaching the maximum stress can also be detected. The maximum strength is called *peak strength*, and the final strength, at very large strains, is called the *residual strength*. For certain types of soils the residual strength is much lower than the peak strength, for instance the calcareous sands that occur in offshore coastal zones of Western Australia and Brazil. An example of such a result is also shown in Figure 23-3. In this type of material the peak strength is so high, with respect to the residual strength, because the sand particles have been cemented together. The sand will become very stiff, but brittle. It appears to be very strong, and it is, but as soon as the structure has been broken, the strength falls down to a much lower value. In the construction of two offshore platforms near the coast of Western Australia this has caused large

problems, because the shear strength of the soil was reduced very severely after the driving of the foundation piles through the soil.

Very often it is not sufficient to just determine the maximum strength of the soil, at failure. Then only part of the available information is being used, and there is a definite risk of overestimating the strength. In general it is much better to determine the relation between stresses and strains over the entire range of possible strains. This also enables to take into account the reduction of the shear strength after a possible peak, depending on the strain of the soil.

Problems

23.1 In a cell test a sample of dry sand, of diameter 6.5 cm is loaded by a weight of 5 kg. Some water is carefully drained from the cell, until the cell pressure remains constant, indicating failure. The cell pressure then is 5.2 kPa. If it is assumed that the cohesion $c = 0$, then what is the friction angle ϕ of the material?

23.2 In a modified cell test a sample of dry sand ($c = 0, \phi = 40°$) is being tested. The diameter of the sample is 6.5 cm. The cell pressure is kept constant at 10 kPa. What can be the vertical load before failure of the sample occurs?

24 Shear test

The notion that failure of a soil occurs by sliding along a plane on which the shear stress reaches a certain maximum value has lead to the development of shear tests. In such tests a sample is loaded such that it is expected that one part of the sample slides over another part, along a given sliding plane. It is often assumed that the sliding plane is fixed and given by the geometry of the equipment used, but it will appear that the deformation mode may be more complicated.

24.1 Direct shear test

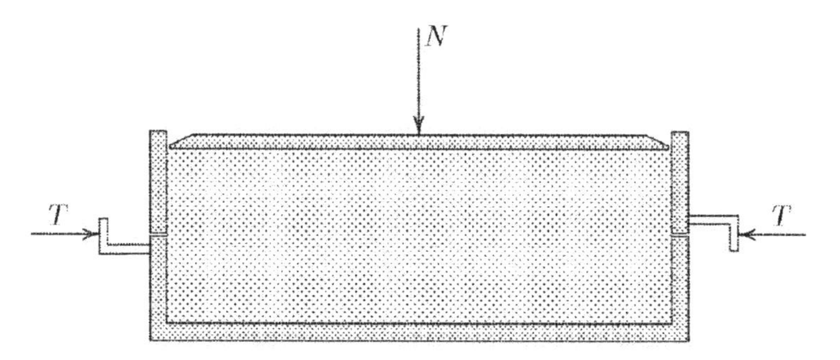

Figure 24-1. Direct shear test.

For dry sand, with $c = 0$, it is possible to determine the friction angle ϕ from one test. In this chapter it is assumed for simplicity that the soil is dry sand, with $c = 0$. The simplest apparatus is shown in Figure 24-1. It consists of a box (*the shear box*) of which the upper half can be moved with respect to the lower half, by means of a motor which pushes the lower part away from the upper part, which is fixed in horizontal direction. The cross section of the container usually is rectangular, but circular versions also exist. The soil sample is loaded initially by a vertical force only, applied by the dead weight of a loading plate and some additional weights on it, through the intermediary of a small steel plate on top of the sample. Because of this plate the sample is free to deform in vertical direction during the test. The actual test consists of the lateral movement of the lower half of the box with respect to the upper half, at a constant (small) speed, with a horizontal force acting in the plane between the two halves. This test is called the *direct shear test*. The horizontal force gradually increases, as the box moves, and is measured by a pressure ring or a strain gauge. The horizontal force reaches a maximum value after some time, and the force remains more or less constant afterwards, or it may slowly increase or decrease. It seems logical to assume that the maximum value of the horizontal force (T_f) is related to the vertical force N by a relation in analogy with Coulomb

$$\tau_f = c + \sigma_n \tan\delta \quad \text{with} \quad \tau_f = \frac{T_f}{A}, \sigma_n = \frac{N}{A} \quad \text{and} \quad \delta = \phi. \tag{24.1}$$

where A is the area of the sample, c is the cohesion of the material. The last assumption ($\delta = \phi$) however is not correct. The shear relation of Coulomb and the corresponding friction angle δ are only based on the normal stress (σ_n), so only the horizontal shear plane is checked for failure. The shear relation of Morh-Coulomb and corresponding angle of internal friction ϕ however, are based on two principle stresses (σ_1 and σ_3), so the soil is checked for failure in all direction. This means that the shear force T not only depends on the vertical stress σ_n, but also on the horizontal stress σ_h. This last one however is unknown during this test and is strongly related with the margins around the sample, which is installed in the shear box.

If the shear relations of Coulomb and Mohr-Coulomb for cohesionless materials are equalized, a relation between the friction angle δ and the angle of internal friction ϕ is found

$$\tan\delta = \frac{1}{2}\sqrt{(1+\frac{\sigma_h}{\sigma_n})^2 \sin^2\phi - (1-\frac{\sigma_h}{\sigma_n})^2}. \tag{24.2}$$

If the horizontal stress is assumed to be equal to the vertical (normal) stress, it follows

$$\tan\delta = \sin\phi. \tag{24.3}$$

Since the horizontal stress during a direct shear test is almost always smaller than the vertical stress, it can be concluded that the friction angle δ is smaller than the angle of internal friction ϕ. Therefore many investigators have found that the test results of shear tests lead to values for the shear strength that are considerably lower than the values obtained from triaxial tests. Furthermore, it has sometimes been found that the reproducibility of the results of shear tests is not so good.

Because of this, the direct shear test is seldom applied to clay- and sand samples, the triaxial test is preferred for these materials. For peat the situation is different. Because peat often contains a lot of old leaves and reed, the horizontal shear strength can be much lower than measured with a triaxial test, because of the diagonal shear plane. Because of this the direct shear test gives better results than the triaxial test for peat, despite the problems mentioned earlier.

24.2 Simple shear test

Apart from the difficulty that the state of stress is not completely given in a shear test, the direct shear test suffers from the disadvantage that the deformation is strongly inhomogeneous, because the deformations are concentrated in a zone in the centre of the shear box. An improved shear box has been developed by Roscoe in

Cambridge (England), in which the deformation is practically homogeneous. The apparatus has been constructed with rotating side walls, so that a uniform shear deformation can be imposed on the sample, see Figure 24-2. This is denoted as the *simple shear* apparatus.

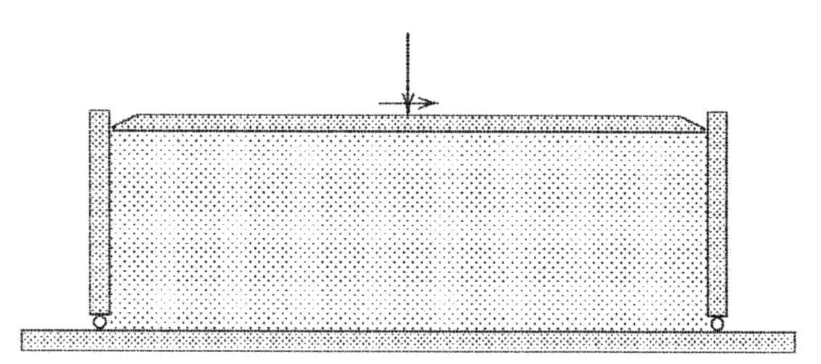

Figure 24-2. Simple shear test.

As in the direct shear box, the cross section in the horizontal plane is rectangular. The improvement is that the hinges at the top and the bottom of the side walls enable a uniform shear deformation of the sample. In Norway a variant of this apparatus has been developed, with a circular cross section. A uniform deformation is then ensured by constructing the box using a system of stiff metal rings, that can slide over each other.

Although the simple shear test is a definite improvement with respect to the direct shear test, because the deformations are much more homogeneous, the problem of the unknown horizontal stress is not solved.

It may be concluded that the shear test is not very well suited for an accurate determination of the shear strength parameters of a soil, because the state of stress is not fully known. The scatter in the results, and the relatively low values that are sometimes obtained, may well be a result of the unknown horizontal stress. The triaxial test does not suffer from this defect, as in this test the horizontal stress and the vertical stress can both be measured accurately.

It may be mentioned that in soil mechanics practice laboratory tests can sometimes be considered as scale tests of the behaviour in the field. The oedometer test can be considered as such, when the initial stresses and the incremental stresses are taken equal to those in the field. For the problem of the shearing resistance of a large concrete offshore caisson, loaded by wave forces on the caisson, a shear test may be used if the vertical normal stress and the shear stress on the sample simulate the stresses to be expected in the field, and the sample has been carefully taken from the field to the shear box. Possible errors or inaccuracies may have the same effect in the laboratory and in the field, so that they do not invalidate the applicability of the test. But in this case it is also important to ensure that the horizontal stress in the sample is of the same order of magnitude as the horizontal stress in the field.

Problems

24.1 A shear test is performed on a sample of sand, with $c = 0$ and $\phi = 40^\circ$, as determined in triaxial tests. The sand has been poured very carefully into the shear box, so that it can be expected that the horizontal stress in the sample is half the value of the vertical stress. The vertical normal stress is 100 kPa. What is the maximum allowable shear stress on the sample?

24.2 If the test described in the previous problem is interpreted in the classical way, with $\delta = \phi$, what would be the value of ϕ?

25 Pore pressures

In the previous chapters the main principles of triaxial tests, and the very similar cell test, have been presented. For simplicity it was assumed that the material was dry soil, so that there were no pore pressures, and the effective stresses were equal to the applied stresses. In reality, especially for clay soils, the sample usually contains water in its pores, and loading the soil may give rise to the development of additional pore pressures. The influence of these pore pressures will be described in this chapter.

25.1 Measuring the pore pressure

There are two possibilities to control the pore pressures in a triaxial test: either execute the test, on a drained sample, at a very low deformation rate, so that no pore pressures are developed at all, or measure the pore pressures during the test. In the first case the drainage of the sample can be ensured by filter paper applied at the top and/or bottom ends of the sample, together with a drainage connection to a water reservoir, and taking care that the duration of the test is so long that consolidation has been completed during the test. The consolidation time should be estimated, using estimated values of the permeability and the compressibility of the sample, and the duration of the test should be large compared to that consolidation time. This may mean that the test will take very long, and usually it is impractical. A better option is to measure the pore pressures in the sample, for instance by means of an electric pore pressure meter. This is a pressure deducer in which the pressure is measured on the basis of the deflection of a thin steel membrane, using a strain gauge on the membrane. The pore pressure meter is connected to the top or the bottom of the sample, see Figure 25-1. An alternative is to measure the pore pressure in the interior of the sample, using a thin needle. Whatever the precise system is, care should be taken that the measuring device is very stiff, i.e. that it requires only a very small amount of water to record a pore pressure increment. Otherwise a considerable time lag between the sample and the measuring device would occur, and the measurements would be unreliable, as they may not be representative of the pore pressures in the sample. An electrical pore pressure meter usually is very stiff: it may require only 1 mm^3 of water to record a pressure increment of 100 kPa. The response of such a stiff instrument is very fast, but it is very sensitive to the inclusion of air bubbles, because air is very compressible. Great care should be taken to avoid the presence of air in the system.

If the pore pressures during the test are known it is simple to determine the effective stresses from the measured total stresses, by subtraction of the pore pressures. Because failure of the soil is determined by the critical values of the effective stresses the shear strength parameters c and ϕ can then be determined. In the stress diagrams the effective stresses must be plotted, and the envelope of the Mohr circles yields the cohesion c and the friction angle ϕ. The procedure is illustrated in Figure 25-2. The test results are recorded in Table 25-1. The table refers to two tests, performed at cell pressures of 40 kPa and 95 kPa. In both tests the pore pressures developed in the first stage of loading, by the application of the cell pressure, have been reduced to zero, by waiting sufficiently long for complete drainage to have taken place. In the second stage of the tests the vertical force has been increased, at a fairly rapid rate, measuring the pore pressures during the test. The total stresses have all been represented in Figure 25-2 by the Mohr circles. For the last circles, corresponding to the maximum values of the vertical load, the effective stress circles have also been drawn, indicated by the dotted circles. On the basis of the two critical effective stress circles the Mohr-Coulomb envelope can be drawn, and the values of the cohesion c and the friction angle ϕ can be determined. In the case of Figure 25-2 the result is $c = 9$ kPa and $\phi = 30°$.

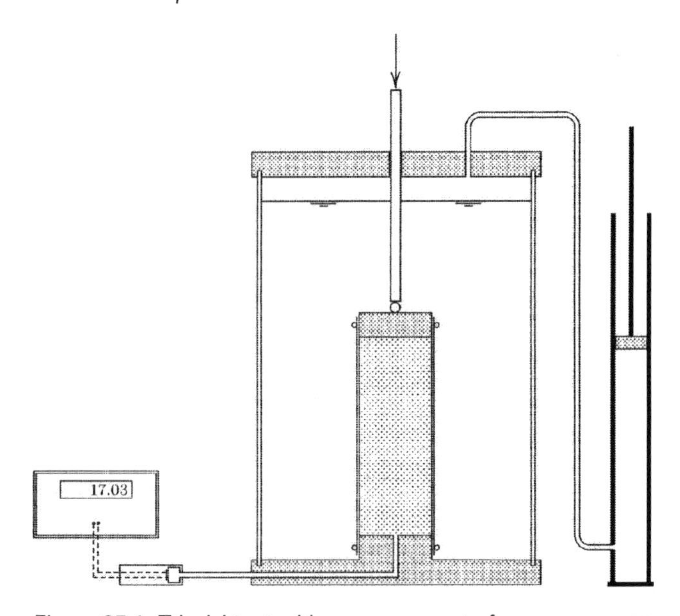

Figure 25-1. Triaxial test with measurement of pore pressures.

It should be emphasized that the strength parameters c and ϕ should be determined on the basis of critical states of stress for the *effective stresses*. If in the test described above some drainage would have occurred, and the pore pressures would have been lower, the critical total stresses would also have been different (lower). Only if the test results are represented in terms of effective stresses they will lead to the same values of c and ϕ, as they should.

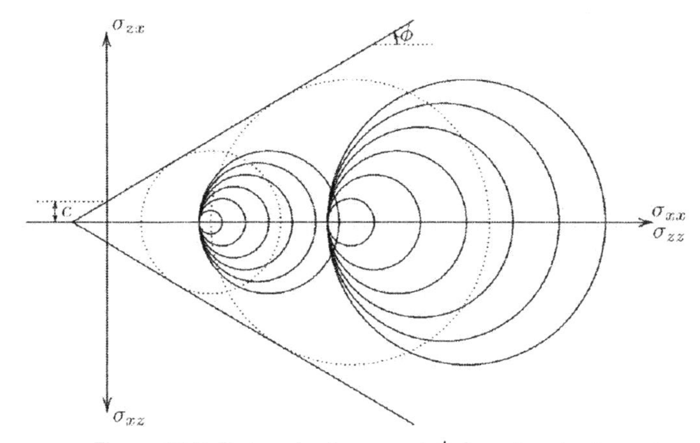

Figure 25-2. Determination c and ϕ from two tests.

Test	σ_3	$\sigma_1-\sigma_3$	p	σ_3'	σ_1'
1	40	0	0	40	40
	40	10	4	36	46
	40	20	9	31	51
	40	30	13	27	57
	40	40	17	23	63
	40	50	21	19	69
	40	60	25	15	75
2	95	0	0	95	95
	95	20	8	87	107
	95	40	17	78	118
	95	60	25	70	130
	95	80	33	62	142
	95	100	42	53	153
	95	120	50	45	165

Table 25-1: Test results.

25.2 Types of triaxial tests

The procedure in the tests described above, with the results given in Table 25-1, is that in the first stage of the tests, the application of the cell pressure, the soil is free to consolidate, and sufficient time is taken to allow for complete consolidation, i.e. the excess pore pressures are reduced to zero. In the second stage of the test, however, no consolidation is allowed, by closing the tap to the drainage reservoir. Such a test is denoted as a *Consolidated Undrained* test, or a CU-test. This is a common procedure, but several other procedures exist.

If in the second stage, the vertical loading of the sample, pore pressures are again avoided by allowing for drainage, and by a very slow execution of the test (a very small loading rate), the test is denoted as a *Consolidated Drained* test, or a CD-test. Such a test takes a rather long time, which is expensive, and sometimes impractical.

A further possibility is to never allow for drainage in the test, not even in the first stage of the test, by sealing off the sample. This is an *Unconsolidated Undrained* test, or a UU-test.

25.3 Elastic response

It may be illustrative to try to predict the pore pressures developed in a triaxial test using basic theory. This will appear to be not very accurate and reliable, but it may give some insight into the various mechanisms that govern the generation of pore pressures.

The basic notion is that the presence of water in the pores obstructs a volume change of the sample. The presence of water in no way hinders the shear deformation of a soil element, but a volume change is possible only if water is drained from the sample or if the water itself is compressed. The particles are assumed to be so stiff that their volume is constant. At the moment of loading drainage can not yet have lead to a volume change, and thus the only possibility for an immediate volume change is a compression of the fluid itself. This can be described by

$$\Delta V = n\beta V \Delta p, \tag{25.1}$$

where V is the volume of the sample, Δp is the increment of the pore pressure, and β is the compressibility of the water, see also Chapter 16. The instantaneous volume strain is

$$\varepsilon_{vol} = \frac{\Delta V}{V} = n\beta\Delta p. \tag{25.2}$$

Because the compressibility of the water (β) is very small, this is a very small quantity.

On the other hand, if the soil skeleton is assumed to deform elastically, the volume strain can be expressed as

$$\varepsilon_{vol} = \frac{\Delta\sigma'}{K}, \tag{25.3}$$

where K is the compression modulus of the soil, and $\Delta\sigma'$ is the increment of the isotropic effective stress. Because the volume strain is so small, the increment of the isotropic effective stress will also be very small. It can be expressed as the increment of the average of the three principal stresses,

$$\Delta\sigma' = \tfrac{1}{3}(\Delta\sigma_1' + \Delta\sigma_2' + \Delta\sigma_3'). \tag{25.4}$$

It follows from Eq. (25.3), with $\sigma' = \sigma - p$, that

$$\varepsilon_{vol} = \frac{\Delta\sigma - \Delta p}{K}. \tag{25.5}$$

From (25.2) and (25.5) it finally follows that

$$\Delta p = \frac{\Delta\sigma}{1+n\beta K}. \tag{25.6}$$

This formula expresses the increment of the pore water pressure into the increment of the isotropic total stress. If the water is incompressible ($\beta = 0$), the increment of the pore pressure is equal to the increment of the isotropic total stress. All this is in complete agreement with the considerations in Chapter 16 on consolidation. The relation (25.6), with $\beta = 0$ can directly be obtained by noting that in a very short time there can be no volume change if the water is incompressible. Hence there can be no change in the isotropic effective stress, and thus the pore pressure must be equal to the isotropic total stress. Only if the water is somewhat compressible there can be a small instantaneous volume change, so that there can be a small increment of the effective stress, and thus the pore pressure is somewhat smaller than the isotropic total stress.
In general equation (25.6) can also be written as

$$\Delta p = \frac{\Delta\sigma_1 + \Delta\sigma_2 + \Delta\sigma_3}{3(1+n\beta K)}, \tag{25.7}$$

In a triaxial test $\Delta\sigma_2 = \Delta\sigma_3$, and in such tests the basic stress parameters are the cell pressure $\Delta\sigma_3$ and the additional vertical stress, produced by the axial load, $\Delta\sigma_1 - \Delta\sigma_3$. This suggests to write Eq. (25.7) in the form

$$\Delta p = \frac{1}{1+n\beta K}[\Delta\sigma_3 + \tfrac{1}{3}(\Delta\sigma_1 - \Delta\sigma_3)]. \tag{25.8}$$

In an undrained triaxial test it can be expected that increasing the cell pressure leads to an increment of the pore pressure practically equal to the increment of the cell pressure, assuming that $n\beta K \ll 1$. Furthermore, if the cell pressure remains constant, and the vertical load increases, the increment of the pore pressure will be about $\frac{1}{3}$ of the additional vertical stress. Indeed, such values are sometimes measured, approximately, see for instance the test results given in Table 25-1. Very often the results show considerable deviations from these theoretical results, because the water may not be incompressible (perhaps due to the presence of air bubbles in the soil), or because the sample is not isotropic, or because the sample exhibits non-linear properties, such as dilatancy. Furthermore, the measurements may be disturbed by inaccuracies in the measurement system, such as air bubbles in the pore pressure meter.

25.4 Dilatancy

The analysis of the previous section may be generalized by taking dilatancy into account. The basic idea remains that at the moment of loading there can not yet have been any drainage, so that the only possibility for a volume change is the compression of the water in the pores. This can be expressed by Eq. (25.2),

$$\varepsilon_{vol} = \frac{\Delta V}{V} = n\beta\Delta p. \tag{25.9}$$

It is now postulated that the volume change of the pore skeleton is related to the stress changes by

$$\varepsilon_{vol} = \frac{\Delta\sigma'}{K} - \frac{\Delta\tau}{M}. \tag{25.10}$$

The first term is the volume change due to the average compressive stress, which is determined by the isotropic effective stress σ'. The second term in Eq. (25.10) is the volume change caused by the shear stresses. It has been assumed that this is determined by some measure for the deviatoric stresses, indicated as τ, and as a first approximation it has been assumed that this volume change is proportional to the increment of τ, with a modulus M. That is a simplification of the real behaviour, but at least it gives the possibility to investigate the effect of dilatancy, because this term expresses that shear stresses lead to a volume increase, if $M > 0$, which indicates a densely packed soil. If $M < 0$ there would be a volume decrease due to an increment of the shear stresses. Such a behaviour can be expected in a loose material.
Because $\sigma' = \sigma - p$ it follows that

$$\Delta p = \frac{1}{1 + n\beta K}(\Delta\sigma - \frac{K}{M}\Delta\tau). \tag{25.11}$$

This is a generalization of the expression (25.6). For the conditions in a triaxial test one may write

$$\Delta\sigma = \tfrac{1}{3}(\Delta\sigma_1 + \Delta\sigma_2 + \Delta\sigma_3) = \Delta\sigma_3 + \tfrac{1}{3}(\Delta\sigma_1 - \Delta\sigma_3). \tag{25.12}$$

The deviator stress τ is assumed to be

$$\Delta\tau = \tfrac{1}{2}(\Delta\sigma_1 - \Delta\sigma_3). \tag{25.13}$$

This means that the radius of the Mohr circle is used as the measure for the deviator stress τ.
The final result is

$$\Delta p = \frac{1}{1 + n\beta K}[\Delta\sigma_3 + (\tfrac{1}{3} - \tfrac{1}{2}\frac{K}{M})(\Delta\sigma_1 - \Delta\sigma_3)]. \tag{25.14}$$

This is a generalization of equation (25.8). Dilatancy does not appear to have any influence in the first stage of a triaxial test, when the isotropic stress is increased. In the second stage of a triaxial test, during the application of the vertical load, the generation of pore pressures is determined by the factor $\frac{1}{3} - \frac{1}{2}\frac{K}{M}$. The first term is a result of compression, the second term is a consequence of the dilatancy (or contractancy, when $M < 0$).
In a dilatant material, with $M > 0$, the pore water pressure will be larger than in a material without dilatancy. This is caused by the tendency of the densely packed

material to expand, which reduces the compression due to the isotropic loading. If the dilatancy effect (here expressed by the parameter M) is very large, the pore pressure may even become negative. In a very dense material the tendency for expansion will lead to a suction of water.
In a contractant material, with $M < 0$, the pore pressures will become larger due to the tendency of the material to contract. The loosely packed soil will tend to contract as a result of shear stresses, thus enlarging the volume decrease due to the isotropic stress increment. The water in the pores opposes such a volume change.

25.5 Skempton's coefficients

Skempton has suggested to write the relation between the incremental pore water pressure and the increments of the total stress in the form

$$\Delta p = B[\Delta\sigma_3 + A(\Delta\sigma_1 - \Delta\sigma_3)]. \qquad (25.15)$$

The coefficients A and B should be measured in an undrained triaxial test.
The relations given in this section would mean that

$$B = \frac{1}{1 + n\beta K}, \qquad (25.16)$$

and

$$A = \frac{1}{3} - \frac{1}{2}\frac{K}{M}. \qquad (25.17)$$

Indeed, the values of B observed in tests are usually somewhat smaller than 1, and for the coefficient A various values, usually between 0 and $\frac{1}{2}$ are found, but sometimes even negative values have been obtained.
Skempton's coefficients A and B have been found to be useful in many practical problems, but it should be noted that they have limited physical significance, because they are based upon a rather special description of the deformation process of a soil, see Eq. (25.10). When their values are measured in a triaxial test, they may be influenced by partial saturation, by anisotropy, and by the stiffness of the pore pressure meter. It should also be noted that the values of the coefficients depends upon the stress level. It is therefore suggested to determine the values of A and B in tests in which the stress changes simulate the real stress changes in the field.

Problems

Test	σ_3	$\sigma_1 - \sigma_3$	p
1	20	40.94	8.19
2	40	69.52	13.90
3	60	98.09	19.62

Table 25-2: Test results.

25.1 On a number of identical soil samples CU-triaxial tests are being performed. The cell pressure is applied, then consolidation is allowed to reduce the pore water pressures to zero, and in the second stage the sample is very quickly brought to failure, undrained. The pore pressures are measured. The results are given in the table (all stresses in kPa). Determine the values of the cohesion c and the friction angle ϕ.

25.2 What can you say about the coefficients A and B in this case?

25.3 Dense soils tend to expand in shear (dilatancy). Loose soils tend to contract (contractancy). Do you think that the soil in problem 25.1 is dense, or loose?

25.4 A completely saturated clay sample is loaded in a cell test by a vertical stress of 80 kPa. Due to this load the cell pressure increases by 20 kPa. If the soil were perfectly elastic, what would then be the increment of the pore pressure?

26 Undrained behaviour

If no drainage is possible from a soil, because the soil has been sealed off, or because the load is applied so quickly and the permeability is so small that there is no time for outflow of water, there will be no consolidation of the soil. This is the undrained behaviour of a soil. This chapter contains an introduction to the description of this undrained behaviour.

26.1 Undrained tests

In an undrained triaxial test on a saturated clay each increase of the cell pressure will lead to an increase of the pore water pressure. As discussed in the previous chapter this can be described by Skempton's formula

$$\Delta p = B[\Delta\sigma_3 + A(\Delta\sigma_1 - \Delta\sigma_3)]. \tag{26.1}$$

The coefficient B can be expected to be about

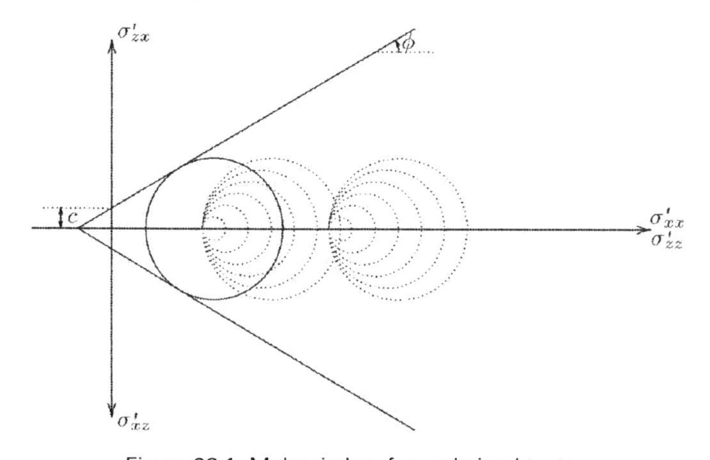

Figure 26-1. Mohr circles for undrained tests.

$$B = \frac{1}{1 + n\beta K}, \tag{26.2}$$

where β is the compressibility of the pore fluid (including possible air bubbles) and K is the compression modulus of the grain skeleton. The value of the coefficient B will be close to 1, as the water is practically incompressible.

Increasing the cell pressure can be expected to result in an increment of the pore pressure by the same amount as the increment of the cell pressure, or slightly less, and thus there will be very little change in the effective stresses. If there is a possibility for drainage, and there is sufficient time for the soil to drain, the pore pressures will be gradually reduced, with a simultaneous increase of the effective stresses. This is the consolidation process. If there is no possibility for drainage, because the sample has been completely sealed off, or because the test is done so quickly that there is no time for consolidation, the test is called *unconsolidated.* In the second stage of a triaxial test, in which only the vertical stress is increased, distinction can also be made in drained or undrained tests. If in this stage no drainage can take place, the test is called *unconsolidated undrained* (a UU-test). If a second UU-test is done at a higher cell pressure, the only difference with the first test will be that the pore pressures are higher. The effective stresses in both tests will be practically the same. If the test results are plotted in a Mohr diagram, there would be just one critical circle for the effective stresses, but in terms of total stresses there will be two clearly distinct circles, of practically the same magnitude, see Figure 26-1. In this figure the critical Mohr circles for the total stresses in the two tests have been dotted. The critical circles for the effective stresses can be obtained by subtracting the pore pressure, and these are represented by full lines. The two circles practically coincide, if the sample is saturated with water.

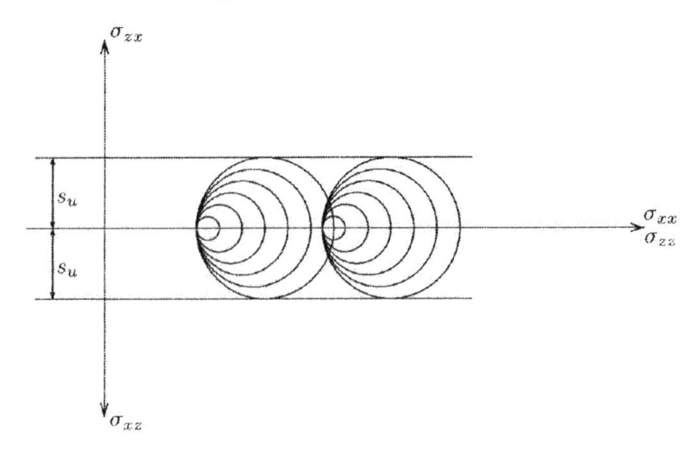

Figure 26-2. Mohr circles for total stresses.

These test results appear to be insufficient to determine the shear strength parameters c and ϕ, because only one critical circle for the effective stresses is available. In order to determine the values of c and ϕ in at least one of the tests the sample should be allowed to consolidate after the first loading stage, so that the isotropic effective stress at the beginning of the second stage, the vertical loading, is different in the two tests. This would mean that this test should be a *Consolidated Undrained* test, or a CU-test.

Admitting that undrained tests can not be used to determine the correct values of the shear strength parameters c and ϕ, they may still be very useful, because in engineering practice there are many situations in which no (or very little) drainage will occur, for instance in case of loading of a soil of very low permeability (clay) for a short time. Examples are a temporary loading for some building operation, or a temporary excavation for the construction of a pipe line. In order to predict the behaviour of the clay in these circumstances it makes sense to just consider the total stresses, and to make use of the results of an undrained test, analysing the test results in terms of total stresses also. That there may be considerable pore pressures in the test as well as in the field, is perhaps interesting, but irrelevant if the period of loading is so short that no consolidation can occur.

The analysis of the tests in terms of total stresses is illustrated in Figure 26-2. As explained above, all critical stress circles will be of the same magnitude, and when the results are interpreted in terms of total stresses only it seems that the friction angle ϕ is practically zero. The strength of the soil can be characterized by a cohesion only, which is then usually denoted as s_u, the *undrained shear strength* of the soil. The analysis, in which the friction of the material and the pore pressures are neglected, is called an *undrained analysis*. Because the analysis of the safety of a structure on a purely cohesive material (with $\phi = 0$) is much simpler that the analysis for a material with internal friction, an undrained analysis is often used in engineering practice.

The applicability of undrained tests, and the use of undrained strength parameters is also justified if it can be expected that the most critical situation will be the undrained state immediately after loading. In many cases of loading of a soil by a constant load, it can be expected that the largest pore pressures will be developed immediately after loading, and that these pore pressures will gradually dissipate during consolidation of the soil, with the effective stresses increasing.

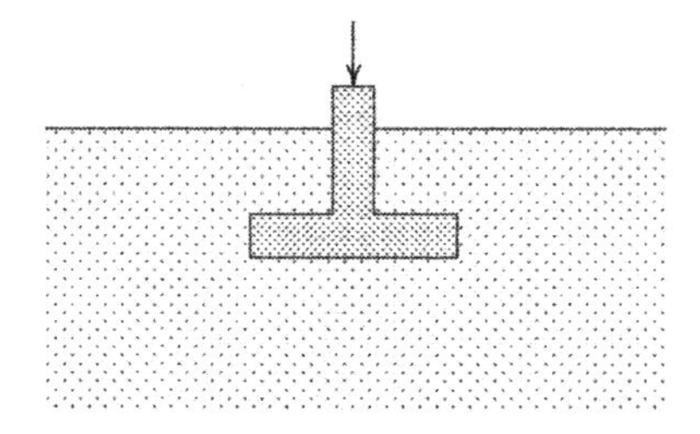

Figure 26-3. Shallow foundation.

For instance, in the case of a permanent load applied to a shallow foundation slab, see Figure 26-3, it can be expected that pore pressures will be developed below the foundation, and that these pore pressures will dissipate in course of time due to consolidation. If the load remains constant, it can be expected that the pore pressures are highest, and thus the effective stresses are smallest, just after the application of the load.

Later, after consolidation, the effective stresses will be higher, so that the Mohr circle will be shifted to the right. This means that the most critical situation occurs immediately after application of the load, in the undrained state. If the structure is safe immediately after application of the load it will certainly be safe at later times, when the pore pressures have been dissipated, the effective stresses have increased, and thus the strength of the soil has been further developed.

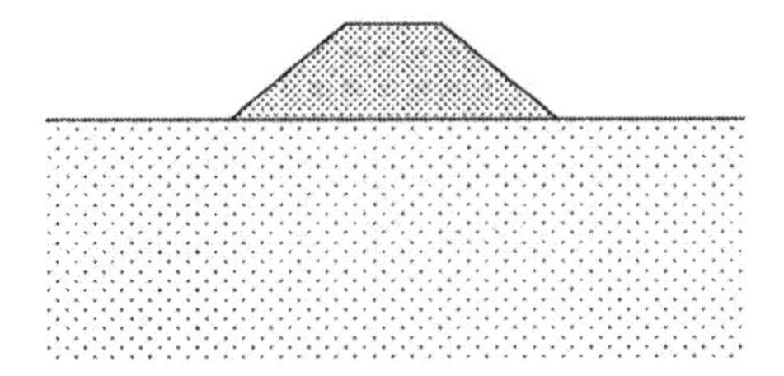

Figure 26-4. Embankment.

In the case of the construction of an embankment, for a dike or a road, an undrained analysis may also be sufficient for the analysis of the stability of the embankment itself, see Figure 26-4. In many cases it can be assumed that the construction of the embankment is one of the most critical phases in its lifetime. If the embankment "survives" the construction, then it will probably will be stable forever. The pore pressures are largest during the construction of the embankment. Later these will be reduced, the effective stresses will increase, and therefore the shear strength will increase. In many cases this additional strength is sufficient to even accept future additional loadings, for instance by water pressures against the slope of the dike, or by traffic, in case of a highway. In some exceptional cases, of very soft soils, with a very low permeability, there may be additional undrained creep deformations, prior to the effect of consolidation, so that the pore pressures may increase in the first few days or weeks after construction. In one case, a dike near Streefkerk in The Netherlands, this has resulted in failure of the dike a few days after its construction.
Of course it is not sufficient to assume, without further proof, that the reduction of the pore pressures, caused by consolidation, will be sufficient to accommodate the additional pore pressures due to the additional loading. A dike is built to withstand the forces of the water during a storm with high water levels, and the behaviour of the dike under these conditions needs careful analysis. Immediately after application of the load, in this case the water pressure against the slope of the dike, the soil may be considered as undrained, but after some days of high water the dike must still be stable. During prolonged periods of high water, the pore pressures in the dike may gradually increase, because of inflow of water into the dike body, and an unsafe situation may be created by the reduction of the effective stresses in the dike. An undrained analysis of the dike stability may be one element in its design, but an effective stress analysis, considering various combinations of loading and drainage must also be performed.
An undrained analysis is unsafe if it is to be expected that the pore pressures will increase after the construction. As an example one may consider the case of an excavation, see Figure 26-5. The excavation can be considered as a negative load, which will result in decreasing total stresses, and therefore decreasing pore pressures immediately after the excavation. Due to consolidation, however, the pore pressures

later will gradually increase, and they will ultimately be reduced to their original value, as determined by the hydrologic conditions. Thus the effective stresses will be reduced in the consolidation process, so that the shear strength of the soil is reduced. This means that in the course of time the risk of a sliding failure may increase. A trench may be stable for a short time, especially because of the increased strength due to the negative pore pressures created by the excavation, but after some time there may be a collapse of the slopes. This may be very dangerous for the people at work in the excavation, of course.

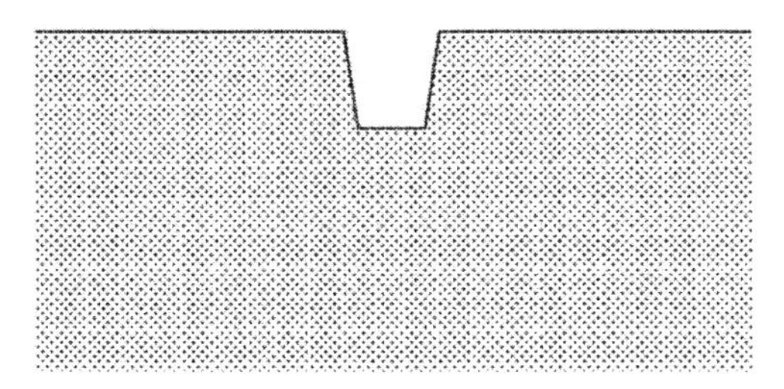

Figure 26-5. Excavation.

26.2 Undrained shear strength

For the comparison of drained and undrained calculations, and for the actual calculation in an undrained analysis, it is often necessary to determine the undrained shear strength s_u of a soil, from the basic shear strength parameters c and ϕ. This can be done by noting that in a saturated soil there can be practically no volume change in undrained conditions, so that the isotropic effective stress remains constant. Thus the average effective stress remains constant, and this means that the location of the Mohr circle is constrained. Usually the state of stress in the soil is such that the vertical stresses are reasonably well known, because of the weight of the soil and a possible load. If the pore water pressure is also known, this means that it can be assumed that the vertical effective stress σ'_{zz} is known. Usually the two horizontal stresses will be equal, and their magnitude may be estimated (or perhaps measured), even while that is not always very easy. Here it is assumed that the horizontal effective stress σ'_{xx} is also known. Thus the average effective stress, $\frac{1}{3}(\sigma'_{zz}+2\sigma'_{xx})$ is known. If the soil is loaded this average effective stress will remain constant,

$$\sigma'_0 = \tfrac{1}{3}(\sigma'_{zz}+2\sigma'_{xx}) = \text{constant}. \tag{26.3}$$

In case of failure of the soil the combination of the major principal stress σ'_1 and the minor principal stress σ'_3 must be such that the Mohr-Coulomb failure criterion is satisfied, i.e., with (21.12),

$$(\frac{\sigma'_1-\sigma'_3}{2}) - (\frac{\sigma'_1+\sigma'_3}{2})\sin\phi - c\cos\phi = 0. \tag{26.4}$$

Because $\sigma'_1+\sigma'_3 = \frac{2}{3}(\sigma'_1+2\sigma'_3)+\frac{1}{3}(\sigma'_1-\sigma'_3)$ this can also be written as

$$(1-\tfrac{1}{3}\sin\phi)(\frac{\sigma_1'-\sigma_3'}{2})-(\frac{\sigma_1'+\sigma_3'}{3})\sin\phi-c\cos\phi=0. \tag{26.5}$$

Because the average effective stress can not change in undrained conditions, we have, before and after the application of the load,

$$\tfrac{1}{3}(\sigma_1'+\sigma_3')=\sigma_0', \tag{26.6}$$

where σ_0' is a given value, determined by the initial stresses, see Eq. (26.3).
From (26.5) and (26.6) the undrained shear strength s_u is found to be

$$s_u=\frac{\sigma_1'-\sigma_1'}{2}=c\frac{\cos\phi}{1-\frac{1}{3}\sin\phi}+\sigma_0'\frac{\sin\phi}{1-\frac{1}{3}\sin\phi}, \tag{26.7}$$

This formula enables to estimate the undrained shear strength if the drained shear strength parameters c and ϕ are known, as well as the initial average effective stress σ_0'. The relation is illustrated in Figure 26-6. In this figure a number of Mohr circles for the effective stresses are shown, on the basis of the assumption that the average effective stress σ_0' remains constant. The total stresses always differ from the effective stresses by the (unknown) value of the pore water pressure. The location of the total stress circles is not known, and not relevant. Their magnitude is always equal to the magnitude of the corresponding effective stress circle, as the pore pressure increases all normal stresses, both σ_{xx}' and σ_{zz}'. The undrained shear strength s_u is also called the undrained cohesion c_u.

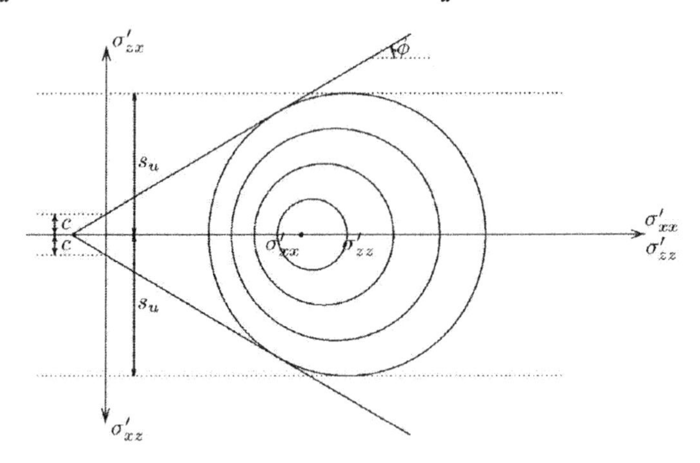

Figure 26-6. Mohr circles for undrained behaviour.

Equation (26.7) indicates that $s_u=c$ if $\phi=0$, as could be expected. If $\phi>0$ the undrained shear strength s_u increases with the average effective stress σ_0'. This means that a preload, followed by consolidation, has a positive effect on the undrained strength of the soil.
It should be noted that in the derivation of equation (26.7) it has been assumed that a volume change can be produced only by a change of the average effective stress.

This means that effects such as anisotropy, dilatancy and contractancy have been disregarded. That is an important restriction, and it means that the formula is a first approximation only.

Problems

26.1 A consolidated undrained triaxial test is done on a clay sample. The cell pressure is 50 kPa, and the sample is found to fail when the additional axial stress is 170 kPa. What is s_u?

26.2 For a certain soil it is known that $c = 20$ kPa and $\phi = 30°$. An undrained analysis must be made for a case of a soil in which the original vertical effective stresses are 80 kPa, and the horizontal effective stresses are estimated to be 40 kPa. What is the value of s_u?

26.3 What would be the answer of the previous problem if the initial horizontal effective stresses would be 80 kPa?

26.4 Check the answers of the previous two problems, as they can be found by using two methods: analytically by using Eq. (26.7), or graphically using the Mohr circle diagram, see Figure 26-6.

27 Stress paths

A convenient way to represent test results, and their correspondence with the stresses in the field, is to use a *stress path*. In this technique the stresses in a point are represented by two (perhaps three) characteristic parameters, and they are plotted in a diagram. This diagram is called a *stress path*.

27.1 Parameters

It is assumed that the state of stress in a point can be characterized by the average stress (the isotropic stress), $\frac{1}{3}(\sigma_1+\sigma_2+\sigma_3)$, and the difference of the major and minor principal stresses, $\sigma_1-\sigma_3$. By doing so it is assumed that the behaviour of a soil depends only upon these two parameters. This means that it is assumed that other parameters, such as the intermediate principal stress σ_2, or the direction of the major principal stress, are unimportant.
Alternatively, the average value of the major and minor principal stresses, $\frac{1}{2}(\sigma_1+\sigma_3)$, may be used rather than the average stress. The two variables will be denoted by σ and τ,

$$\sigma=\tfrac{1}{2}(\sigma_1+\sigma_3), \tag{27.1}$$

$$\tau=\tfrac{1}{2}(\sigma_1-\sigma_3). \tag{27.2}$$

The introduction of the factor $\frac{1}{2}$ in the two definitions results in σ and τ being the location of the centre, and the magnitude of the circle in Mohr's diagram, see Figure 27-1.

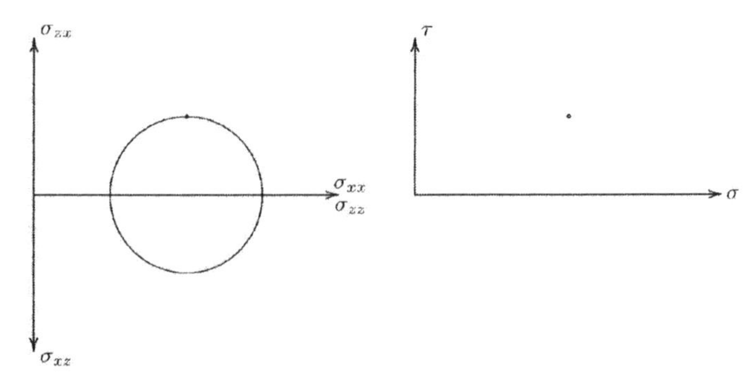

Figure 27-1. Mohr's circle and stress point.

By choosing these parameters it is implicitly assumed that other parameters are unimportant for the description of the material behaviour of the soil. It is assumed, for instance, that the intermediate principal stress is unimportant, as is the orientation of the principal stresses. This is approximately correct for the failure state of a soil, because the Mohr-Coulomb failure criterion can be formulated in σ and τ, but for smaller stresses it may be a first approximation only. Actually, even the failure criterion of a soil is often found to be dependent on other parameters (such as the value of σ_2) too, so that the Mohr-Coulomb failure criterion should be considered as merely a first approximation of real soil behaviour.
In many publications the symbols p and q are used, rather than σ and τ, and the diagram is denoted as a p,q-diagram. This will not be done here, as the notation p is reserved for the pore pressure.
The state of stress is represented in the right half of Figure 27-1 in the σ,τ-diagram. The basic principle is that the Mohr circle is characterized by the location of its top only. If the state of stress changes, the values of σ and τ will be different, so that the location of the stress point changes. The path of the stress point is called the *stress path*. Such a stress path can be drawn for the total stresses as well as the effective stresses, in the same diagram. The difference is the pore pressure, see Figure 27-2. The total stress path will be indicated by TSP, and the effective stress path by ESP.

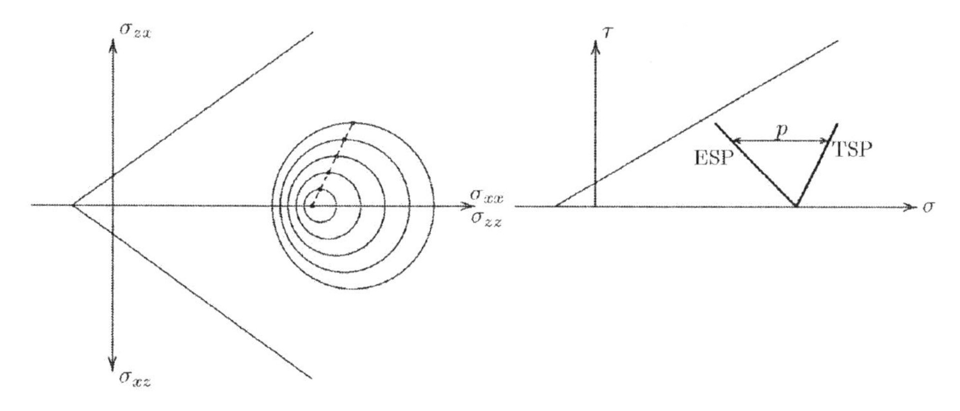

Figure 27-2. Stress paths.

The possible states of stress are limited by the Mohr-Coulomb failure criterion, see equation (21.12). In a diagram of Mohr circles this is a straight line, limiting the stress circles, see the left half of Figure 27-2. This limit is described by

$$(\frac{\sigma_1' - \sigma_3'}{2}) - (\frac{\sigma_1' + \sigma_3'}{2})\sin\phi - c\cos\phi = 0. \tag{27.3}$$

Expressed in terms of σ and τ this is

$$\tau' = \sigma'\sin\phi + c\cos\phi. \tag{27.4}$$

This describes a straight line in the σ,τ-diagram. This straight line has been indicated in the right half of Figure 27-2. The slope of this line is $\sin\phi$, which is slightly less steep than the envelope in the diagram of Mohr circles. The intersection with the vertical axis is $c\cos\phi$.
It may be noted that some researchers use different parameters to characterize the stresses in soils, because they are claimed to provide a better approximation of the behaviour of soils in certain tests. Actually, any combination of stress invariants may be used, for instance the three principal stresses. The parameters σ and τ used here are convenient because the Mohr-Coulomb failure criterion can so easily be formulated in terms of σ and τ. This criterion is not a basic physical principle, however, but rather a simple way to represent some test results. Other failure criteria, perhaps involving more parameters (such as the intermediate principal stress), may be formulated, and these may give a better approximation of a wider class of test results. In conclusion, the choice of stress path parameters is based upon considerations of convenience and experience as well as pure science.

27.2 Triaxial test

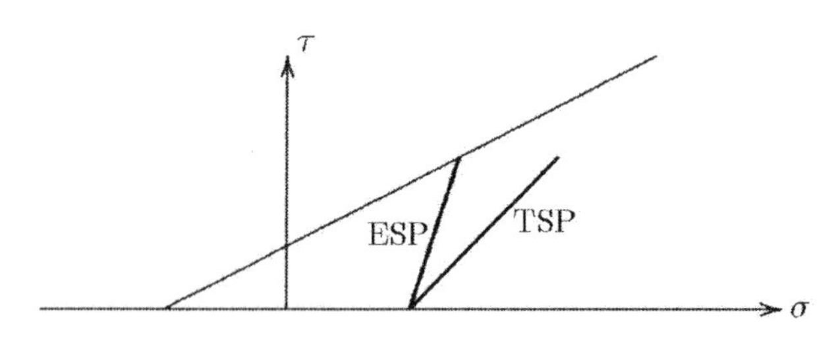

Figure 27-3. Stress path in triaxial test.

In the usual triaxial test the cell pressure is kept constant, and this is the minor principal stress. This means that σ_3 is constant. During the test the value of σ_1 increases. The total stress path is a straight line, with a slope of 45°, see Figure 27-3. Its mathematical description is

$$\Delta\sigma_3 = 0: \qquad \Delta\tau = \Delta\sigma. \tag{27.5}$$

The course of the effective stress path depends upon the pore pressures. In Chapter 25 it was postulated that these may be expressed by Skempton's formula,

$$\Delta p = B[\Delta\sigma_3 + A(\Delta\sigma_1 - \Delta\sigma_3)]. \tag{27.6}$$

This formula can also be written as

$$\Delta p = B[\Delta\sigma - (1-2A)\Delta\tau]. \tag{27.7}$$

In case of a triaxial test the pore pressure is

$$\Delta\sigma_3 = 0: \qquad \Delta p = 2BA\Delta\sigma. \tag{27.8}$$

For a completely saturated isotropically elastic material the values of the coefficients A and B are, if the compressibility of the water is neglected (see Chapter 25): $B=1$ and $A=\frac{1}{3}$. It then follows from (27.8) that the pore pressure increment will be $\frac{2}{3}$ of the increment of σ, $\Delta p=\frac{2}{3}\Delta\sigma$, see also Eq. (25.7). For such an idealized material behaviour the effective stress path will be a straight line at a slope of 3 : 1, see Figure 27-3.

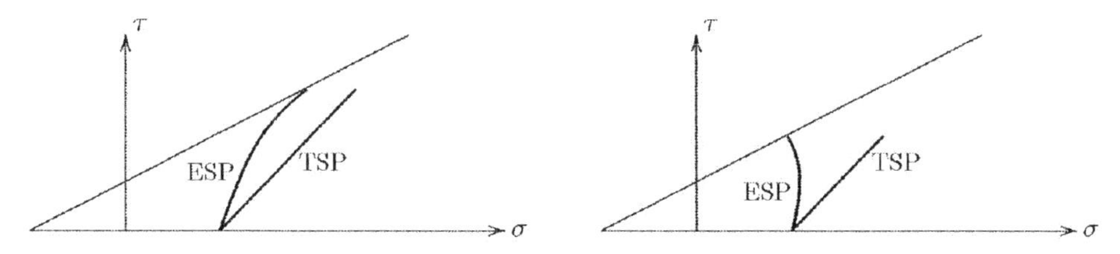

Figure 27-4. Stress paths in triaxial test on dilatant and contractant material.

Figure 27-4 shows the stress paths for a dilatant material and for a contractant material. When the material is dilatant, it will tend to expand during shear, so that the pore pressures will be reduced (the volume expansion results in suction). In a contractant material the volume will tend to decrease, so that the pore pressures are increased. It can be seen from the figure that in a contractant material failure will be reached much faster than in a non-contractant or dilatant material. The two mechanisms of pore pressure development, increasing the isotropic total stress (i.e. compression) and shear deformation, add up to a relatively large pore pressure increase, so that the isotropic effective stress σ' decreases, and this may result in rapid failure. In a dilatant material the two phenomena (compression and shear) counteract. The compression tends to increase the pore pressure, whereas the shear tends to decrease the pore pressure. The effective stress path will be located to the right of the path for a non-dilatant material. In a triaxial test this will result in a large apparent strength, as the vertical load can be very high before failure is reached.

27.3 Example

As a further illustration the example given in Chapter 25 will be further elaborated, using stress paths. The test results have been taken from Table 25-1, but they have been elaborated some more, to calculate the values of σ, σ' and τ, see Table 27-1.

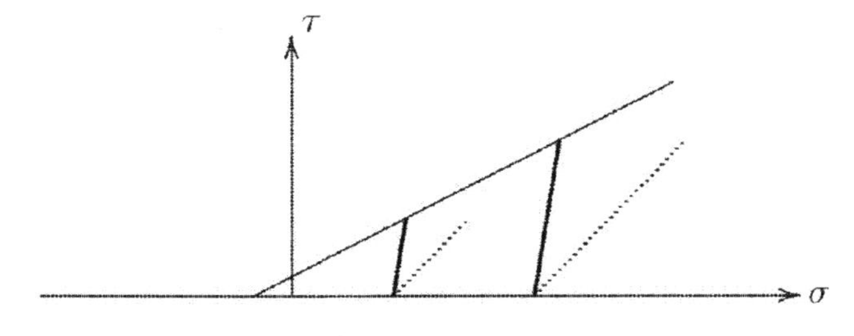

Figure 27-5. Stress paths in triaxial tests.

Test	σ_3	$\sigma_1-\sigma_3$	p	σ_1	σ	σ'	τ
1	40	0	0	40	40	40	0
	40	10	4	50	46	41	5
	40	20	9	60	51	41	10
	40	30	13	70	57	42	15
	40	40	17	80	63	43	20
	40	50	21	90	69	44	25
	40	60	25	100	75	45	30
2	95	0	0	95	95	95	0
	95	20	8	115	107	97	10
	95	40	17	135	118	98	20
	95	60	25	155	130	100	30
	95	80	33	175	142	102	40
	95	100	42	195	153	103	50
	95	120	50	215	165	105	60

Table 27-1: Test results.

The stress paths for the two tests are shown in Figure 27-5. The paths for the total stresses have been indicated by dotted lines, the effective stress paths have been indicated by fully drawn lines. The two end points of the effective stress paths determine the critical envelope.
According to Eq. (27.4) the critical points of the effective stress paths are located on the straight line

$$\tau' = a\sigma' + b, \tag{27.9}$$

where $a = \sin\phi$ and $b = c\cos\phi$. In this case there are two critical points: $\sigma' = 45$ kPa, $\tau = 30$ kPa and $\sigma' = 105$ kPa, $\tau = 60$ kPa. Substitution of these two pairs of values into Eq. (27.9) leads to two equations with two unknowns, a and b. This gives $a = 0.5$ and $b = 7.5$ kPa. This means that $\phi = 30°$ and $c = 8.7$ kPa. These results are in agreement with the values obtained in Chapter 25.

Problems

27.1 In a triaxial apparatus it is also possible to apply a negative value of the axial force (by pulling on the steel rod), at constant cell pressure. This is called a triaxial extension test. Draw the total stress path for such a test.

27.2 Also draw the effective stress path, for an isotropic elastic material, for a contractant material, and for a dilatant material.

Part VI
Stress distributions

28 Elastic stresses and deformations

An important class of soil mechanics problems is the determination of the stresses and deformations in a soil body, by the application of a certain load. The load may be the result of construction of a road, a dike, or the foundation of a building. The actual load may be the weight of the structure, but it may also consist of the forces due to traffic, wave loads, or the weight of the goods stored in a building. The stresses in the soils must be calculated in order to verify whether these stresses can be withstood by the soil (i.e. whether the stresses remain below the failure criterion), or in order to determine the deformations of the soil, which must remain limited.

28.1 Stresses and deformations

A three-dimensional computation of stresses and deformations in general involves three types of equations : equilibrium, constitutive relations, and compatibility. For soils the main difficulty is that the constitutive relations are rather complicated, and that their accurate description and formulation requires a large number of parameters, which are not so easy to determine, and which must be determined for every soil anew. In principle this should include the non-linear behaviour of soils, both in compression and in shear, and possible effects such as time dependence (creep), dilatancy, contractancy and anisotropy. The calculation of the real stresses and deformations in a soil is a well nigh impossible task, for which advanced numerical models are being developed. Such models, usually based upon the finite element method, are applied very often in engineering practice, and it can be expected that their use will be further expanded.
As an introduction into the methods of analysis the problem will be severely schematised here, and will be kept as simple as possible, by assuming that the material is isotropic linear elastic. This means that it is assumed that the relation between stresses and strains is described by Hooke's law. This is a severe approximation, but it may still be useful, as it contains all the necessary elements of a continuum analysis. Also, it will appear that in many cases some of the results, in particular the calculation of the vertical stresses, may be reasonably accurate. The idea then is that for the stresses in a soil body caused by the application of a certain load, see Figure 28-1, it is perhaps not so important what the precise properties of the materials are. A complete linear elastic computation at least ensures that equilibrium is satisfied, whereas the field of deformations and displacements satisfies the

compatibility equations. For the vertical stresses on the soft layer in Figure 28-1 it probably does not matter so much what the precise stiffness of al the layers is, as long as equilibrium, the geometry, and the distribution of the load have been taken into account. For other quantities, such as the vertical deformations, the stiffness of the layers may be very important, and the results of an elastic analysis may not be so relevant. It will be shown, however, that an analysis on the basis of linear elasticity may still be used as the first step in a reliable computation for the deformations as well.

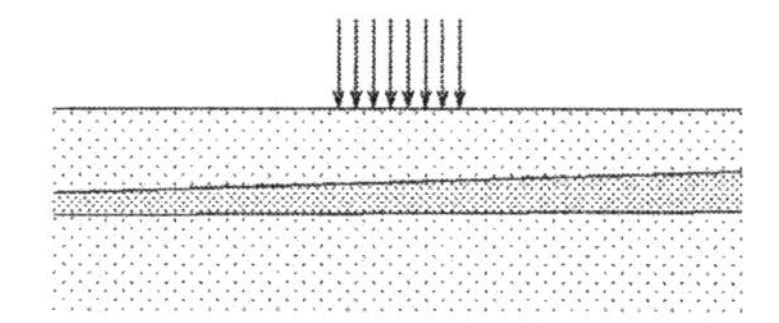

Figure 28-1. Load.

It may also be mentioned that the applicability of a linear elastic analysis has been verified for several problems by comparison with more complex computations. For instance, the results of computations for anisotropic materials, or layered materials, have been compared with the solutions for the linear elastic approximation. This confirms that the errors in the vertical normal stresses often are very small. On the other hand, the horizontal stresses, and the displacements, are very sensitive to the description of the material properties. It is fortunate that the vertical normal stresses often are the most interesting quantities, and these appear to be least sensitive for the material properties.

A useful procedure is to determine the stresses from a linear elastic analysis, and then, in a next step, the deformations are calculated from these stresses, using the best known relations between stresses and strains. From a theoretical or scientific viewpoint this is not justified, as the compatibility relations are ignored in the second step, and the coupling between stresses and the real deformations is also disregarded, but for engineering it appears to be a powerful and useful method. For instance, for a layered soil the stresses may be calculated assuming that the soil is homogeneous and linearly elastic, completely ignoring the difference in properties of the various layers, and then in a second step the deformations of each layer are calculated using Terzaghi's logarithmic compression formula, or some other realistic formula. The vertical deformations of the layers are finally added to determine the settlement of the soil surface. This procedure will be elaborated in Chapter 32.

28.2 Elasticity

For the analysis of stresses and strains in a homogenous, isotropic linear elastic material various methods have been developed. The general theory can be found in many textbooks on the theory of elasticity. Here, only the basic equations will be given, without giving the details of the derivations. Some of these details, and some derivations of solutions are given in Appendix B.

In this chapter, and in the next chapters, the analysis always concerns the calculation of stresses and deformations caused by some applied load. This means that the stresses and deformations in each case are incremental quantities. The initial stresses should be added in order to determine the actual stresses. It is assumed that these initial stresses also account for the weight of the soil, so that for the analysis of incremental stresses the weight of the soil itself may be disregarded. Thus there will be no body forces due to gravity.

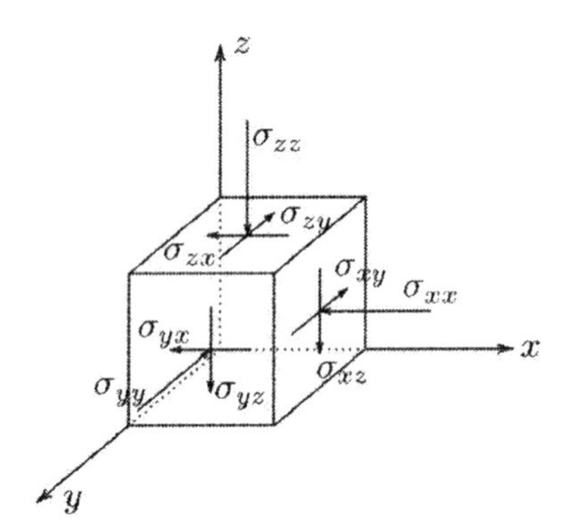

Figure 28-2. Stresses on element.

For a small element the stresses on the three visible faces are shown in Figure 28-2. The equations of equilibrium in the three coordinate directions are

$$\begin{aligned}
\frac{\partial \sigma_{xx}}{\partial x} + \frac{\partial \sigma_{yx}}{\partial y} + \frac{\partial \sigma_{zx}}{\partial z} &= 0, \\
\frac{\partial \sigma_{xy}}{\partial x} + \frac{\partial \sigma_{yy}}{\partial y} + \frac{\partial \sigma_{zy}}{\partial z} &= 0, \\
\frac{\partial \sigma_{xz}}{\partial x} + \frac{\partial \sigma_{yz}}{\partial y} + \frac{\partial \sigma_{zz}}{\partial z} &= 0.
\end{aligned} \tag{28.1}$$

Because of equilibrium of moments the stress tensor must be symmetric,

$$\begin{aligned}
\sigma_{xy} &= \sigma_{yx}, \\
\sigma_{yz} &= \sigma_{zy}, \\
\sigma_{zx} &= \sigma_{xz}.
\end{aligned} \tag{28.2}$$

The equations of equilibrium constitute a set of six equations involving nine stress components. In itself this can never be sufficient for a mathematical solution. The deformations must also be considered before a solution can be contemplated.

For a linear elastic material the relation between stresses and strains is given by Hooke's law,

$$\varepsilon_{xx} = \frac{1}{E}[\sigma_{xx} - \nu(\sigma_{yy} + \sigma_{zz})],$$
$$\varepsilon_{yy} = \frac{1}{E}[\sigma_{yy} - \nu(\sigma_{zz} + \sigma_{xx})], \tag{28.3}$$
$$\varepsilon_{zz} = \frac{1}{E}[\sigma_{zz} - \nu(\sigma_{xx} + \sigma_{yy})],$$

$$\varepsilon_{xy} = \frac{1+\nu}{E}\sigma_{xy},$$
$$\varepsilon_{yz} = \frac{1+\nu}{E}\sigma_{yz}, \tag{28.4}$$
$$\varepsilon_{zx} = \frac{1+\nu}{E}\sigma_{zx}.$$

where E is the modulus of elasticity (Young's modulus), and ν is Poisson's ratio. The equations (28.3) and (28.4) add six equations to the system, at the same time introducing six additional variables.
The six strains can be related to the three components of the displacement vector,

$$\varepsilon_{xx} = -\frac{\partial u_x}{\partial x},$$
$$\varepsilon_{yy} = -\frac{\partial u_y}{\partial y}, \tag{28.5}$$
$$\varepsilon_{zz} = -\frac{\partial u_z}{\partial z},$$

$$\varepsilon_{xy} = -\tfrac{1}{2}(\frac{\partial u_x}{\partial y} + \frac{\partial u_y}{\partial x}),$$
$$\varepsilon_{yz} = -\tfrac{1}{2}(\frac{\partial u_y}{\partial z} + \frac{\partial u_z}{\partial y}), \tag{28.6}$$
$$\varepsilon_{zx} = -\tfrac{1}{2}(\frac{\partial u_z}{\partial x} + \frac{\partial u_x}{\partial z}).$$

These are the *compatibility equations*. In total there are now just as many equations as there are variables, so that the system may be solvable, at least if there are a sufficient number of boundary conditions.
The minus sign in the equations has been introduced because a volume decrease leads to a positive strain and a volume increase to a negative strain.
For a number of problems solutions of the system of equations can be found in the literature on the theory of elasticity. In soil mechanics the solutions for a half space or a half plane, with a horizontal upper surface, are of special interest. The solutions for some problems are given in Appendix B. In order to conform to the sign conventions used in the extensive literature about continuum mechanics, the stresses

in that appendix are denoted by τ, and the sign convention is that tensile stresses and increase of volume are considered positive. In the next chapters some of the most important solutions for soil mechanics are further discussed.
For the solved elastic problems the strains and effective stresses can be calculated for drained or consolidated loads, with the help of two stiffness parameters $(E,\nu$ or $K,G)$. If these parameters are replaced by two undrained parameters (E_u,ν_u) the strains and total stresses can be calculated for undrained problems (short loads on clay or peat). When calculating these two parameters, one should realize that an undrained load, because of the incompressibility of water, causes no volume change, so

$$\nu_u = 0.5. \tag{28.7}$$

The undrained compression modulus becomes (see Chapter 14)

$$K_u = \frac{E_u}{3(1-2\nu_u)} = \infty. \tag{28.8}$$

Besides this the water has no shear stresses, both in the drained as in the undrained state. From this it follows that

$$G_u = G = \frac{E}{2(1+\nu)} = \frac{E_u}{2(1+\nu_u)}. \tag{28.9}$$

From which can be derived that

$$E_u = E\frac{(1+\nu_u)}{(1+\nu)}. \tag{28.10}$$

Replacing the drained stiffness parameters (E,ν) from the previous chapters by the undrained stiffness parameters (E_u,ν_u) and calculating these values in the way demonstrated above, the solutions can be used not only for the drained, but also for the undrained case.

29 Boussinesq

In 1885 the French scientist Boussinesq obtained a solution for the stresses and strains in a homogeneous isotropic linear elastic half space, loaded by a vertical point force on the surface, see Figure 29-1. A derivation of this solution is given in Appendix B, see also any textbook on the theory of elasticity (for instance S.P. Timoshenko, *Theory of Elasticity*, paragraph 123). The stresses are found to be

$$\sigma_{zz} = \frac{3P}{2\pi}\frac{z^3}{R^5}, \tag{29.1}$$

$$\sigma_{rr} = \frac{P}{2\pi}[\frac{3r^2 z}{R^5} - (1-2\nu)\frac{1}{R(R+z)}], \tag{29.2}$$

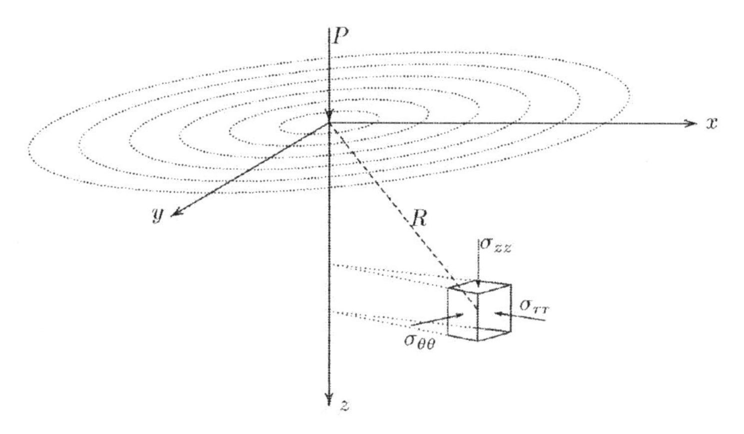

Figure 29-1. Point load on half space.

$$\sigma_{\theta\theta} = \frac{P}{2\pi}\frac{1-2\nu}{R^2}(\frac{R}{R+z} - \frac{z}{R}), \tag{29.3}$$

$$\sigma_{rz} = \frac{3P}{2\pi}\frac{rz^2}{R^5}. \tag{29.4}$$

In these equations r is the cylindrical coordinate,

$$r = \sqrt{x^2 + y^2}, \tag{29.5}$$

and R is the spherical coordinate,

$$R=\sqrt{x^2+y^2+z^2}. \tag{29.6}$$

The solution for the displacements is

$$u_r=\frac{P(1+\nu)}{2\pi ER}\left[\frac{r^2z}{R^3}-(1-2\nu)(1-\frac{z}{R})\right], \tag{29.7}$$

$$u_\theta=0, \tag{29.8}$$

$$u_z=\frac{P(1+\nu)}{2\pi ER}[2(1-\nu)+\frac{z^2}{R^2})]. \tag{29.9}$$

The vertical displacement of the surface is particularly interesting. This is

$$z=0 \quad : \qquad u_z=\frac{P(1-\nu^2)}{\pi Er}. \tag{29.10}$$

For $r\rightarrow 0$ this tends to infinity. At the point of application of the point load the displacement is infinitely large. This singular behaviour is a consequence of the singularity in the surface load, as in the origin the stress is infinitely large. That the displacement in that point is also infinitely large may not be so surprising.
Another interesting quantity is the distribution of the stresses as a function of depth, just below the point load, i.e. for $r=0$. This is found to be

$$r=0 \quad : \qquad \sigma_{zz}=\frac{3P}{2\pi z^2}, \tag{29.11}$$

$$r=0 \quad : \qquad \sigma_{rr}=\sigma_{\theta\theta}=-(1-2\nu)\frac{P}{4\pi z^2}. \tag{29.12}$$

These stresses decrease with depth, of course. In engineering practice, it is sometimes assumed, as a first approximation, that at a certain depth the stresses are spread over an area that can be found by drawing a line from the load under an angle of about 45°. That would mean that the vertical normal stress at a depth z would be $P/\pi z^2$, homogeneously over a circle of radius z. That appears to be incorrect (the error is 50 % if $r=0$), but the trend is correct, as the stresses indeed decrease with $1/z^2$. In Figure 29-2 the distribution of the vertical normal stress σ_{zz} is represented as a function of the cylindrical coordinate r, for two values of the depth z.
The assumption of linear elastic material behaviour means that the entire problem is linear, as the equations of equilibrium and compatibility are also linear. This implies that the principle of superposition of solutions can be applied. Boussinesq's solution can be used as the starting point of more general types of loading, such as a system of point loads, or a uniform load over a certain given area.
As an example consider the case of a uniform load of magnitude p over a circular area, of radius a. The solution for this case can be found by integration over a

circular area (S.P. Timoshenko, *Theory of Elasticity*, paragraph 124), see Figure 29-3.

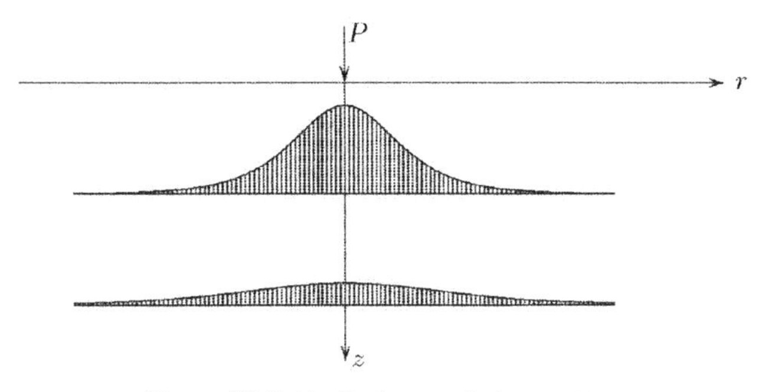

Figure 29-2. Vertical normal stress σ_{zz}.

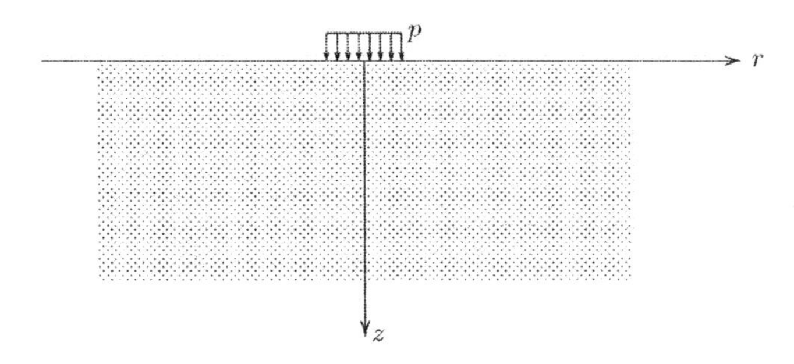

Figure 29-3. Uniform load over circular area.

The stresses along the axis $r = 0$, i.e. just below the load, are found to be

$$r = 0 \quad : \qquad \sigma_{zz} = p(1 - \frac{z^3}{b^3}), \tag{29.13}$$

$$r = 0 \quad : \qquad \sigma_{rr} = p[(1+\nu)\frac{z}{b} - \tfrac{1}{2}(1 - \frac{z^3}{b^3})], \tag{29.14}$$

in which $b = \sqrt{z^2 + a^2}$.

The displacement of the origin is

$$r = 0, z = 0 \quad : \qquad u_z = 2(1-\nu^2)\frac{pa}{E}. \tag{29.15}$$

This solution will be used as the basis of a more general case in the next chapter.

Another important problem, which was already solved by Boussinesq (see also Timoshenko) is the problem of a half space loaded by a vertical force on a rigid plate. The force is represented by $P = \pi a^2 p$, see Figure 29-4. The distribution of the normal stresses below the plate is found to be

$$z=0,\ 0<r<a \qquad : \qquad \sigma_{zz}=\frac{\frac{1}{2}\overline{p}}{\sqrt{1-r^2/a^2}}. \tag{29.16}$$

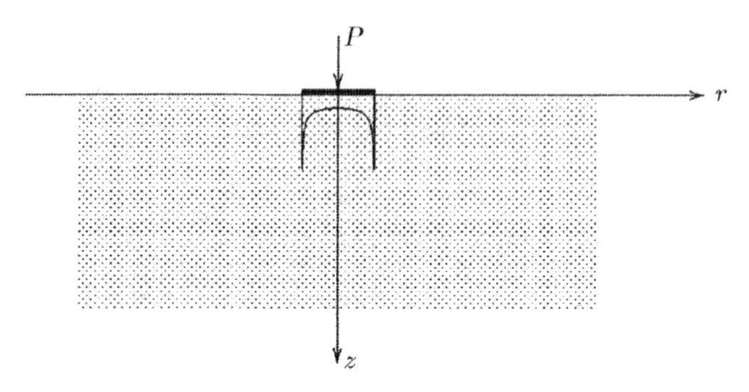

Figure 29-4. Rigid plate on half space.

This stress distribution is shown in Figure 29-4. At the edge of the plate the stresses are infinitely large, as a consequence of the constant displacement of the rigid plate. In reality the material near the edge of the plate will probably deform plastically. It can be expected, however, that the real distribution of the stresses below the plate will be of the
form shown in the figure, with the largest stresses near the edge. The centre of the plate will subside without much load.
The displacement of the plate is

$$z=0,\ 0<r<a \qquad : \qquad u_z=\frac{\pi}{2}(1-\nu^2)\frac{\overline{p}a}{E}. \tag{29.17}$$

When this is compared with the displacement below a uniform load, see (29.15), it appears that the displacement of the rigid plate is somewhat smaller, as could be expected.

Problems

29.1 A circular area is loaded by a uniform vertical load p. The stress distribution is approximated by assuming that the vertical normal stress at a depth z is uniform, over an area πz^2. Sketch the distribution of the stresses just below the load, as a function of depth, and compare the result with the exact solution (29.13).

29.2 From the expression of the previous problem, and using the relations $\varepsilon=\sigma/E$ and $\varepsilon=\partial u_z/\partial z$, derive an expression for the displacement of the surface, and compare the result with Eq. (29.15).

29.3 In engineering practice the displacement due to a loaded plate is often expressed by a subgrade constant, by writing $u_z=p/c$. Derive a relation between the subgrade constant c and the elastic modulus E.

30 Newmark

An ingenious method for the determination of the vertical normal stresses at the certain depth, caused by some arbitrary load distribution on the surface was developed by Newmark. The basis of the method is equation (29.13), which gives the vertical normal stress at a depth z below the centre of a uniform load p on ac circular area, see Figure 30-1. This equation can also be written as

$$r=0 \quad : \qquad \frac{\sigma_{zz}}{p}=1-\frac{1}{\sqrt{(1+a^2/z^2)^3}}. \tag{30.1}$$

a/z	0.0000	0.2698	0.4005	0.5181	0.6370	0.7664	0.9176	1.1097	1.3871	1.9083	∞
p_{zz}/p	0.0	0.1	0.2	0.3	0.4	0.5	0.6	0.7	0.8	0.9	1.0

Table 30-1: Vertical stresses below a circular load.

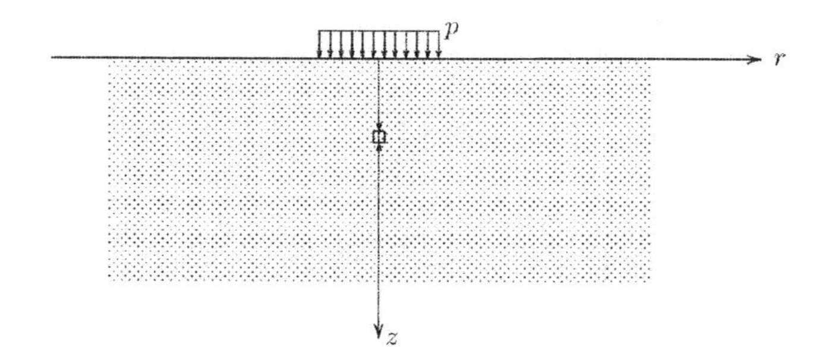

Figure 30-1. Uniform load on circular area.

This formula gives the value of the vertical normal stress σ_{zz} for a given value of a/z. Conversely, it is also possible, of course, to calculate the value of a/z for which a certain value of σ_{zz}/p occurs, in multiples of $0.1p$. For instance, by taking $\sigma_{zz}/p=0.5$, it follows from Eq. (30.1) that a/z should be equal to 0.7664. In this way the values in Table 30-1 have been calculated.

All the values given in the table can easily be verified using Eq. (30.1). The first value expresses that a ring of radius 0 leads to a stress $\sigma_{zz}=0$, and the last value expresses that in case of a uniform load over the entire surface the vertical stress σ_{zz} at a depth z always is p.

It also follows from the table that for a load on a circle of radius $0.7664z$ the stress σ_{zz} at a depth is $0.5p$, and that for a load on a circle of radius $0.9176z$ the stress is $0.6p$. The increase by $0.1p$ must be due to the additional load, which is a uniform

load over a circular ring, between the circles of radius $0.7664z$ and the circle of radius $0.9176z$, see Figure 30-2. It can be concluded that each ring shaped load, between two successive circles from Table 30-1 leads to a stress $\sigma_{zz} = 0.1p$ in a point at depth z just between the centre of the circles and rings. If each ring is further subdivided into 100 equal sectors it follows that the load on every individual segment leads to a contribution to the stress σ_{zz} of magnitude $0.001p$.
Using these properties, Newmark's diagram can be constructed, see Figure 30-3. A load of magnitude p on each of the 1000 elementary rectangles in this diagram gives rise to a stress $\sigma_{zz} = 0.001p$, in the point at a depth z below the origin.

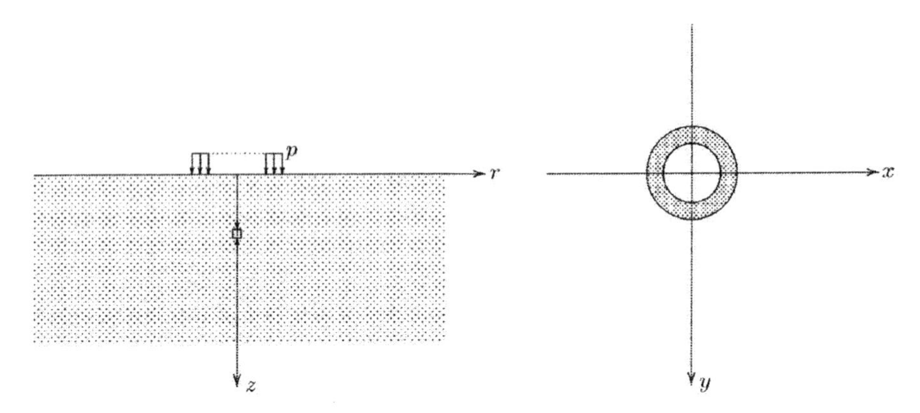

Figure 30-2. Uniform load on a ring.

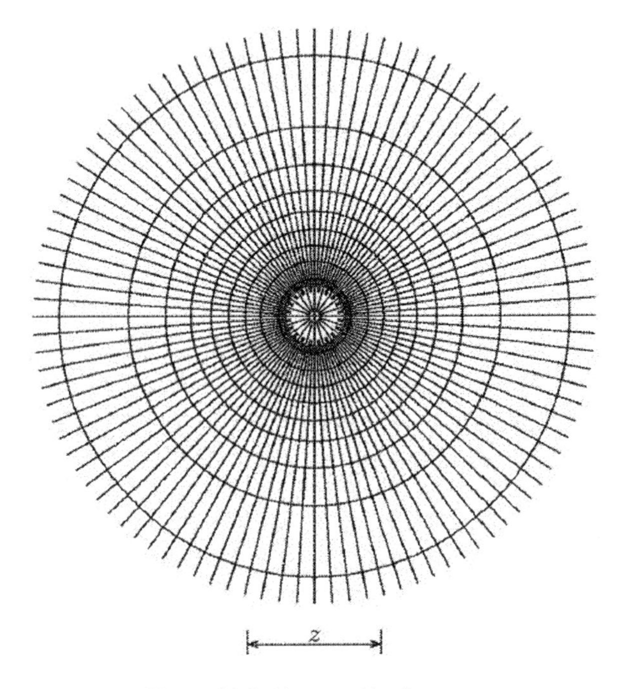

Figure 30-3. Newmark's diagram.

For clarity, not all the radial lines inside the smallest circle in Figure 30-3 have been fully continued to the centre, because they all intersect in that point, and drawing all lines would result in a big blue spot. It should be remembered that in reality the inner ring, inside the smallest circle, contains 100 elementary rectangles, as each of the other rings. In the outer ring all radial lines should be extended towards infinity.
The principle of the application of Newmark's diagram is that a load of magnitude p on any one of the 1000 elementary rectangles leads to a stress $\sigma_{zz} = 0.001p$ in a point at depth z just below the centre of the circles. This is valid for each of the rectangles, and because the problem is linear, the stress caused by various loads may be superimposed. This means that a load of magnitude p on a surface that covers n rectangles, leads to a stress σ_{zz} at a depth z below the origin of magnitude $\sigma_{zz} = n \times 0.001p$.
The value of the depth z plays an important role in the method. It actually determines the scale of the problem in the horizontal plane. To determine the stress at a deeper point the value of z is somewhat larger, and the size of the loaded area in the diagram will be somewhat smaller. This smaller area then covers a smaller number of rectangles, so that the stress will be smaller, as expected. This will be illustrated by an example.
The method can also be used for non-uniform loads. As a load p on any rectangle leads to a stress $\sigma_{zz} = 0.001p$, it follows that a load kp on a rectangle results in a contribution to the stress of magnitude $\sigma_{zz} = k \times 0.001p$. This will also be illustrated in the example.

30.1 Example

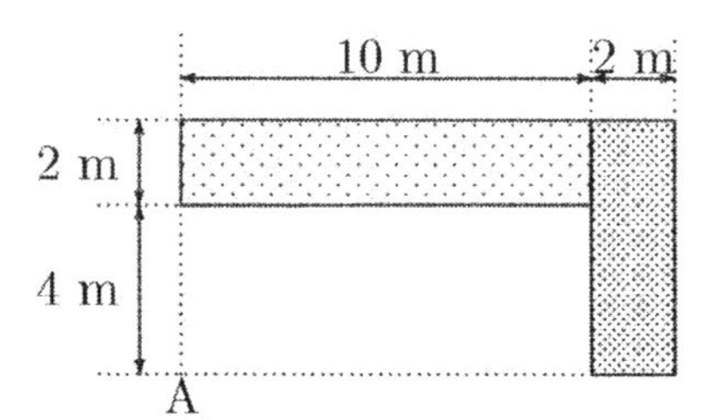

Figure 30-4. Example.

As an example consider the problem of a load on an L-shaped region, see Figure 30-4. On the short leg the load is 15 kPa, on the larger leg the load is 5 kPa. The problem is to determine the vertical normal stress at a depth of 8 m below the point A. The first step in the solution is to draw the loaded area on such a scale that the reference length z in Newmark's diagram corresponds to 8 m, see Figure 30-5, and such that the point A is located in the origin of the circles. The short leg of the loaded area now covers about 7 rectangles, and the long leg covers about 34 rectangles. This means that the stress is

$$\sigma_{zz} = 7 \times 0.001 \times 15 \text{ kPa} + 34 \times 0.001 \times 5 \text{ kPa} = 0.275 \text{ kPa}.$$

In order to determine the stress at a greater depth, say at a depth of 16 m, the loaded area should be drawn half as large. This then covers a smaller number of rectangles, so that the stress will be smaller.

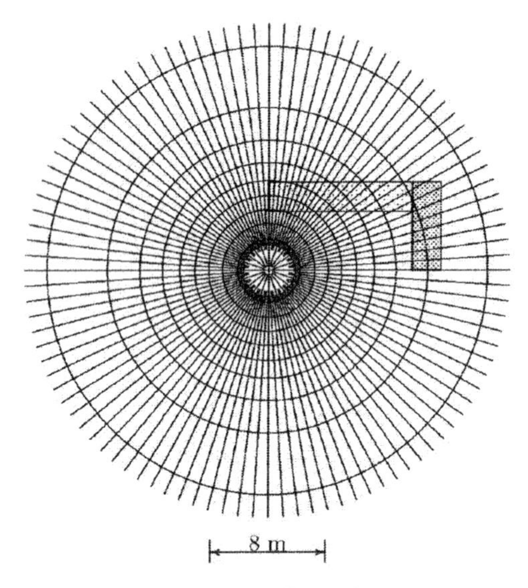

Figure 30-5. Example.

Problems

30.1 In the example considered above, determine the vertical stress σ_{zz} at a depth of 8 m below the corner point in the upper right corner in the plan.

30.2 Determine the stress in that corner point at the surface (for $z = 0$). Also determine the stress in the point A at the surface. And finally, also calculate the stress at a depth of 8000 m.

30.3 A square region of dimensions 4 m times 4 m is loaded by a uniform load 10 kPa. Determine the vertical normal stress at a depth of 4 m in a number of characteristic points: the centre, one of the corners, and a point in the centre of one of the sides.

30.4 Starting form the (incorrect, but approximate) assumption that the vertical stresses below a uniform load are uniformly distributed over an area that can be found by drawing a line under 45° from the loaded area, it may also be possible to construct a diagram similar to Newmark's. Construct this diagram. Are the stress overestimated or underestimated by this method?

30.5 For the construction of Newmark's diagram equation (29.13) for the stress σ_{zz} below the centre of a uniform circular load was the starting point. The radial stress σ_{rr} for this basic problem is given by Eq. (29.14). Investigate whether for this case a Newmark type diagram can be constructed, that gives the horizontal stress σ_{rr} for a load on an arbitrary area.

31 Flamant

In 1892 Flamant obtained the solution for a vertical line load on a homogeneous isotropic linear-elastic half space, see Figure 31-1. This is the two-dimensional equivalent of Boussinesq's basic problem. It can be considered as the superposition of an infinite number of point loads, uniformly distributed along the *y*-axis. A derivation is given in Appendix B.

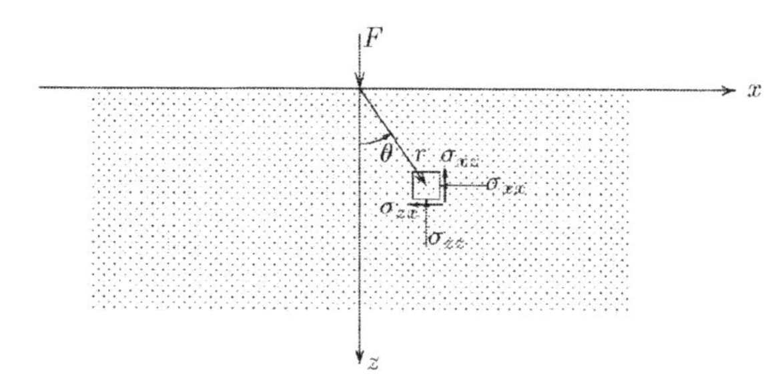

Figure 31-1. Flamant's Problem.

In this case the stresses in the *x,z*-plane, resulting from the line load, are

$$\sigma_{zz} = \frac{2F}{\pi}\frac{z^3}{r^4} = \frac{2F}{\pi r}\cos^3\theta, \tag{31.1}$$

$$\sigma_{xx} = \frac{2F}{\pi}\frac{x^2 z}{r^4} = \frac{2F}{\pi r}\sin^2\theta\cos\theta, \tag{31.2}$$

$$\sigma_{xz} = \frac{2F}{\pi}\frac{xz^2}{r^4} = \frac{2F}{\pi r}\sin\theta\cos^2\theta. \tag{31.3}$$

In these equations $r = \sqrt{x^2 + z^2}$. The quantity F has the dimension of a force per unit length, so that F/r has the dimension of a stress.

Expressions for the displacements are also known, but these contain singular terms, with a factor $\ln r$. This factor is infinitely large in the origin and at infinity. Therefore these expressions are not so useful.

On the basis of Flamant's solution several other solutions may be obtained using the principle of superposition. An example is the case of a uniform load of magnitude p on a strip of width $2a$, see Figure 31-2. In this case the stresses are

$$\sigma_{zz} = \frac{p}{\pi}[(\theta_1 - \theta_2) + \sin\theta_1 \cos\theta_1 - \sin\theta_2 \cos\theta_2], \tag{31.4}$$

$$\sigma_{xx} = \frac{p}{\pi}[(\theta_1 - \theta_2) - \sin\theta_1 \cos\theta_1 + \sin\theta_2 \cos\theta_2], \tag{31.5}$$

$$\sigma_{xz} = \frac{p}{\pi}[\cos^2\theta_2 - \cos^2\theta_1]. \tag{31.6}$$

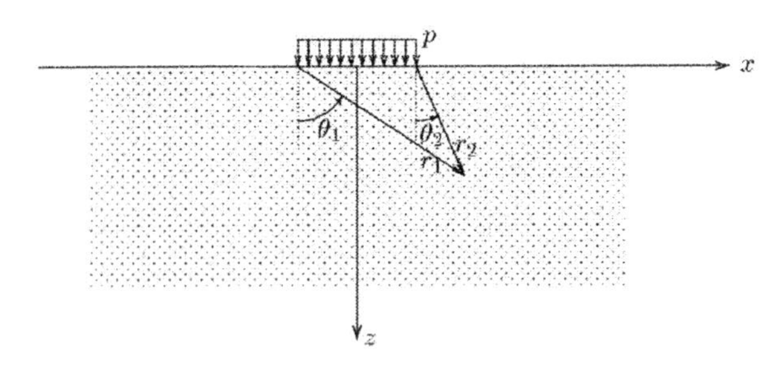

Figure 31-2. Strip load.

In the centre of the plane, for $x = 0$, $\theta_2 = -\theta_1$. Then the stresses are

$$x = 0 \quad : \qquad \sigma_{zz} = \frac{2p}{\pi}[\theta_1 + \sin\theta_1 \cos\theta_1], \tag{31.7}$$

$$x = 0 \quad : \qquad \sigma_{xx} = \frac{2p}{\pi}[\theta_1 - \sin\theta_1 \cos\theta_1], \tag{31.8}$$

$$x = 0 \quad : \qquad \sigma_{xz} = 0. \tag{31.9}$$

That the shear stress $\sigma_{xz} = 0$ for $x = 0$ is a consequence of the symmetry of this case. The stresses σ_{xx} and σ_{zz} are shown in Figure 31-3, as functions of the depth z. Both stresses tend towards zero for $z \to \infty$, of course, but the horizontal normal stress appears to tend towards zero much faster than the vertical normal stress. It also appears that at the surface the horizontal stress is equal to the vertical stress. At the surface this vertical stress is equal to the load p, of course, because that is a boundary condition of the problem. Actually, in every point of the surface below the strip load the normal stresses are $\sigma_{xx} = \sigma_{zz} = p$.

It may be interesting to further explore the result that the shear stress $\sigma_{xz} = 0$ along the axis of symmetry $x = 0$ in the case of a strip load, see Figure 31-2. It can be expected that this symmetry also holds for the horizontal displacement, so that $u_x = 0$ along the axis $x = 0$. This means that this solution can also be used as the solution of the problem that is obtained by considering the right half of the strip problem only, see Figure 31-4. In this problem the quarter plane $x > 0$, $x > 0$ is supposed to be loaded by a strip load of width a on the surface $z = 0$, and the

boundary conditions on the boundary $x=0$ are that the displacement $u_x=0$ and the shear stress $\sigma_{xz}=0$, representing a perfectly smooth and rigid vertical wall. The wall is supposed to extend to an infinite depth, which is impractical. For a smooth rigid wall of finite depth the solution may be considered as a first approximation.
The formulas (31.7) and (31.8) can also be written as

$$x=0 \quad : \qquad \sigma_{zz}=\frac{2p}{\pi}[\arctan(\frac{a}{z})+\frac{az}{a^2+z^2}], \tag{31.10}$$

$$x=0 \quad : \qquad \sigma_{xx}=\frac{2p}{\pi}[\arctan(\frac{a}{z})-\frac{az}{a^2+z^2}]. \tag{31.11}$$

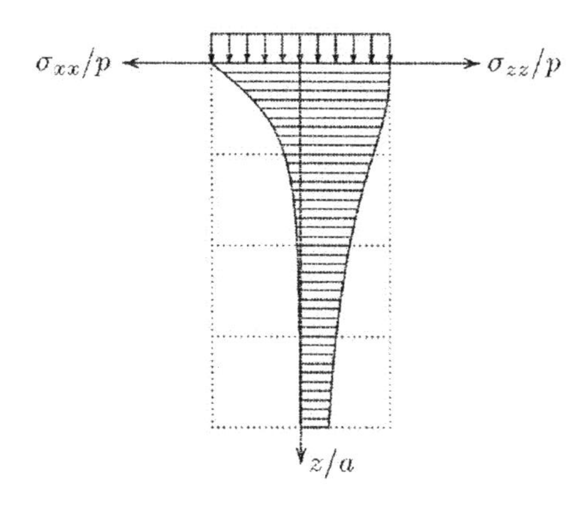

Figure 31-3. Stresses for x=0.

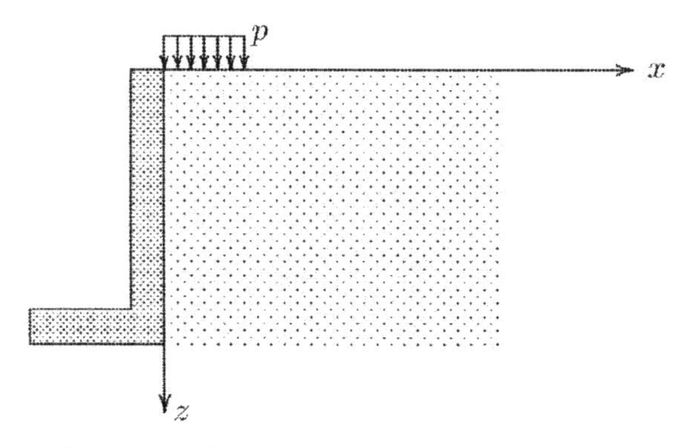

Figure 31-4. Strip load next to a smooth rigid wall.

Integration of the horizontal stress σ_{xx} from $z=0$ to $z=h$ gives the total force on a wall of height h,

$$Q=\frac{2}{\pi}ph\arctan(\frac{a}{h}). \tag{31.12}$$

For a very deep wall ($h \ll a$) this becomes, because then $\arctan(a/h) \approx a/h$,

$$h \to \infty: \qquad Q = \frac{2}{\pi} pa = 0.637 pa. \tag{31.13}$$

The quantity *pa* is the total vertical load *F* (per unit length perpendicular to the plane of the drawing). It appears that the horizontal reaction in an elastic material is 0.637*F*.
For a very shallow wall ($h \ll a$) the total lateral force will be, because then $\arctan(a/h) \approx \pi/2$,

$$h \to 0: \qquad Q = ph. \tag{31.14}$$

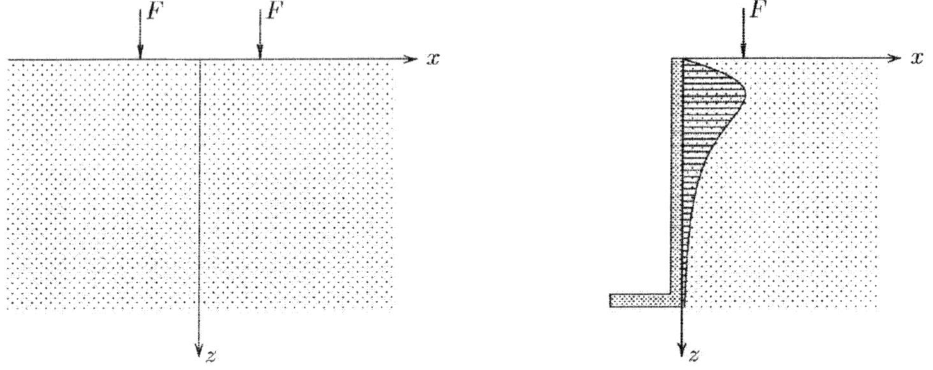

Figure 31-5. Line load next to a smooth rigid wall.

This is in agreement with the observation made earlier that the value of the horizontal stress at the surface, just below the load, is $\sigma_{xx} = p$. For a very short wall the horizontal force will be that horizontal stress, multiplied by the length of the wall.
Another interesting application of Flamant's solution is shown in Figure 31-5. In the case of a surface load by two parallel line loads it can be expected that at the axis of symmetry ($x = 0$) the horizontal displacement and the shear stress will be zero, because of symmetry. This means that the solution of that problem can also be used as the solution of the problem of a line load at a certain distance from a smooth rigid wall, because the boundary conditions along the wall are that $u_x = 0$ and $\sigma_{zx} = 0$ for $x = 0$. By the symmetry of the problem shown in the left half of Figure 31-5, these conditions are satisfied by the solution of that problem.
The horizontal stresses against the wall are given by equation (31.2) for Flamant's basic problem, multiplied by 2, because there are two line loads and each gives the same stress. In this formula the value of x should be taken as $x = a$, where a is the distance of the force to the wall. The horizontal stress against the wall in this case is

$$\sigma_{xx} = \frac{4}{\pi} \frac{Fa^2 z}{(a^2 + z^2)^2}. \tag{31.15}$$

The distribution of the horizontal stresses against the wall are also shown in Figure 31-5. The maximum value occurs for $z=0.577a$, and that maximum stress is

$$\sigma_{xx-\max}=0.4135\frac{F}{a}.$$

The total force on a wall of depth h can be found again by integration of the horizontal stress over the depth of the wall. This gives

$$Q=\frac{2}{\pi}\frac{F}{1+a^2/h^2}. \tag{31.16}$$

If $a=0$ this is $Q=0.637F$. If a increases the value of Q will gradually become smaller. A force at a larger distance from the wall will give smaller stresses against the wall.

It should be kept in mind that only the extra forces and stresses resulting from the loads are mentioned. The weight of the soil itself also causes stresses, both in vertical and horizontal direction.

Problems

31.1 Is it a coincidence that in each of the formulas (31.12) and (15.16) the same factor $2/\pi$ appears?

31.2 Transform the stresses in Flamant's solution into polar coordinates, or in other words, derive expressions for the stress components σ_{rr}, $\sigma_{\theta\theta}$ and $\sigma_{r\theta}$. Note that the result is somewhat peculiar.

32 Deformation of layered soil

An important problem of soil mechanics practice is the prediction of the settlements of a structure built on the soil. For a homogeneous isotropic linear elastic material the deformations could be calculated using the theory of elasticity. That is a completely consistent theory, leading to expressions for the stresses and the displacements. However, solutions are available only for a half space and a half plane, not for a layered material (at least not in closed form). Moreover, soils exhibit non-linear properties (such as a stiffness increasing with the actual stress), often have anisotropic properties, and in many cases the soil consists of layers of different properties. For such materials the description of the material properties is already a complex problem, let alone the analysis of stresses and deformations.
For these reasons an approximation is often used, based on a semi-elastic analysis. In this approximation it is first postulated that the vertical stresses in the soil, whatever it true properties are, can be approximated by the stresses that can be calculated from linear elastic theory. On the basis of these stresses the deformations are then determined, using the best available description of the relation between stress and strain, which may be non-linear. If the soil is layered, the deformations of each layer are calculated using its own properties, and then the surface displacements are determined by a summation of the deformations of all layers. In this way the different properties of the layers can be taken into account, including a possible increase of the stiffness with depth.

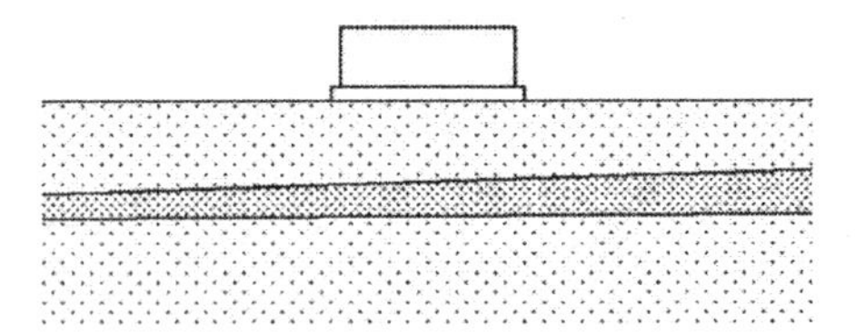

Figure 32-1. Load on layered soil.

The procedure is not completely consistent, because in a soil consisting of layers of different stiffness, the stress distribution will not be the same as in a homogeneous linear elastic material. A partial justification may be that the stresses following from an elastic computation at least satisfy the equilibrium conditions. Also it has been found, by comparison of solutions for layered materials with the solution for a homogeneous material, that the distribution of the vertical stress σ_{zz} is not very sensitive to the material properties, provided that the differences in material properties are not very large, i.e. excepting extreme cases such as a very stiff layer on a very soft subsoil.

32.1 Example

The computation method can best be illustrated by considering an example. This example concerns a circular fluid reservoir, having a diameter of 20 meter, which is being constructed on a foundation plate on a layer of fairly soft soil, of 20 meter thickness, see Figure 32-2. Below the soft soil the soil is a hard layer of sand or rock. The compressibility of the soft soil is about $C_{10} = 50$. The pressure of the foundation plate on the soil is 20 kPa, and the additional load by the fluid in the reservoir is 100 kPa. The problem is to determine the settlement caused by the load, in the centre of the reservoir.

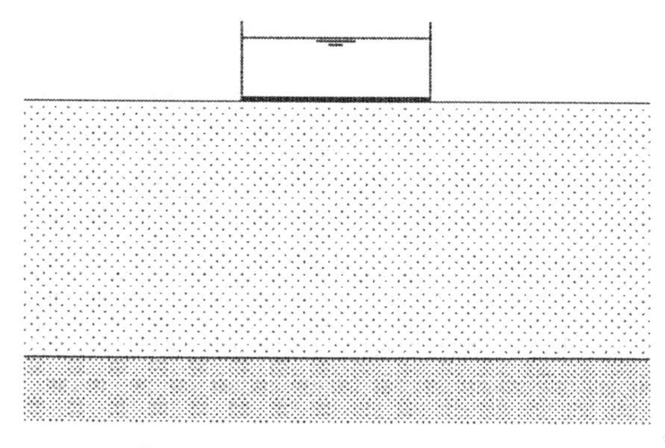

Figure 32-2. Reservoir on soft soil.

depth (m)	weight (kPa)	found. (kPa)	σ_1 (kPa)	load (kPa)	σ kPa	$\log(\sigma/\sigma_1)$	Δh (m)
1	10.00	19.98	29.99	99.90	129.89	0.6366	0.025
3	30.00	19.52	49.52	97.63	147.15	0.4730	0.019
4	50.00	18.21	68.21	91.06	159.27	0.3683	0.015
5	70.00	16.23	86.23	81.14	167.37	0.2880	0.011
6	90.00	14.01	104.04	70.06	174.07	0.2236	0.009
11	110.00	11.90	121.90	59.49	181.39	0.1726	0.007
13	130.00	10.04	140.04	50.20	190.24	0.1330	0.005
15	150.00	8.48	158.48	42.40	200.88	0.1030	0.004
17	170.00	7.19	177.19	35.96	213.15	0.0802	0.003
19	190.00	6.14	196.14	30.70	226.84	0.0631	0.002

Table 32-1: Computation of settlement.

The example has been elaborated in Table 32-1. The soil has been subdivided into 10 layers, of 2 meter thickness each. The first column of the table gives the average depth of each layer. The second column gives the effective stress due to the weight of the soil, assuming that the effective unit weight of the soil is 10 kN/m^3, so that for each meter depth the stress increases by 10 kPa. The third column gives the additional stress due to the weight of the foundation plate. These stresses have been calculated using the formula for the stresses below a uniform circular load, equation (29.13),

$$\sigma_{zz} = p(1 - \frac{z^3}{b^3}), \tag{32.1}$$

in which $b = \sqrt{z^2 + a^2}$, and a is the radius of the circular area. The fourth column is the sum of the second and third columns. This is considered as the initial stress, before the application of the load, but after the construction of the foundation plate. The fifth column gives the stresses caused by the load, the weight of the fluid in the reservoir. These stresses have also been calculated by the formula (32.1). The sixth column is the sum of the fourth and the fifth column. These are the final effective stresses. The seventh and eight columns contain the actual computation of the deformations of each layer, using Terzaghi's logarithmic formula, see Chapter 15, and the value $C_{10} = 50$. By adding the deformations of the layers the total settlement of the reservoir is obtained, which is found to be 0.10 m. That is not very small, and it might mean that the construction of the reservoir on such a soft soil is not feasible. The procedure described above can easily be extended. It is, for instance, simple to account for different properties in each layer, by using a variable compressibility. The method is also not restricted to circular loads. The method can easily be combined with Newmark's method to calculate the stresses below a load of arbitrary magnitude on an area of arbitrary shape. It is also possible to incorporate creep by adding the formula of Koppejan in the calculation. The method can also be elaborated with little difficulty to a computer program. Such a program may use a numerical form of Newmark's method to determine the stresses, and then calculate the settlements of the loaded area by the method illustrated above. The formula to compute the deformation of each layer may be Terzaghi's formula, but it may also include a time dependent term, to account for creep and consolidation.

Problems

32.1 The method presented in this chapter can easily be executed by a spreadsheet program on a computer. Write such a program and repeat the calculations from the table. The program will be more flexible if a separate column is used for the compressibility of the layers.

32.2 Repeat the computations of the example for the case of a layered soil, with $C_{10} = 20$ in the top 10 meter, and $C_{10} = 80$ in the lower 10 meter. On the average that is 50, just as in the example, but the settlement will appear to be different. Do you expect the settlement to be more than 0.10 m or less?

32.3 Calculate the settlement of a point on the boundary of the reservoir. Use Newmark's method to determine the stresses at the various depths.

33 Lateral stresses in soils

In the previous chapters some elastic solutions of soil mechanics problems have been given. It was argued that elastic solutions may provide a reasonable approximation of the vertical stresses in a soil body loaded at its surface by a vertical load. Also, an approximate procedure for the prediction of settlements has been presented. In this chapter, and the next chapters, the analysis of the horizontal stresses will be discussed. This is of particular interest for the forces on a retaining structure, such as a retaining wall or a sheet pile wall.

33.1 Coefficient of lateral earth pressure

As stated before, see Chapter 6, even in the simplest case of a semi-infinite soil body, without surface loading, see Figure 33-1, it is impossible to determine all stresses caused by the weight of the soil. It seems reasonable to assume that in a homogeneous soil body with a horizontal top surface the shear stresses σ_{zx}, σ_{zy} and σ_{xy} are zero, and it also seems natural to assume that the vertical normal stress σ_{zz} increases linearly with depth, $\sigma_{zx} = \gamma z$. These assumptions ensure that the condition of vertical equilibrium is satisfied. The horizontal stresses σ_{xx} and σ_{yy}, however, can not be determined unequivocally from the equilibrium conditions.

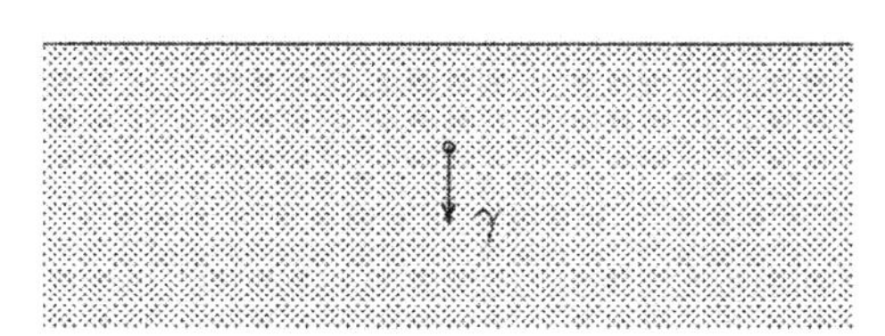

Figure 33-1. Half space.

Actually, it can be stated that the stresses must satisfy the equations of equilibrium,

$$\frac{\partial \sigma_{xx}}{\partial x} + \frac{\partial \sigma_{yx}}{\partial y} + \frac{\partial \sigma_{zx}}{\partial z} = 0, \tag{33.1}$$

$$\frac{\partial \sigma_{xy}}{\partial x} + \frac{\partial \sigma_{yy}}{\partial y} + \frac{\partial \sigma_{zy}}{\partial z} = 0, \tag{33.2}$$

$$\frac{\partial \sigma_{xz}}{\partial x} + \frac{\partial \sigma_{yz}}{\partial y} + \frac{\partial \sigma_{zz}}{\partial z} - \gamma = 0, \tag{33.3}$$

$$\sigma_{yz} = \sigma_{zy}, \tag{33.4}$$

$$\sigma_{zx} = \sigma_{xz}, \tag{33.5}$$

$$\sigma_{xy} = \sigma_{yx}. \tag{33.6}$$

These equations constitute a set of six conditions for the nine stress components, at every point of the soil body. It seems probable that many solutions of these equations are possible, and it can not be decided, without further analysis, what the best solution is. It seems natural to assume, at least for a homogenous material, or a material consisting of horizontal layers, that the stress state may be such that vertical normal stress increases linearly with depth, in proportion to the unit weight of the soil. More precisely, it is assumed that the stresses can be written as

$$\sigma_{zz} = \gamma z, \tag{33.7}$$

$$\sigma_{xx} = \sigma_{yy} = f(z), \tag{33.8}$$

$$\sigma_{yz} = \sigma_{zy} = 0, \tag{33.9}$$

$$\sigma_{zx} = \sigma_{xz} = 0, \tag{33.10}$$

$$\sigma_{xy} = \sigma_{yx} = 0. \tag{33.11}$$

This field of stresses satisfies all the equilibrium conditions, and the boundary conditions on the upper surface of the soil body, i.e. for $z = 0$ the stresses on a horizontal plane are zero, $\sigma_{zz} = 0$, $\sigma_{zx} = 0$, and $\sigma_{zy} = 0$. To assume that all shear stresses in the soil body seems a realistic assumption if all vertical columns of the soil have the same properties. There will probably be no shear stress transfer between these columns.

The function $f(z)$ in equation (33.8) remains arbitrary, and in principle the stresses σ_{xx} and σ_{yy} need not be equal. It has been assumed that the horizontal stress in any horizontal plane is the same in all directions, so that there are no preferential directions in the horizontal plane. Theoretically speaking, the function $f(z)$, which describes the horizontal stresses, need not be continuous. Discontinuities in this function are allowed, and may occur especially if there are discontinuities in the soil properties. It may be remarked that even the expressions for the vertical normal stress σ_{zz} and for the shear stresses do not follow necessarily from the equilibrium conditions. It may well be that these stresses depend upon x and y, if the soil stiffness is not constant in horizontal planes. In case of a very soft inclusion in a rather stiff soil body, the stresses may be concentrated in a region around the soft inclusion. This is called arching, as the stiffer soil may form a certain arch to transmit the load from upper layers to the subsoil. In homogeneous soil, however, or in soils without large differences in stiffness, the stress distribution given above can be considered as a reasonable approximation. Such a soil body has often been created, in its geological history, by gradual sedimentation, often under water. In such conditions the gradual increase of the thickness of the soil body will normally lead to a stress state of the form given above.

The stress state described by equations (33.7) – (33.11) can be made somewhat more practical by writing $f(z) = K\sigma_{zz}$, where K is an unknown coefficient, that may depend upon the vertical coordinate z. The horizontal stresses then are

$$\sigma_{xx} = \sigma_{yy} = K\sigma_{zz} = K\gamma z, \tag{33.12}$$

where K is the *coefficient of lateral earth pressure*. It gives the ratio of the lateral normal (effective) stress to the vertical (effective) stress. Theoretically speaking, the problem has not be cleared, because the value of K is still unknown, but it seems to make sense to assume that the horizontal stresses will also increase with depth, if the vertical stresses do so. Thus, it can be assumed that the coefficient K will not vary too much, at least compared to the original function $f(z)$.

It may be mentioned that for historical reasons the coefficient K is denoted as a coefficient of earth pressure, in agreement with most soil mechanics literature. This is one of the few instances where the word earth is used in soil mechanics, rather than the word soil, or ground. No special meaning should be attached to this terminology. In this book the coefficient will sometimes also be denoted as the horizontal (or lateral) stress coefficient.

The value of the lateral earth pressure coefficient K depends upon the material, and also on the geological history of the soil. In this chapter some examples of possible values, or the possible range of values, will be given, for certain simple materials. It will appear to be illustrative to describe the relations between vertical and horizontal stresses in a stress path. In Chapter 27 the quantities σ and τ have been introduced for that purpose, being the location of the centre and the radius of Mohr's circle. In this case these quantities are

$$\sigma = \tfrac{1}{2}(\sigma_{zz} + \sigma_{xx}), \tag{33.13}$$

$$\tau = \tfrac{1}{2}\,|\,\sigma_{zz} - \sigma_{xx}\,|. \tag{33.14}$$

It now follows, with eq.(33.12), and assuming that $K < 1$,

$$\frac{\tau}{\sigma} = \frac{1-K}{1+K}. \tag{33.15}$$

Often the horizontal stress will indeed be smaller than the vertical stress, so that $K < 1$, but this is not absolutely necessary.

33.2 Fluid

In a fluid the shear stresses can be neglected, compared to the pressure. This means that the normal stress is equal in all directions. This means that

$$K = 1. \tag{33.16}$$

If $K = 1$ the horizontal stress is equal to the vertical stress. With (33.15) this gives

$$\frac{\tau}{\sigma} = 0. \tag{33.17}$$

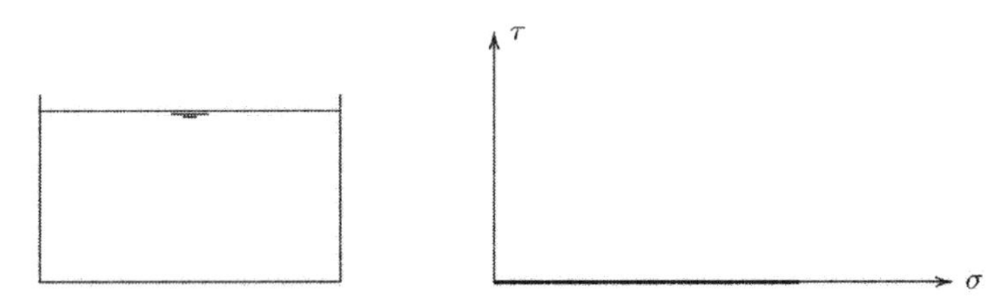

Figure 33-2. Stress path for a fluid.

The stress path is shown in Figure 33-2. This stress path refers to the case that a container is gradually filled with water. It would also apply if gravity would gradually develop in a fluid.

Soil is not a fluid, but certain very soft soils come close: the mud collected by dredging often is similar to a thick fluid. Very soft clay, with a high water content, also behaves similar to a fluid. When spread out it will flow until an almost horizontal surface has been formed. For such soils the value of K will be close to 1, and the stress path of Figure 33-2 is realistic.

33.3 Elastic material

A possible approach to the behaviour of soils is to consider it as an elastic material. In such a material the stresses and strains satisfy Hooke's law. In a situation in which there can be no lateral deformation, the stresses must satisfy the condition

$$\varepsilon_{xx} = \frac{1}{E}[\sigma_{xx} - \nu(\sigma_{yy} + \sigma_{zz})] = 0,$$

$$\varepsilon_{yy} = \frac{1}{E}[\sigma_{yy} - \nu(\sigma_{zz} + \sigma_{xx})] = 0,$$

if the z-direction is vertical. In a medium of large horizontal extent it can be expected that $\sigma_{xx} = \sigma_{yy}$. Then

$$\varepsilon_{xx} = \varepsilon_{xx} = o \quad : \qquad \sigma_{xx} = \sigma_{yy} = \frac{\nu}{1-\nu}\sigma_{zz},$$

or

$$K = \frac{\nu}{1-\nu}. \tag{33.18}$$

If Poisson's ratio varies between 0 and 0.5, the value of K varies from 0 to 1.
It follows from (33.15) and (33.18) that in this case

$$\frac{\tau}{\sigma} = 1 - 2\nu. \tag{33.19}$$

For a number of values of Poisson's ratio ν, between 0 and $\frac{1}{2}$, the stress path is shown in Figure 33-3. If $\nu = \frac{1}{2}$ the horizontal stresses are equal to the vertical stresses. In that case there are no volume changes, just as in a fluid. The stress path then is equal to the stress path in a fluid. If $\nu = 0$ the stress path has a slope of 45°. If the horizontal strains are not zero, but it is still assumed that the two horizontal stresses, σ_{xx} and σ_{yy}, are equal, these stresses are

$$\sigma_{xx} = \sigma_{yy} = \frac{\nu}{1-\nu}\sigma_{zz} + \frac{E}{1-\nu}\varepsilon_{xx}. \tag{33.20}$$

In case of a negative horizontal strain, the horizontal stress decreases, and then K is getting smaller. A positive horizontal strain, for instance due to some lateral compression, will result in a larger horizontal stress. The value of K then will seem to increase. These are general tendencies, with a validity beyond elasticity.

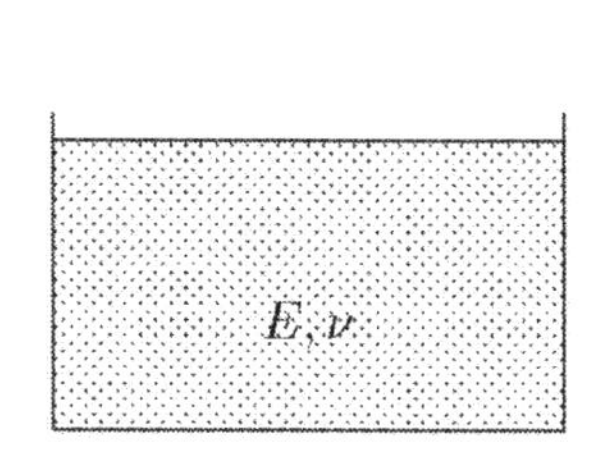

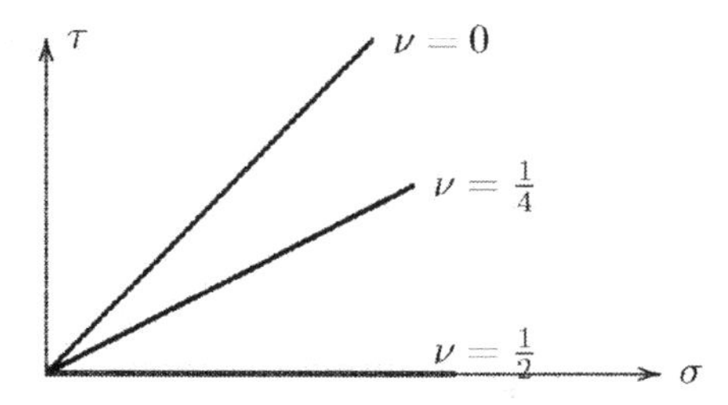

Figure 33-3. Stress path for elastic material.

In some older publications equation (33.18) has been proposed as a generally applicable relation for soil and rock. That is not true. An elastic analysis supposes that the stresses are being developed gradually, by gravity being applied gradually, on an existing soil in an unstressed state. And during this entire process the relation between stress and strain should be linear, and no horizontal deformations should occur. Geological history usually is much more complex, and the material behaviour is non-linear. This means that the value of the lateral stress coefficient K in general can not be predicted with any accuracy. It can be expected that in a region between two deep rivers the value of K will be relatively small, whereas in a valley between two mountain ridges that are moving towards each other due to tectonic motion, the stress coefficient K will be relatively large.

33.4 Elastic material under water

In order to take groundwater into account, the soil may be schematised as a linear elastic material, that is being deposited under water, see Figure 33-4.

Figure 33-4. Elastic material under water.

If the weight of the material is again carried by the vertical stresses, the vertical total stress will increase linearly with depth,

$$\sigma_{zz} = \gamma_s z, \tag{33.21}$$

in which γ_s is the total volumetric weight of the soil, including the water in the pores. The pore pressures are assumed to be hydrostatic,

$$p = \gamma_w z, \tag{33.22}$$

so that the vertical effective stresses are

$$\sigma'_{zz} = \sigma_{zz} - p = (\gamma_s - \gamma_w)z. \tag{33.23}$$

It is now postulated that in the process of the development of these stresses no horizontal deformations of the soil skeleton can occur. The deformation of this soil skeleton is determined by the effective stresses, and in this case, for a linear elastic material, it follows that

$$\sigma'_{xx} = \sigma'_{yy} = \frac{\nu}{1-\nu}\sigma'_{zz} = \frac{\nu}{1-\nu}(\gamma_s - \gamma_w)z. \tag{33.24}$$

This means that

$$K = \frac{\nu}{1-\nu}. \tag{33.25}$$

The coefficient of horizontal earth pressure K is in general the ratio between the horizontal and vertical effective stresses

$$\sigma'_{xx} = K\sigma'_{zz}. \tag{33.26}$$

Therefore it can be deduced that the horizontal total stress becomes

$$\sigma_{xx} = \sigma'_{xx} + p = K(\gamma_s - \gamma_w)z + \gamma_w z. \tag{33.27}$$

The lateral stress coefficient K should be used for the effective stresses only. The horizontal total stresses should be determined by adding the pore pressure to the horizontal effective stress.

Problems

33.1 Make a graph of the effective stress path (ESP) at a certain depth, if an elastic material is being built up, as shown in Figure 33-4, assuming that $\nu = 1/4$.

33.2 Also show the total stress path, in the same graph, and in the same point.

34 Rankine

The possible stresses in a soil are limited by the Mohr-Coulomb failure criterion. Following Rankine (1857) this condition will be used in this chapter to determine limiting values for the horizontal stresses, and for the lateral stress coefficient K.
For reasons of simplicity the considerations will be restricted to dry soils at first. The influence of pore water will be investigated later.

34.1 Mohr-Coulomb

As seen before, see Chapter 21, the stress states in a soil can be limited, with a good approximation by the Mohr-Coulomb failure criterion. This criterion is that the shear stresses on any plane are limited by the condition

$$\tau < \tau_f = c + \sigma \tan\phi, \tag{34.1}$$

where c is the cohesion, and ϕ is the angle of internal friction. The criterion can be illustrated using Mohr's circle, see Figure 34-1.

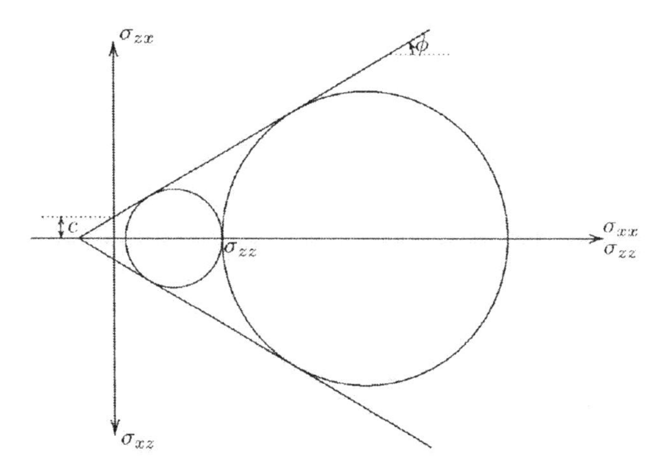

Figure 34-1. Mohr-Coulomb.

If it is assumed that σ_{zz} and σ_{xx} are principal stresses, and that σ_{zz} is known (by the weight of the load and the soil), it follows that the value of the horizontal stress σ_{xx} can not be smaller than indicated by the small circle, and not larger than defined by the large circle. The ratio between the minor and the major principal stress can be determined by noting, see Figure 34-2, that the radius of Mohr's circle is

$\frac{1}{2}(\sigma_1-\sigma_3)$, and that the location of the centre is at a distance $\frac{1}{2}(\sigma_1+\sigma_3)$ from the origin. It follows that for a circle touching the envelope,

$$\sin\phi=\frac{\frac{1}{2}(\sigma_1-\sigma_3)}{\frac{1}{2}(\sigma_1+\sigma_3)+c\cot\phi}.$$

So that

$$\sigma_3=\frac{1-\sin\phi}{1+\sin\phi}\sigma_1-2c\frac{\cos\phi}{1+\sin\phi}. \tag{34.2}$$

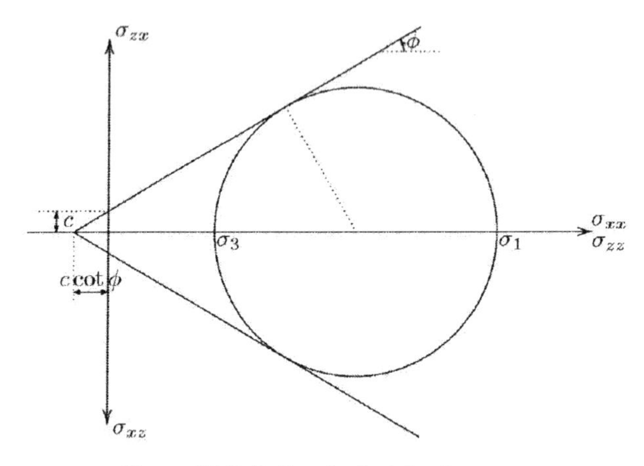

Figure 34-2. Ratio of principle stresses.

This relation has been derived before, in Chapter 21. The two coefficients in this equation can be related by noting that

$$\frac{\cos\phi}{1+\sin\phi}=\frac{\sqrt{1-\sin^2\phi}}{1+\sin\phi}=\frac{\sqrt{(1+\sin\phi)(1-\sin\phi)}}{1+\sin\phi}=\sqrt{\frac{1-\sin\phi}{1+\sin\phi}}.$$

This means that equation (34.2) can be written as

$$\sigma_3=K_a\sigma_1-2c\sqrt{K_a}, \tag{34.3}$$

Apart from the constant term $2c\sqrt{K_a}$ there appears to be a given ratio of the minor and the major principal stress.
Formula (34.3) can be written in inverse form as

$$\sigma_1=K_p\sigma_3+2c\sqrt{K_p}, \tag{34.4}$$

where

$$K_p=\frac{1+\sin\phi}{1-\sin\phi}. \tag{34.5}$$

The coefficients K_a and K_p, which give the smallest and the largest ratio of the two principal stresses (apart from a constant term), are denoted as the coefficients of *active earth pressure* (K_a) and *passive earth pressure* (K_p), respectively.
If the cohesion is zero ($c = 0$) it can be seen that

$$c = 0 \quad : \qquad K_a < K < K_p. \tag{34.6}$$

If $\phi = 30^\circ$ (this is a common value for sand, on the small side), it follows that

$$c = 0 \text{ and } \phi = 30^\circ : \quad \tfrac{1}{3} < K < 3. \tag{34.7}$$

The lateral stress coefficient K appears to be limited by values about $\frac{1}{3}$, and about 3. The precise limits depend upon the angle of internal friction ϕ.
As seen in the previous chapter for the elastic case, lateral extension of the soil leads to a smaller value of the lateral stress coefficient K, whereas lateral compression leads to a larger value of the coefficient K. The extreme situations are denoted as *active earth pressure* and *passive earth pressure*, respectively. The case of active earth pressure is supposed to occur when a retaining structure is being pushed away by the soil stresses. Passive earth pressure denotes that the structure is being pushed into the ground, in which a reaction is being developed.
The large difference between the minimum and the maximum lateral stress is characteristic for frictional materials such as soils.

34.2 Active earth pressure

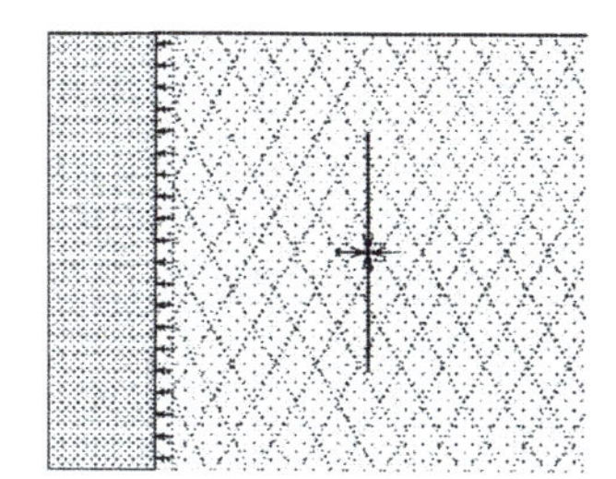

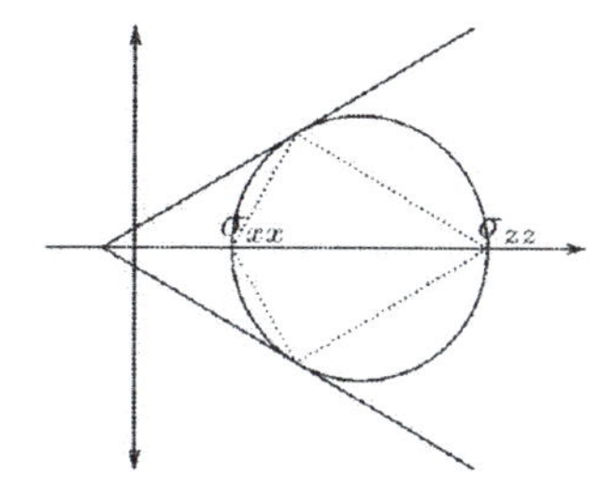

Figure 34-3. Active earth pressure.

It can be expected that the smallest value of the horizontal stress occurs in the case of a retaining wall that is moving away from the soil, see Figure 34-3. The Mohr circle for that case is also shown in the figure. The pole for the vectors normal to the planes is the rightmost point of the circle. This means that the critical shear stress acts on planes making an angle $\frac{1}{4}\pi + \frac{1}{2}\phi$ with the horizontal direction, that is an angle of $\frac{1}{4}\pi - \frac{1}{2}\phi$ with the vertical direction. These planes have been indicated in the left half of Figure 34-3. It is sometimes imagined that the soil indeed slides along these planes in case of failure.

The vertical stresses along the wall are

$$\sigma_{zz} = \gamma z, \tag{34.8}$$

in which γ is the volumetric weight of the soil, and z is the depth below soil surface. The horizontal stresses now are, with (34.3),

$$\sigma_{xx} = K_a \gamma z - 2c\sqrt{K_a}. \tag{34.9}$$

The total horizontal force on a wall of height h is obtained by integration from $z = 0$ to $z = h$. This gives

$$Q = \tfrac{1}{2} K_a \gamma h^2 - 2ch\sqrt{K_a}. \tag{34.10}$$

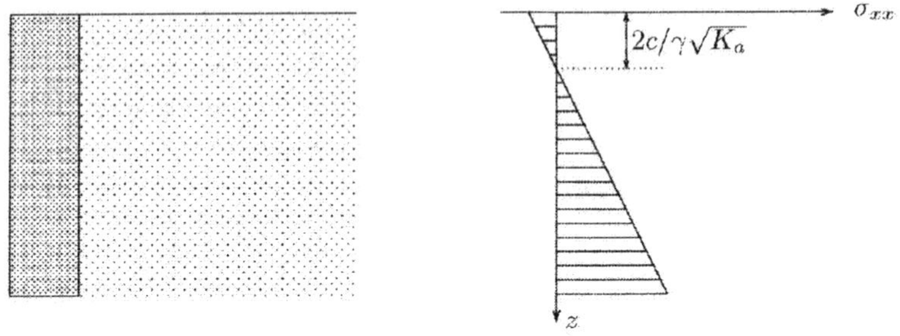

Figure 34-4. Horizontal stress in case of active earth pressure.

The distribution of the horizontal normal stress σ_{xx} against the wall is shown in Figure 34-4. It appears that at the top of the wall tensile stresses are generated, over a depth of $2c/\gamma\sqrt{K_a}$. That may be possible in the soil for a short while, in undrained conditions, with negative stresses in the water. In fully drained conditions this is not possible, because then there should be tensile stresses between the particles, or between the particles and the wall. Therefore, it is usually assumed that in a top layer of the soil, of height $2c/\gamma\sqrt{K_a}$, cracks will appear in the soil, and between the soil and the wall. For that case the stress distribution is shown in Figure 34-5. For the vertical stresses the cracked zone acts as a surcharge.

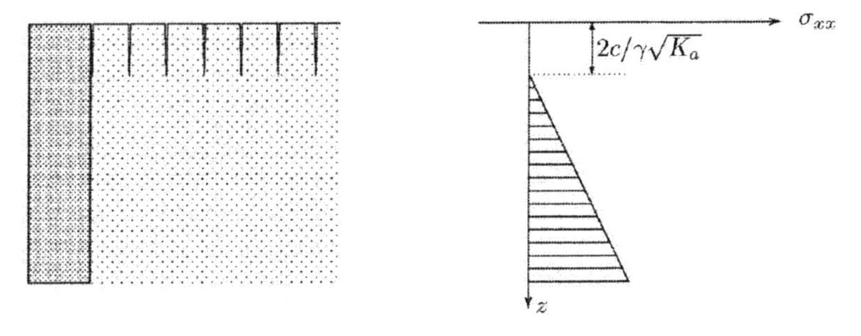

Figure 34-5. Horizontal stresses, with cracks in the soil.

The total horizontal force now is

$$Q=\tfrac{1}{2}K_a\gamma h_r^{\,2}, \tag{34.11}$$

in which h_r is the reduced height of the wall,

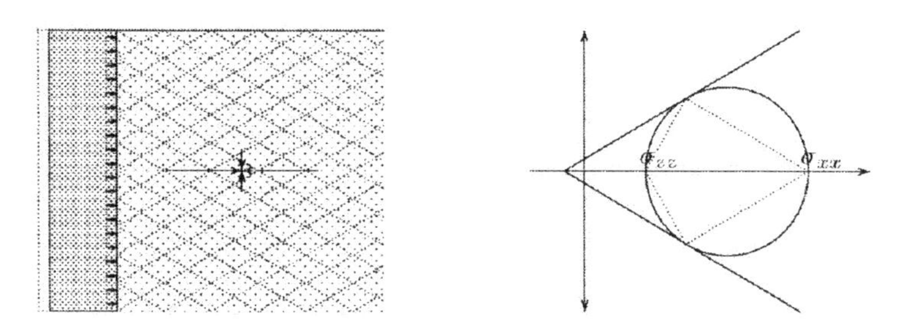

Figure 34-6. Passive earth pressure.

$$h_r=h-2c/\gamma\sqrt{Ka}. \tag{34.12}$$

34.3 Passive earth pressure

The case of passive earth pressure, in which the horizontal soil stress has its maximum value, can be considered to correspond to a smooth vertical wall that is being pushed in horizontal direction into the soil, see Figure 34-6. Again the Mohr circle has been shown in the figure as well, with the pole in this case being located in the leftmost point of the circle. The critical shear stress $\tau=\tau_f=c+\sigma\tan\phi$ occurs on planes making an angle $\frac{1}{4}\pi-\frac{1}{2}\phi$ with the horizontal direction. These planes have also been indicated in the left half of the figure. In this case the potential sliding planes are shallower than 45°.
The horizontal stresses against the wall in this case are

$$\sigma_{xx}=K_p\gamma z+2c\sqrt{K_p}. \tag{34.13}$$

They are shown in Figure 34-7.

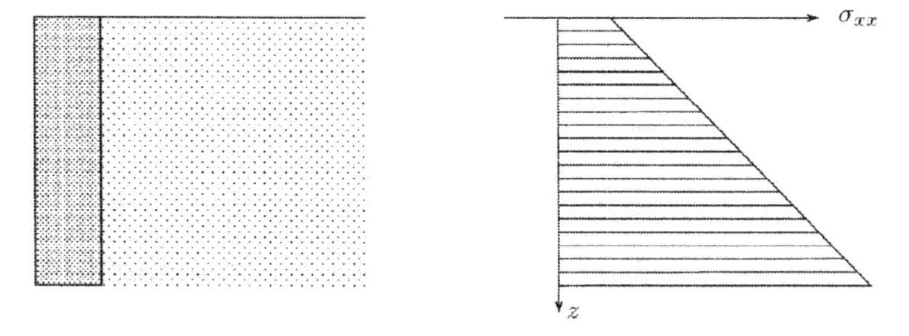

Figure 34-7. Horizontal stresses in case of passive earth pressure.

The total horizontal force on a wall of height h is obtained by integration of the horizontal stresses from $z=0$ to $z=h$. This gives

$$Q = \tfrac{1}{2} K_p \gamma h^2 + 2ch\sqrt{K_p}. \tag{34.14}$$

In the passive case the cohesion c appears to lead to a constant factor in the expression for the horizontal stresses. There is no reason for cracks to appear, as there are no tensile stresses in this case.

The two extreme states of stress considered here are often denoted as the Rankine states, after the English scientist Rankine (1857), who indicated that these stress states are the limiting conditions. In the case of a solid retaining wall, on a good foundation, the actual horizontal stresses will be somewhere between these two extremes. As the limits are so far apart (there may be a factor about 9 between them), this leaves the horizontal stress σ_{xx} undetermined to a high degree.

34.4 Neutral earth pressure

It has been found that the possible states of stress in a soil may vary between fairly wide limits, especially if the friction angle is large. For a normal sand, with $\phi = 30°$, the smallest value of the horizontal stress is $\frac{1}{3}$ of the vertical stress (which usually is known from the surcharge and the weight of the overlying soil), and the largest value is 3 times the vertical stress. The laterall stress in a non preloaded horizontal terrain, which is a totally different case than a moving retaining wall, is unknown, at least from a strictly scientific viewpoint. In reality there may be some additional information that may help to determine the probable range of the horizontal stress. If the horizontal displacements of the wall are practically zero, the ratio of the horizontal stress to the vertical stress is denoted as the *neutral earth pressure coefficient* K_0. In an elastic material this value would be $K_0 = \nu/(1-\nu)$, but this is not a very good estimate, as soil is not an elastic material, and the history of the development of stresses in the soil may be more important than the condition of zero lateral deformation. Nevertheless, in many cases it is unlikely that the coefficient K_0 would be larger than 1, as this would require some form of motion of the wall towards the soil mass. Also, the active state of stress (say $\frac{1}{3}$) may also be unlikely, if the wall is rather stiff and strong. All this suggests that the neutral stress coefficient may perhaps vary in the range from 0.5 to 1.0. For soft clay the value may be close to 1, and for sands values of about 0.6 or 0.7 have been found to give reasonable results. Lacking any better information the value may be estimated by the formula proposed by Jaky,

$$K_0 \approx 1 - \sin\phi, \tag{34.15}$$

but there is no real basis for this formula, except that it gives values between 0 and 1, with the value being close to 1 if the friction angle ϕ is very small (as it is for soft clays).

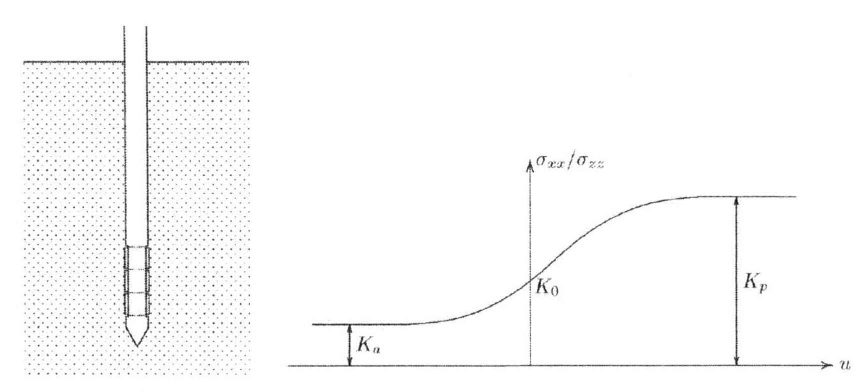

Figure 34-8. CAMKO-meter

Figure 34-9. Horizontal stress as a function of the displacement.

The best procedure is to try to measure the value of K_0, using an instrument of which the response is determined by the horizontal stress. For example, in the CAMKO-meter, developed in Cambridge (UK), a rubber membrane around a pipe is being pressurized, and the resulting deformation is measured. The idea is that the soil response will be different for lateral pressures below and above the original neutral stress. The membrane consists of three parts, with the central part being the measuring cell. A similar instrument is Marchetti's dilatometer, which consists of a hollow circular plate that is pushed into the soil, in a vertical position. By expanding the plate the lateral response is measured, and from this response the lateral stiffness and the neutral stress coefficient may be estimated. Another method is to inject water into the soil from a tubular instrument. The idea is that a vertical crack may be produced in the soil if the water pressure exceeds the horizontal total stress, because the soil skeleton can not transfer tensile stresses. In petroleum engineering this process is called *hydraulic fracturing*, and it is used to improve the permeability of porous rock.

A possible relation between the horizontal stress against a retaining structure and its horizontal displacement is shown in Figure 34-9. If the displacement is zero, the lateral stress coefficient will be K_0. If the structure now is being pressed towards the soil, the lateral stress will gradually increase, until finally the passive coefficient K_p is reached. On the other hand, if the structure moves away from the soil, the lateral stress will decrease, until its lowest value is reached, as defined by the active coefficient K_a. In advanced computations a relation as shown in Figure 34-9 may be used to determine the displacement of the structure and the stresses against it. In many cases it can be argued that one of the two limiting values can be considered as appropriate, see the next chapter.

34.5 Groundwater

In the case of a soil saturated with water it should be noted that the Mohr-Coulomb criterion describes the limiting states of effective stress in the soil. The correct procedure should be that the smallest and the largest horizontal stress can be deduced from the vertical effective stress, using the active or passive stress coefficient. The horizontal total stress can be obtained by adding the pore water pressure to the horizontal effective stress.

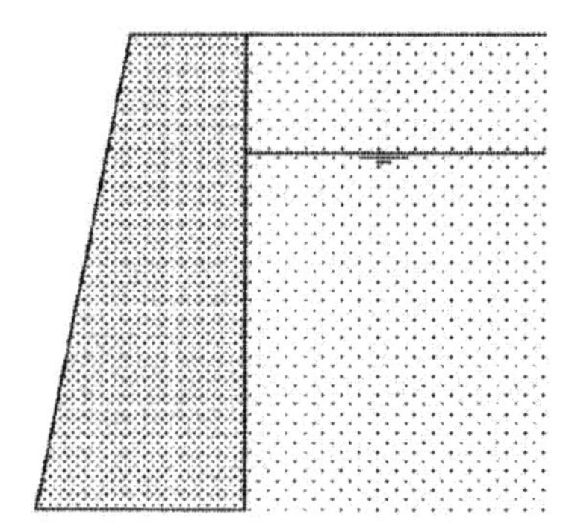

Figure 34-10. Groundwater in the soil.

As an example a retaining wall is shown in Figure 34-10. The wall is assumed to be 8 meter high, with the groundwater level at 2 meter below the soil surface. The question is to determine the horizontal total stress at a depth of 8 m, assuming that the soil is sand, with $c = 0$ and $\phi = 30°$, for the case of active earth pressure. The volumetric weight of the soil is 16 kN/m^3 when dry, and 20 kN/m^3 when saturated with water. It is assumed that there is no capillary rise in the sand. The vertical total stress at a depth of 8 m is the weight of 2 m dry soil and 6 m of saturated soil, which gives $\sigma_{zz} = 152$ kPa. Because the pore pressure at that depth is 60 kPa is, the vertical effective stress is $\sigma'_{zz} = 92$ kPa. The active stress coefficient is $K_a = \frac{1}{3}$, so that the horizontal effective stress is $\sigma'_{xx} = 31$ kPa. The total stress is found by adding the pore pressure, i.e. $\sigma_{xx} = 91$ kPa. It is interesting to note that this consists for 66 % of water pressure, and for only 34 % of effective stress. This illustrates that the contribution of the pore water pressure may be very large. This is especially true in the case of active earth pressure.

Problems

34.1 Compose a simple table for the active and passive earth pressure coefficients, as a function of the angle of internal friction ϕ.

34.2 If the cohesion c is unequal to zero, it follows from eq. (34.3) that the minor principal stress σ_3 can be zero, while the major principal stress σ_1 is unequal to zero. This means that in a cohesive material an excavation can be made with vertical sides. What is the maximum depth of such an excavation, on the basis of this formula?

34.3 Why is the instrument shown in Figure 34-8 known as the CAMKO-meter?

34.4 A bulldozer, having a blade of 4 m width and 1 m height, is used to remove an amount of dry sand, of 1 m height. Estimate the total force necessary to move the sand.

34.5 A concrete wall that retained a mass of gravel, of 5 m height, has collapsed by overturning. Estimate the total horizontal force on the wall at the moment of failure, per meter width.

34.6 Make a graph of the horizontal total stresses against the retaining wall shown in Figure 34-10, as a function of the depth, and calculate the total horizontal force, per unit meter width.

35 Coulomb

Long before the analysis of Rankine the French scientist Coulomb presented a theory on limiting states of stress in soils (in 1776), which is still of great value. The theory enables to determine the stresses on a retaining structure for the cases of active and passive earth pressure. The method is based upon the assumption that the soil fails along straight slip planes.

35.1 Active earth pressure

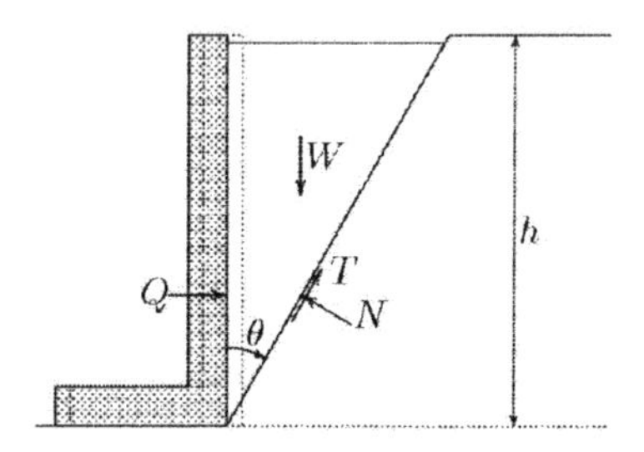

Figure 35-1. Active earth pressure.

For the active case (a retreating wall) the procedure is illustrated in Figure 35-1. It is assumed that in case of a displacement of the wall towards the left, a triangular wedge of soil will slide down, along a straight slip plane. The angle of the slope with the vertical direction is denoted by θ. It is also assumed that at the moment of sliding, the weight of the soil wedge is just in equilibrium with the forces on the slip surface and the forces on the wall. For reasons of simplicity it is assumed, at least initially, that the force between the soil and the wall (Q) is directed normal to the surface of the wall, i.e. shear stresses along the wall are initially neglected. In later chapters such shear stresses will be taken into account as well. The purpose of the analysis is to determine the magnitude of the force Q. The principle of Coulomb's method is that it is stated that the wall must be capable of withstanding the force Q for all possible slip planes. Therefore the slip plane that leads to the largest value of Q is to be determined. The various slip planes are characterized by the angle θ, and this angle will be determined such that the maximum value of Q is obtained. The starting point of the analysis is the weight of the soil wedge (W), per unit width perpendicular to the plane shown in the figure,

$$W = \tfrac{1}{2}\gamma h^2 \tan\theta. \tag{35.1}$$

This weight must be balanced by the horizontal force Q (horizontal because the wall has been assumed to be perfectly smooth), and the forces N and T on the slip plane. The direction of the shear force T is determined by the assumed sliding direction, with the soil body moving down, in order to follow the motion of the wall to the left. Furthermore, because the length of the slip plane is $h/\cos\theta$,

$$T = \frac{ch}{\cos\theta} + N\tan\phi. \tag{35.2}$$

The equations of equilibrium of the soil body, in horizontal and vertical direction, are

$$Q + T\sin\theta - N\cos\theta = 0, \tag{35.3}$$

$$W - N\sin\theta - T\cos\theta = 0. \tag{35.4}$$

With eq. (35.2) the shear force T can be eliminated. This gives

$$Q = \frac{N}{\cos\phi}\cos(\theta+\phi) - ch\tan\theta, \tag{35.5}$$

$$W = \frac{N}{\cos\phi}\sin(\theta+\phi) + ch. \tag{35.6}$$

From these two equations the normal force N can be eliminated,

$$Q = W\frac{\cos(\theta+\phi)}{\sin(\theta+\phi)} - ch\frac{\cos\phi}{\cos\theta\sin(\theta+\phi)}. \tag{35.7}$$

With eq. (35.1) this gives

$$Q = \tfrac{1}{2}\gamma h^2\frac{\sin\theta\cos(\theta+\phi)}{\cos\theta\sin(\theta+\phi)} - ch\frac{\cos\phi}{\cos\theta\sin(\theta+\phi)}. \tag{35.8}$$

This equation expresses the force Q as a function of the angle θ. The relation is rather complex (the angle θ appears in 6 places), so that it does not seem to be very simple to determine the maximum value. However, the expression can be simplified by using various trigonometric relations, such as $\sin\theta\cos(\theta+\phi) = \cos\theta\sin(\theta+\phi) - \sin\phi$. This gives

$$Q = \tfrac{1}{2}\gamma h^2 - \frac{\tfrac{1}{2}\gamma h^2\sin\phi + ch\cos\phi}{\cos\theta\sin(\theta+\phi)}. \tag{35.9}$$

Now the angle θ appears in 2 places only, in the denominator of the second term. The maximum value of Q can be determined by the maximum value of the function

$$f(\theta) = \cos\theta\sin(\theta+\phi).$$

The maximum of this function occurs when its derivative with respect to θ is zero. Differentiation gives

$$\frac{df}{d\theta}=\cos(2\theta+\phi),$$

and a second differentiation gives

$$\frac{d^2f}{d\theta^2}=-2\sin(2\theta+\phi).$$

It now follows that $df/d\theta=0$ if $2\theta+\phi=\frac{1}{2}\pi$ or

$$\frac{df}{d\theta}=0: \qquad \theta=\tfrac{1}{4}\pi-\tfrac{1}{2}\phi. \tag{35.10}$$

Then $d^2f/d\theta^2=-2$, so that the function f indeed has a maximum for this value of θ. This means that the horizontal force Q also has a maximum for $\theta=\frac{1}{4}\pi-\frac{1}{2}\phi$ This maximum value is, after some elaboration,

$$\theta=\tfrac{1}{4}\pi-\tfrac{1}{2}\phi \quad : \qquad Q=\tfrac{1}{2}\gamma h^2K_a-2ch\sqrt{K_a}, \tag{35.11}$$

in which K_a is the coefficient of active earth pressure, defined before,

$$K_a=\frac{1-\sin\phi}{1+\sin\phi}. \tag{35.12}$$

These results are in full agreement with the results obtained in the previous chapter on active earth pressure, see equation (34.11). The value for the horizontal force Q corresponds to the sliding of a wedge of soil along a slip plane making an angle $\frac{1}{4}\pi-\frac{1}{2}\phi$ with the vertical direction. These are just the planes shown in Figure 34-3. In the previous chapter it was found that along these planes the stresses first reach the Mohr-Coulomb envelope. It might be noted that in this analysis possible tension cracks in the soil have been ignored.

Coulomb's method contains a possible confusing step, in the procedure of *maximizing* the force Q to determine the appropriate value of the angle θ. This might suggest that the procedure gives a high value for Q, whereas in reality the value of Q indicates the smallest possible value of the horizontal force against a retaining wall, as can be seen from Rankine's analysis. The confusion is caused by the assumption in Coulomb's analysis that the soil slides along a slope defined by an angle θ with the vertical, and not along any other plane. For a value of θ other than the critical value $\theta=\frac{1}{4}\pi-\frac{1}{2}\phi$, the force Q may be smaller, but in that case there will be other planes on which the maximum shear stress exceeds Coulomb's maximum $\tau_f=c+\sigma_n\tan\phi$. In the analysis it ought to be investigated whether the assumed slip plane, at an angle θ, is indeed the most critical plane. This is the case only if $\theta=\frac{1}{4}\pi-\frac{1}{2}\phi$, as can be seen from Rankine's analysis. In that analysis the stresses on all planes are considered, by using Mohr's circle. In Coulomb's analysis the stresses on planes other than the assumed sliding plane are not considered at all.

In engineering practice, the horizontal stress against a retaining wall, or a sheet pile wall is often calculated using the active stress coefficient K_a. This may seem on the unsafe side, because it gives the smallest possible value of the horizontal stress, and will occur only in case of failure of the soil. The application is based upon the following argumentation. It is admitted that the analysis following Rankine or Coulomb, for the active stress state, yields the smallest possible value for the lateral force. In reality the lateral force may be higher, especially if the foundation of the retaining wall is stiff and strong. If the lateral force is so large that the wall's foundation can not withstand that force, it will deform, away from the soil. During that deformation the lateral force will decrease. Eventually this deformation may be so large that the active state of stress is attained. If the foundation and the structure are strong enough to withstand the active state of stress, the deformations will stop as soon as this active state is reached. These deformations may be large, but the structure will not fail. Thus, the structure will be safe if it can withstand active earth pressure, provided that there is no objection to a considerable deformation. For instance, the pile foundation of a quay wall in a harbour can be designed on the basis of active earth pressure against the quay wall, if it is accepted that considerable lateral deformations (say 1 % or 2 % of the height of the wall) of the quay wall may occur. If this is undesirable, for esthetic reasons or because other structures (the cranes) might be damaged by such large deformations, the foundation must be designed for larger lateral forces. This will mean that many more piles are needed.

35.2 Passive earth pressure

For the case of passive earth pressure (i.e. the case of a wall that is being pushed towards the soil mass, by some external cause) Coulomb's procedure is as follows, see Figure 35-2. Because the wedge of soil in this case is being pushed upwards, the shear force T will be acting in downward direction. The weight of the wedge is, as in the active case,

$$W = \tfrac{1}{2}\gamma h^2 \tan\theta. \tag{35.13}$$

The equations of equilibrium in x- and z-direction now are

$$Q - T\sin\theta - N\cos\theta = 0, \tag{35.14}$$

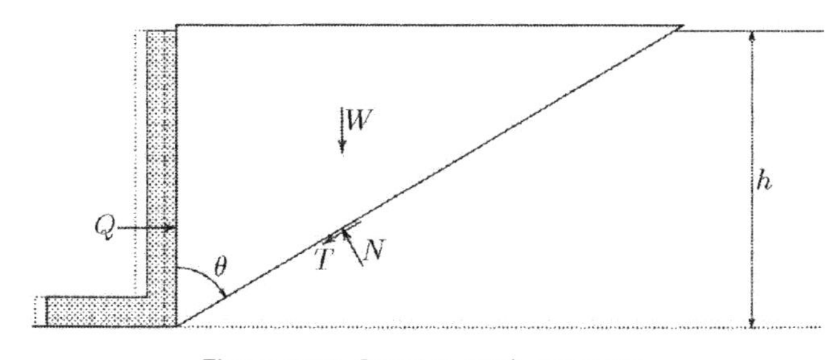

Figure 35-2. Passive earth pressure.

$$W - N\sin\theta + T\cos\theta = 0. \tag{35.15}$$

After elimination of T and N from the equations, and some trigonometric manipulations, the force Q is found to be

$$Q = \tfrac{1}{2}\gamma h^2 + \frac{\tfrac{1}{2}\gamma h^2 \sin\phi + ch\cos\phi}{\cos\theta\sin(\theta - \phi)}. \tag{35.16}$$

Again this force appears to be dependent on the angle θ. The minimum value of Q occurs if the function

$$f(\theta) = \cos\theta\sin(\theta - \phi),$$

has its largest value. Differentiation gives

$$\frac{df}{d\theta} = \cos(2\theta - \phi),$$

It now follows that $df/d\theta = 0$ if $2\theta - \phi = \frac{1}{2}\pi$, or

$$\frac{df}{d\theta} = 0: \qquad \theta = \tfrac{1}{4}\pi + \tfrac{1}{2}\phi. \tag{35.17}$$

Then $d^2f/d\theta^2 = -2$, and the function f indeed has a maximum for that value of θ. That means that the horizontal force Q has a minimum.
This minimum is

$$\theta = \tfrac{1}{4}\pi + \tfrac{1}{2}\phi \quad : \qquad Q = \tfrac{1}{2}\gamma h^2 K_p + 2ch\sqrt{K_p}, \tag{35.18}$$

where K_p is the passive earth pressure coefficient, defined before,

$$K_p = \frac{1+\sin\phi}{1-\sin\phi}. \tag{35.19}$$

Again, the result is in complete agreement with the value obtained in Rankine's analysis. Coulomb's procedure appears to lead to the maximum (passive) earth pressure.
Coulomb's method can easily be extended to more general cases. It is possible, for instance, that the surface of the wall is inclined with respect to the vertical direction, and the soil surface may also be sloping. Also, the soil may carry a given surface load. For all these cases the method can easily be extended. The general procedure is to assume a straight slip plane, consider equilibrium of the sliding wedge, and then maximizing or minimizing the force against the wall. Analytical, graphical and numerical methods have been developed. In the next chapter a number of tables is presented.

Problems

35.1 In some textbooks the coefficients K_a and K_p are defined as
$K_a = \tan^2(\frac{1}{4}\pi - \frac{1}{2}\phi)$, $\quad K_p = \tan^2(\frac{1}{4}\pi + \frac{1}{2}\phi)$. Is that an error?

35.2 If $K_a = 0.273$, then what is K_p?

35.3 A vertical wall retains a mass of dry sand, of 4 m height. The friction angle of the sand is 30°, and the volumetric weight is 17 kN/m^3. What is the design value of the horizontal force (per meter width) on the wall, if the deformations are not important?

35.4 Investigate the sensitivity of the previous problem for the friction angle, by determining the result for a friction angle that is 10 % higher.

35.5 What should be the design value of the horizontal force if the client wishes that the wall does not deform under any circumstances?

36 Tables for lateral earth pressure

The computation of lateral earth pressure against retaining walls is such an important problem of soil mechanics that tables have been produced for its solution, all on the basis of Coulomb's method. These tables can be found in many handbooks, such as the German "Grundbau Taschenbuch". Following Coulomb these tables apply to soils without cohesion ($c = 0$), that is for sand or gravel. In this chapter some tables are given for the active and the passive earth pressure against a retaining wall, with a surface that is practically vertical, and a sloping soil surface.

36.1 The problem

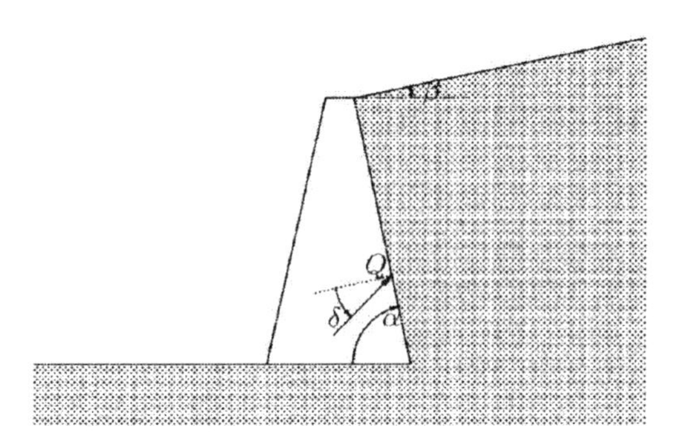

Figure 36-1. Horizontal earth pressure.

The general problem considered in this chapter concerns a retaining wall, having a surface inclined at an angle α with the horizontal direction. The soil surface is horizontal, or it may be sloping at an angle β with the horizontal direction, see Figure 36-1. The wall may be perfectly smooth, or it may have a certain friction, so that the direction of the force Q is at an angle δ with the direction normal to the wall. The friction angle δ is supposed to be given. Because the wall often is rather smooth, its value is often taken somewhat smaller than the friction angle of the soil itself, say $\delta = \frac{2}{3}\phi$. The angle δ is considered positive in the active case, illustrated in Figure 36-1, in which the sliding soil wedge is expected to slide in downward direction, along the surface of the wall. In the case of passive earth pressure it can be expected that the soil will move in upward direction along the surface of the wall. The angle δ then should be given a negative value.

The tables record the values of the coefficient K in the formula

$$Q = \tfrac{1}{2} K \gamma h^2. \tag{36.1}$$

This coefficient would be equal to 1 in the case of a fluid against a vertical wall.
It should be noted that Q is the total force. The angle of this force with the vertical direction is $\alpha - \delta$. The horizontal component of this force is

$$Q_h = Q \sin(\alpha - \delta). \tag{36.2}$$

If the tables are used to determine the horizontal force, the multiplication by the factor $\sin(\alpha - \delta)$ should be performed by the user.
The values of the active coefficient K_a were already calculated by Coulomb. He obtained

$$K_a = \frac{\sin^2(\alpha + \phi)}{\sin^2\alpha \sin(\alpha - \delta)\left[1 + \sqrt{\{\sin(\phi + \delta)\sin(\phi - \beta)\}/\{\sin(\alpha - \delta)\sin(\alpha + \beta)\}}\right]^2}. \tag{36.3}$$

For the passive case the formula is

$$K_p = \frac{\sin^2(\alpha - \phi)}{\sin^2\alpha \sin(\alpha - \delta)\left[1 - \sqrt{\{\sin(\phi - \delta)\sin(\phi + \beta)\}/\{\sin(\alpha - \delta)\sin(\alpha + \beta)\}}\right]^2}. \tag{36.4}$$

It may be mentioned that the active coefficients in the tables may be somewhat too small, and that the passive coefficients may be too large. This may be because in reality the soil may not yet have reached a critical state, but also because in Coulomb's method only straight slip surfaces are considered. In reality a curved slip surface, for instance a circular slip surface may give a higher active earth pressure or a lower passive pressure. This last possibility can easily be imagined: if the soil can fail along a circular slip surface for a force that is smaller than the critical straight sliding plane, there is no reason why the soil would not fail along the circular slip surface. A chain breaks if the weakest link fails.
It has been found that using circular slip surfaces leads to a very small increase of the active coefficients. The passive coefficients, however, may become considerably lower when circular slip surfaces are also taken into account. In particular, all values larger than 10 in the tables are unreliable. This can be very dangerous, for instance when calculating the maximum holding force of an anchor. This may be severely overestimated by using tables based upon straight slip planes only (as in this chapter). More reliable values are given in the tables in "Grundbau Taschenbuch, Teil 1".
It should be noted that in some tables the definition (and the notation) of the angles α, β and δ differs from the definitions used here. Great care should be used when taking values from an unfamiliar table.

36.2 Example

As an example the case of a wall at an inclination of 80° is considered. The slope of the soil is 10°, see Figure 36-2. The soil is sand, with $\phi = 30°$, and the friction angle between the wall and the soil is $\delta = 20°$. The problem is to determine the horizontal component of the force against the wall, in the case of active earth pressure.

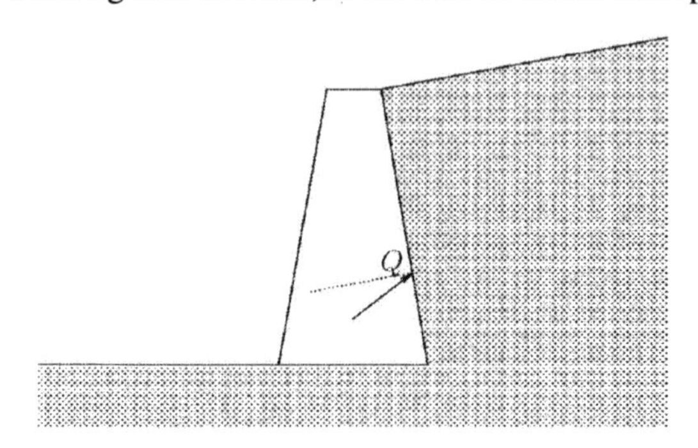

Figure 36-2. Example: Active earth pressure.

In this case Table 36-2 gives $K_a = 0.438$, so that the force on the wall is $Q = 0.219\gamma h^2$. Its horizontal component is, with (36.2), $Q_h = 0.190\gamma h^2$.

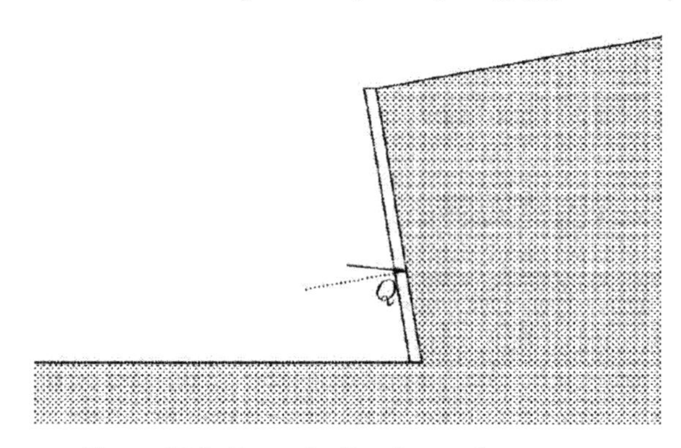

Figure 36-3. Example: Passive earth pressure.

In the case of passive earth pressure, when the wall is moving to the right, it will push the soil wedge up. It can be expected that then the wall will exert a shear force on the wall in downward direction, with the value of δ being negative, $\delta = -20°$, see Figure 36-3. In this case Table 36-3 gives $K_p = 7.162$. The force on the wall then is $Q_h = 3.527\gamma h^2$, because in this case $\alpha - \delta = 100°$.

36.3 Tables

On the following pages some values of K_a and K_p are given in tabular form.

$\alpha = 90°, \beta = 0°$:

$\delta \backslash \varphi$	10°	15°	20°	25°	30°	35°	40°	45°
0°	0.704	0.589	0.49	0.406	0.333	0.271	0.217	0.172
5°	0.662	0.556	0.465	0.387	0.319	0.26	0.21	0.166
10°	0.635	0.533	0.447	0.373	0.308	0.253	0.204	0.163
15°	0.617	0.518	0.434	0.363	0.301	0.248	0.201	0.160
20°	0.607	0.508	0.427	0.357	0.297	0.245	0.199	0.160
25°	0.604	0.505	0.424	0.355	0.296	0.244	0.199	0.160
30°	0.606	0.506	0.424	0.356	0.297	0.246	0.201	0.162

$\alpha = 90°, \beta = 10°$:

$\delta \backslash \varphi$	10°	15°	20°	25°	30°	35°	40°	45°
0°	0.970	0.704	0.569	0.462	0.374	0.300	0.238	0.186
5°	0.974	0.679	0.547	0.444	0.359	0.289	0.230	0.180
10°	0.985	0.664	0.531	0.431	0.350	0.282	0.225	0.177
15°	1.004	0.655	0.522	0.423	0.343	0.277	0.221	0.174
20°	1.032	0.654	0.518	0.419	0.340	0.275	0.220	0.174
25°	1.070	0.658	0.518	0.419	0.340	0.275	0.221	0.175
30°	1.120	0.669	0.524	0.422	0.343	0.278	0.223	0.177

$\alpha = 90°, \beta = 20°$:

$\delta \backslash \varphi$	10°	15°	20°	25°	30°	35°	40°	45°
0°			0.883	0.572	0.441	0.344	0.267	0.204
5°			0.886	0.558	0.428	0.333	0.259	0.199
10°			0.897	0.549	0.42	0.326	0.254	0.195
15°			0.914	0.546	0.415	0.323	0.251	0.194
20°			0.940	0.547	0.414	0.322	0.250	0.193
25°			0.974	0.553	0.417	0.323	0.252	0.195
30°			1.020	0.565	0.424	0.328	0.256	0.198

$\alpha = 90°, \beta = 30°$:

$\delta \backslash \varphi$	10°	15°	20°	25°	30°	35°	40°	45°
0°					0.750	0.436	0.318	0.235
5°					0.753	0.428	0.311	0.229
10°					0.762	0.423	0.306	0.226
15°					0.776	0.422	0.305	0.225
20°					0.798	0.425	0.305	0.225
25°					0.828	0.431	0.309	0.228
30°					0.866	0.442	0.315	0.232

Table 36-1: Active earth pressure coefficient, K_a.

$\alpha = 80°, \beta = 0°$:

$\delta \backslash \varphi$	10°	15°	20°	25°	30°	35°	40°	45°
0°	0.757	0.652	0.559	0.478	0.407	0.343	0.287	0.238
5°	0.720	0.622	0.536	0.460	0.393	0.333	0.280	0.233
10°	0.699	0.603	0.520	0.448	0.384	0.326	0.275	0.229
15°	0.687	0.592	0.511	0.441	0.378	0.323	0.273	0.228
20°	0.684	0.588	0.508	0.438	0.377	0.322	0.273	0.229
25°	0.689	0.591	0.510	0.440	0.379	0.325	0.276	0.232
30°	0.702	0.600	0.517	0.446	0.385	0.330	0.281	0.237

$\alpha = 80°, \beta = 10°$:

$\delta \backslash \varphi$	10°	15°	20°	25°	30°	35°	40°	45°
0°	1.047	0.784	0.654	0.550	0.461	0.384	0.318	0.261
5°	1.067	0.766	0.636	0.534	0.448	0.374	0.311	0.255
10°	1.097	0.759	0.626	0.524	0.440	0.368	0.307	0.253
15°	1.138	0.759	0.622	0.520	0.437	0.366	0.305	0.252
20°	1.191	0.768	0.625	0.521	0.438	0.367	0.306	0.254
25°	1.259	0.785	0.634	0.528	0.443	0.371	0.310	0.257
30°	1.346	0.811	0.650	0.539	0.452	0.379	0.317	0.264

$\alpha = 80°, \beta = 20°$:

$\delta \backslash \varphi$	10°	15°	20°	25°	30°	35°	40°	45°
0°			1.015	0.684	0.548	0.444	0.360	0.291
5°			1.035	0.676	0.538	0.436	0.354	0.286
10°			1.064	0.674	0.534	0.432	0.351	0.283
15°			1.103	0.679	0.535	0.432	0.350	0.284
20°			1.155	0.690	0.540	0.435	0.354	0.286
25°			1.221	0.708	0.551	0.443	0.360	0.292
30°			1.305	0.734	0.568	0.456	0.370	0.300

$\alpha = 80°, \beta = 30°$:

$\delta \backslash \varphi$	10°	15°	20°	25°	30°	35°	40°	45°
0°					0.925	0.566	0.433	0.337
5°					0.943	0.563	0.428	0.333
10°					0.969	0.564	0.427	0.332
15°					1.005	0.570	0.430	0.333
20°					1.051	0.582	0.437	0.338
25°					1.111	0.600	0.448	0.346
30°					1.189	0.624	0.463	0.358

Table 36-2: Active earth pressure coefficient, K_a.

$\alpha = 90°, \beta = 0°$:

$\delta \backslash \varphi$	10°	15°	20°	25°	30°	35°	40°	45°
0°	1.420	1.698	2.040	2.464	3.000	3.690	4.599	5.828
-5°	1.569	1.901	2.313	2.833	3.505	4.391	5.593	7.278
-10°	1.730	2.131	2.635	3.285	4.143	5.309	6.946	9.345
-15°	1.914	2.403	3.029	3.855	4.976	6.555	8.872	12.466
-20°	2.130	2.735	3.525	4.597	6.105	8.324	11.771	17.539
-25°	2.395	3.151	4.169	5.599	7.704	10.980	16.473	26.696
-30°	2.726	3.691	5.036	7.013	10.095	15.273	24.933	46.087

$\alpha = 90°, \beta = 10°$:

$\delta \backslash \varphi$	10°	15°	20°	25°	30°	35°	40°	45°
0°		2.099	2.595	3.235	4.080	5.228	6.841	9.204
-5°		2.467	3.086	3.908	5.028	6.605	8.923	12.518
-10°		2.907	3.700	4.783	6.314	8.569	12.076	17.944
-15°		3.456	4.496	5.969	8.145	11.536	17.225	27.812
-20°		4.166	5.572	7.652	10.903	16.370	26.569	48.891
-25°		5.122	7.093	10.181	15.384	25.117	46.474	108.431
-30°		6.470	9.371	14.274	23.468	43.697	102.545	426.159

$\alpha = 90°, \beta = 20°$:

$\delta \backslash \varphi$	10°	15°	20°	25°	30°	35°	40°	45°
0°	1.363	1.582	1.843	2.156	2.535	3.002	3.587	4.332
-5°	1.480	1.737	2.045	2.418	2.879	3.456	4.193	5.158
-10°	1.600	1.905	2.273	2.725	3.292	4.017	4.966	6.244
-15°	1.732	2.096	2.540	3.094	3.802	4.730	5.981	7.726
-20°	1.883	2.321	2.861	3.549	4.450	5.666	7.363	9.838
-25°	2.060	2.590	3.257	4.127	5.299	6.937	9.329	13.021
-30°	2.274	2.923	3.759	4.881	6.450	8.742	12.286	18.184

$\alpha = 90°, \beta = 30°$:

$\delta \backslash \varphi$	10°	15°	20°	25°	30°	35°	40°	45°
0°		1.935	2.308	2.767	3.343	4.079	5.043	6.340
-5°		2.218	2.668	3.233	3.96	4.914	6.201	7.998
-10°		2.541	3.093	3.805	4.742	6.010	7.783	10.372
-15°		2.922	3.614	4.528	5.767	7.504	10.045	13.969
-20°		3.387	4.272	5.474	7.162	9.636	13.465	19.844
-25°		3.975	5.131	6.759	9.148	12.854	19.039	30.500
-30°		4.740	6.295	8.583	12.137	18.084	29.127	53.188

Table 36-3: Passive earth pressure coefficient, K_p.

Problems

36.1 Check whether the two basic cases of Coulomb (vertical wall, horizontal soil surface) are correctly given in the tables.

36.2 Check that in the example the tables indeed give $K_a = 0.438$ and $K_p = 7.162$. Also verify whether the analytic formulas given in this chapter give these same values.

36.3 Why do the tables not give values for cases with $\phi < \beta$?

36.4 A retaining wall of 5 m height, with a smooth vertical wall is bounded by a soil with a horizontal surface. The angle of internal friction of the soil is $\phi = 35°$, and the volumetric weight of the soil is $\gamma = 17 \text{ kN/m}^3$. Determine the horizontal force against the wall.

36.5 Repeat the previous problem for the case that the wall is not vertical, but inclined at 10° with respect to the vertical direction.

36.6 An anchor in dry soil consists of a square plate, of dimensions 2 m × 2 m. The plate has been pushed into the soil in vertical direction, and its top coincides with the soil surface. Estimate the holding force of the anchor.

37 Sheet pile walls

An effective way to retain a soil mass is by installing a vertical wall consisting of long thin elements (steel, concrete or wood), that are being driven into the ground. The elements are usually connected by joints, consisting of special forms of the element at the two ends. Compared to a massive wall (of concrete or stone), a sheet pile wall is a flexible structure, in which bending moments will be developed by the lateral load, and that should be designed so that they can withstand the largest bending moments. Several methods of analysis have been developed, of different levels of complexity. The simplest methods, that will be discussed in this chapter, are based on convenient assumptions regarding the stress distribution against the sheet pile wall. These methods have been found very useful in engineering practice, even though they contain some rather drastic approximations.

37.1 Homogeneous dry soil

A standard type of sheet pile wall is shown in Figure 37-1. The basic idea is that the pressure of the soil will lead to a tendency of the flexible wall for displacements towards the left. By this mode of deformation the soil pressures on the right side of the wall will become close to the active state. This soil pressure must be equilibrated by forces acting towards the right. A large horizontal force may be developed at the lower end of the wall, embedded into the soil on the left side, by the displacement. In this part passive earth pressure may develop if the displacements are sufficiently large. The usual schematisation is to assume that on the right side of the wall active stresses will be acting, and below the excavated soil level at the left side of the wall passive stresses will develop. Because the resulting force of the passive stresses is below the resulting force of the active stresses, complete equilibrium is not possible by these stresses alone, as the condition of equilibrium of moments can not be satisfied. Equilibrium can be ensured by adding an anchor at the top of the wall, on the right side. This anchor can provide an additional force to the right. Without such an anchor the sheet pile wall would rotate, until at the extreme lower end of the wall passive earth pressures would be developed on the right side. With an anchor equilibrium can be achieved, without the need for very large deformations. It may be noted that the anchoring force can also be provided by a strut between two parallel walls. This is especially practical in case of a narrow excavation trench.
For the sheet pile wall to be in equilibrium the depth of embedment should be sufficiently large, so that a passive zone of sufficient length can be developed. In case of a very small depth, with a thin passive zone at the toe, the lower end of the

wall might be pushed through the soil, with the structure rotating around the anchor point. The determination of the minimum depth of the embedment of the sheet pile wall is an important part of the analysis, which will be considered first. For reasons of simplicity it is assumed that the soil is homogeneous, dry sand. The assumed stress distribution is shown in Figure 37-1.

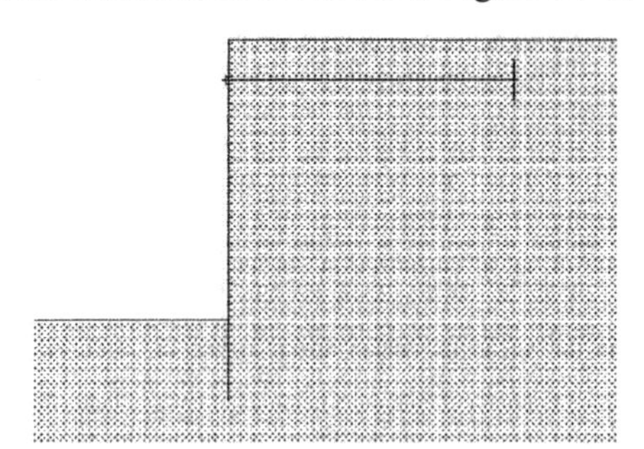

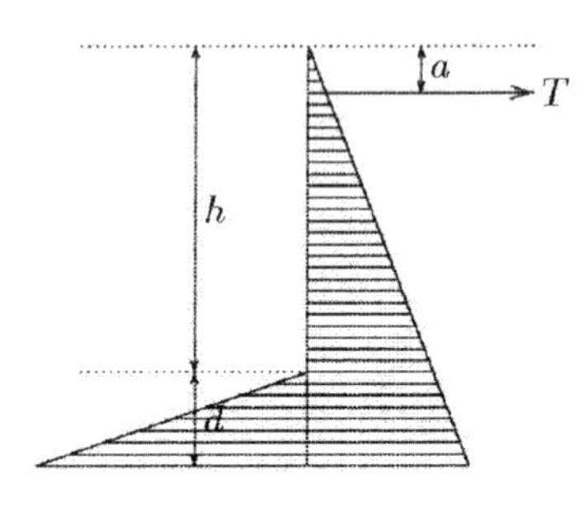

Figure 37-1. Anchored sheet pile wall.

If the retaining height (the difference of the soil levels at the right and left sides of the wall) is h, the length of the toe is d, and the depth of the anchor rod is a, then the condition of equilibrium of moments around the anchor point gives

$$\tfrac{1}{2}K_a\gamma(h+d)^2(\tfrac{2}{3}h+\tfrac{2}{3}d-a)-\tfrac{1}{2}K_p\gamma d^2(h+\tfrac{2}{3}d-a)=0.$$

It follows that

$$(h+d)^2(\tfrac{2}{3}h+\tfrac{2}{3}d-a)=\frac{K_p}{K_a}d^2(h+\tfrac{2}{3}d-a). \tag{37.1}$$

This is an equation of the third degree equation in the variable d. It can be solved iteratively by writing

$$(\frac{d}{h})^2=\frac{2K_a}{3K_p}(1+\frac{d}{h})^2\frac{1+(d/h)-\frac{3}{2}(a/h)}{1+\frac{2}{3}(d/h)-(a/h)}. \tag{37.2}$$

Starting from an initial estimate, for example $d/h=0$, ever better estimates for d/h can be obtained by substituting the estimated value into the right hand of eq. (37.2). This process has been found to iterate fairly rapidly. About 10 iterations may be needed to obtain a relative accuracy of 10^{-6}. The results for a series of values of K_p/K_a and a/h are recorded in Table 37-1.

The magnitude of the anchor force can be determined from the condition of horizontal equilibrium,

$$T=\tfrac{1}{2}\gamma(h+d)^2-\tfrac{1}{2}K_p\gamma d^2. \tag{37.3}$$

The values of T/F_a are given in Table 37-2. The quantity F_a is the total active force,

$$F_a=\tfrac{1}{2}K_a\gamma(h+d)^2. \tag{37.4}$$

a/h	K_p/K_a							
	4	6	8	9	10	12	14	16
0.00	0.793	0.550	0.438	0.401	0.371	0.326	0.294	0.269
0.05	0.785	0.545	0.433	0.396	0.367	0.323	0.290	0.265
0.10	0.777	0.539	0.428	0.392	0.363	0.319	0.287	0.262
0.15	0.768	0.532	0.422	0.386	0.385	0.314	0.282	0.258
0.20	0.759	0.524	0.416	0.380	0.352	0.309	0.278	0.254
0.25	0.749	0.516	0.409	0.374	0.346	0.303	0.273	0.249
0.30	0.737	0.507	0.401	0.366	0.339	0.297	0.267	0.243
0.35	0.724	0.496	0.392	0.358	0.330	0.289	0.260	0.237
0.40	0.710	0.484	0.381	0.348	0.321	0.281	0.252	0.229
0.45	0.693	0.470	0.369	0.336	0.310	0.270	0.242	0.220
0.50	0.674	0.454	0.354	0.322	0.296	0.258	0.230	0.209

Table 37-1: Depth of sheet pile wall, (d/h).

It appears that the anchor carries a substantial part of the total active load, varying from 20 % to more than 50 %. The remaining part is carried by the passive earth pressure, of course.

If the length of the sheet pile wall ($h+d$) and the anchor force are known, the shear force Q and the bending moment M can easily be calculated, in any section of the wall. For the case $K_a = 1/3$, $K_p = 3$ and $a/h = 0.2$ the results are given in Table 37-3. At the location of the anchor the shear force jumps by the magnitude of the anchor force. At the top and at the toe of the wall the shear force and the bending moment are zero.

a/h	K_p/K_a							
	4	6	8	9	10	12	14	16
0.00	0.218	0.244	0.258	0.263	0.267	0.274	0.279	0.283
0.05	0.226	0.254	0.269	0.275	0.279	0.286	0.292	0.296
0.10	0.235	0.265	0.281	0.287	0.292	0.300	0.306	0.310
0.15	0.245	0.277	0.295	0.301	0.306	0.315	0.321	0.326
0.20	0.255	0.290	0.309	0.316	0.322	0.331	0.338	0.344
0.25	0.267	0.305	0.326	0.334	0.340	0.350	0.358	0.364
0.30	0.280	0.321	0.345	0.353	0.360	0.371	0.380	0.387
0.35	0.294	0.340	0.366	0.375	0.383	0.395	0.405	0.413
0.40	0.311	0.361	0.390	0.401	0.409	0.423	0.434	0.443
0.45	0.329	0.386	0.419	0.431	0.441	0.456	0.469	0.478
0.50	0.351	0.415	0.453	0.466	0.478	0.496	0.510	0.521

Table 37-2: Anchor force, (T/F_a).

The largest bending moment, which determines the profile of the pile sheets, is $0.032\gamma h^3$. The results of this example are shown in graphical form in Figure 37-2.

A simple verification of the order of magnitude of the results can be made by considering the sheet pile wall as a beam on two supports, say at $z/h = 0.2$ and at $z/h = 1.2$. The length of the beam then is h, and the average load is $K_a\gamma(0.7h)$. If this load is thought to be distributed homogeneously along the beam, the maximum bending moment would be $M = qh^2/8 = 0.029\gamma h^3$, which is reasonably close to the true value given above.

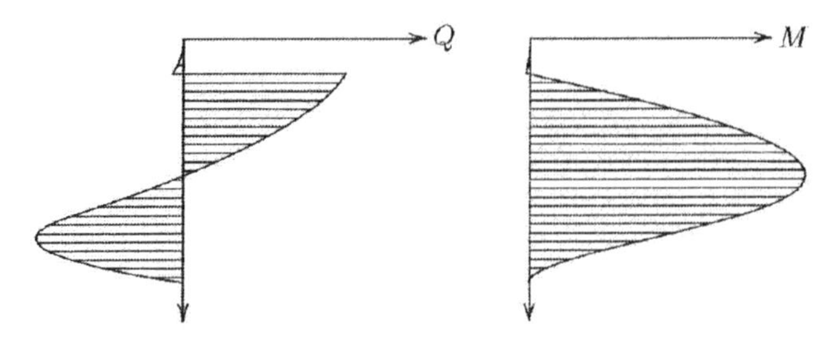

Figure 37-2. Shear force and Bending moment.

If the sheet pile wall is designed on the basis of the maximum bending moment there is no safety against failure. In order to increase the safety of the structure the passive earth pressure is often reduced, by using a conservative value for K_p. The tables then remain valid, but the result will be a somewhat greater length, as can be seen from Table 37-1. If K_p/K_a is taken smaller, the needed value of d/h will be larger. In the next chapter a more advanced method to reduce the risk of failure will be presented.

z/h	$f/\gamma h$	$Q/\gamma h^2$	$M/\gamma h^3$
0.00000	0.00000	0.00000	0.00000
0.10000	-0.03333	-0.00167	-0.00006
0.19999	0.06666	-0.00667	-0.00044
0.20001	0.06667	0.09381	-0.00044
0.30000	0.10000	0.08548	0.00855
0.40000	0.13333	0.07381	0.01654
0.50000	0.16667	0.05881	0.02320
0.60000	0.20000	0.04048	0.02819
0.70000	0.23333	0.01881	0.03119
0.80000	0.26667	-0.00619	0.03184
0.90000	0.30000	-0.03452	0.02984
1.00000	0.33333	-0.06619	0.02483
1.10000	0.06667	-0.08619	0.01699
1.20000	-0.20000	-0.07952	0.00848
1.30000	-0.46667	-0.04619	0.00197
1.38047	-0.68125	0.00000	0.00000

Table 37-3: Sheet pile wall.

An elementary computer program for the calculation of the minimum length of the sheet pile wall, the corresponding anchor force, and the distribution of shear forces and bending moments, is shown as Program 37-1. Input data to this program can be entered interactively. After entering the values of K_a, K_p and a/h the program first calculates the values of the depth of embedment d/h and the anchor force T, and then gives, for an arbitrary value of z/h, to be given by the user, the shear force Q and the bending moment M. The program can be improved in many ways, especially by adding more advanced forms of input and output, such as graphs of the shear force and the bending moment, to be shown on the screen or on a printer. The implementations of such improvements to the program are left as exercises for the reader.

37.2 Pore pressures

In the previous sections the soil was assumed to be dry, for simplicity. In general the soil may consist of soil and water, however, and the excavation may even contain free water. Thus the general problem of a sheet pile wall should take into account the presence of groundwater in the soil. Because the failure of soils, as described by the Mohr-Coulomb criterion, for instance, refers to effective stresses, the relations formulated above for the earth pressure coefficients K_a and K_p, should be applied to effective stresses only. This means that the vertical effective stresses should be calculated first, before the horizontal effective stresses can be determined. The horizontal total stresses can then be determined in the next step by adding the pore pressure.

```
100 CLS:PRINT "Sheet pile wall in homogeneous dry soil"
110 PRINT "Minimal length":PRINT
120 INPUT "Retaining height ................ ";H
130 INPUT "Depth of anchor ................. ";A
140 INPUT "Active stress coefficient ....... ";KA
150 INPUT "Passive stress coefficinet ...... ";KP
160 PA=KP/KA:A=A/H:B=1/(1.5*PA):D=0:A$="& ###.#####"
170 C=B*(1+D)*(1+D)*(1+D-1.5*A)/(1+D/1.5-A)
180 IF C<0 THEN PRINT "No solution":END
190 C=SQR(C):E=ABS(C-D):D=C:IF E>0.000001 THEN 170
200 PRINT USING A$;"d/h = ";D
210 T=KA*(1+D)*(1+D)/2-KP*D*D/2
220 PRINT USING A$;"T/ghh = ";T
230 INPUT "z/h = ";Z
240 IF Z<0 THEN END
250 IF Z>1+D THEN PRINT " Impossible":GOTO 230
260 F=KA*Z:IF Z>1 THEN F=F-KP*(Z-1)
270 Q=-KA*Z*Z/2:IF Z>A THEN Q=Q+T
280 IF Z>1 THEN Q=Q+KP*(Z-1)*(Z-1)/2
290 M=-KA*Z*Z*Z/6:IF Z>A THEN M=M+T*(Z-A)
300 IF Z>1 THEN M=M+KP*(Z-1)*(Z-1)*(Z-1)/6
310 PRINT USING A$;" f/gh = ";F;
320 PRINT USING A$;" Q/ghh = ";Q;
330 PRINT USING A$;" M/ghhh = ";M
340 GOTO 230
```

Program 37-1: Sheet pile wall in homogeneous dry soil.

The general procedure for the determination of the horizontal stresses is as follows.

1. Determine the total vertical stresses, from the surcharge and the weight of the overlying soil layers.
2. Determine the pore water pressures, on the basis of the location of the phreatic surface. If the pore pressures can be assumed to be hydrostatic (if there is no vertical groundwater flow) these can be determined from the depth below the phreatic surface. Above the phreatic surface the pore pressures may be negative in case of a soil with a capillary rise.
3. Determine the value of the vertical effective stress, as the difference of the vertical total stress and the pore pressure. If the result of this computation is negative, it may be assumed that a crack will develop, as tension between the soil particles usually is impossible. The vertical effective stress then is zero.
4. Determine the horizontal effective stress, using the appropriate value of K_a or K_p at the depth considered, and, if applicable, the local value of the cohesion c.
5. Determine the horizontal total stress by adding the pore pressure to the horizontal effective stress.

The algorithm for this procedure can be summarized as

$$\sigma_{xx} = q_z + \sum \gamma dz, \tag{37.5}$$

$$p = \gamma_w(z - z_w), \quad \text{if} \quad z < z_w - h_c \quad \text{then} \quad p = 0, \tag{37.6}$$

$$\sigma'_{zz} = \sigma_{zz} - p, \quad \text{if} \quad \sigma'_{zz} < 0 \quad \text{then} \quad \sigma'_{zz} = 0, \tag{37.7}$$

$$\sigma'_{zz} = K\sigma'_{zz} \pm 2c\sqrt{K}, \tag{37.8}$$

$$\sigma_{xx} = \sigma'_{xx} + p. \tag{37.9}$$

In these equations it has been assumed that the phreatic level is located at a depth $z = z_w$, and that in a zone of thickness hc above that level capillary water is present in the pores. Above the level $z = z_w - h_c$ there is no water in the pores, which can be expressed by $p = 0$. It has also been assumed, in eq. (37.7), that the particles can not transmit tensile forces. It may also be noted that in computations such as these open water, above the soil, may also have to be considered as soil, having a volumetric weight γ_w. The pore pressures in such a water layer will be found as zero, and the horizontal total stress will automatically be found equal to the vertical total stress. For the analysis of the forces on a wall these forces are essential parts of the analysis. For the analysis of a sheet pile wall the stress calculation must be performed for both sides of the wall separately, because on the two sides the soil levels and the groundwater levels may be different.

An example is shown in Figure 37-3. In this case an excavation of 6 m depth is made into a homogeneous soils. On the right side the groundwater level is located at a depth of 1 m below the soil surface, and on the left side the groundwater level coincides with the bottom of the excavation. For simplicity it is assumed that on both sides of the sheet pile wall the groundwater pressures are hydrostatic. This might be possible if the toe of the wall reaches into a clay layer of low permeability. Otherwise the groundwater pressures should include the effect of a groundwater movement from the right side to the left side. That complication is omitted here. An anchor has been installed at a depth of 0.50 m, at the right side. The length of the wall is initially unknown, but is assumed to be 9 m, for the representation of the horizontal stresses. The soil is homogeneous sand, having a dry volumetric weight of 16 kN/m^3, a saturated volumetric weight of 20 kN/m^3. It is assumed that for this sand $K_a = 0.3333$, $K_p = 3.0$, $c = 0$ and $h_c = 0$.

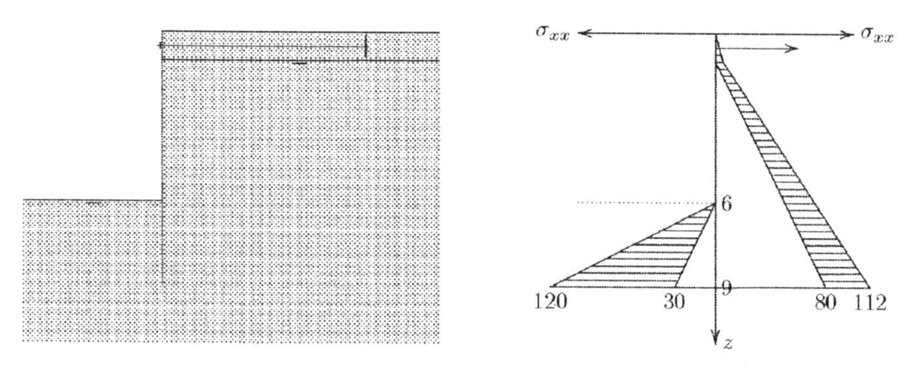

Figure 37-3. Example: The influence of groundwater.

In order to present the stresses against the wand, the simplest procedure is to calculate these stresses in a number of characteristic points. At a depth of 1 m, for instance, at the right side, the vertical total stress is σ_{zz}=16 kPa. Because the pore pressure is zero at that depth the horizontal effective stress is $\sigma'_{xx} = 5.3$ kPa, and the horizontal total stress is equal to that value, because $p = 0$. At a depth of 9 m, the total stress is larger by the weight of 8 m saturated soil, so that $\sigma_{zz} = 176$ kPa. At that depth the pore pressure is $p = 80$ kPa, and the horizontal effective stress is now $\sigma'_{zz} = 96$ kPa. Because $K_a = 0.3333$ the horizontal effective stress is $\sigma'_{xx} = 32$ kPa. Finally, the horizontal total stress is $\sigma_{xx} = 112$ kPa.

At the left side of the wall all stresses are zero down to the level of the bottom of the excavation, at 6 m depth. At a depth of 9 m: $\sigma_{zz} = 60$ kPa and $p = 30$ kPa. This gives $\sigma'_{zz} = 30$ kPa and, because $K_p = 3$, $\sigma'_{xx} = 90$ kPa. The horizontal stress is obtained by adding the pore pressure, i.e. $\sigma_{xx} = 120$ kPa.

Even in this simple case, of a homogeneous soil, the determination of the horizontal loads on the wall is not a trivial problem. In many problems of engineering practice the analysis may be much more complicated, as the soil may consist of layers of different volumetric weight and composition, with variable values of the coefficients

K_a and K_p. This may lead to discontinuities in the distribution of the horizontal stress. The groundwater pressures also need not be hydrostatic. In the case of a permeable soil the determination of the groundwater pressures may be a separate problem.
Another difficulty can be seen in Figure 37-3. Because the water level left of the sheet pile wall is different from the water level on the right, a flow under the sheet pile wall will occur. Because of this the pore pressures around the sheet pile tip will be levelled.
The length of the sheet pile wall is initially unknown. It can be determined by requiring that equilibrium is possible with the toe of the wall being a free end, with $Q=0$ and $M=0$. As in the simple case considered before, see Figure 37-1, the length can be determined from the condition of equilibrium of moments with respect to the anchor point. The simplest procedure is to first assume a certain very short depth of the embedment, with full passive pressures at the left side, then calculating the bending moment at the toe, and then gradually reducing the embedment depth until this bending moment is zero.

The computations can be executed by the Program 37-2. In this program the sheet pile wall is subdivided into a large number of small elements, of length **DZ=H/N**, where **N=NN/3** and **NN=10000**. The horizontal stresses on the right side and the left side are calculated from top to toe, at the same time calculating the moment with respect to the anchor point (this is the variable **MT**). This is done first for the part from to the top to the bottom of the excavation, in lines 220 until 270. The vertical total stresses σ_{zz} to the left and right of the wall are denoted as **TLZ** and **TRZ**, the vertical effective stresses as **SLZ** and **SRZ**, the horizontal effective stresses as **SLX** and **SRX,** and the horizontal total stresses as **TLX** en **TRX**. The quantity **F(I)** is the total distributed load, the sum of the loads from the left and the right. The total length of the wall is gradually increased, from its initial value **HH=H**, in small steps of magnitude **DZ**, until a change of sign of the moment **MT** occurs. Then the length of the wall is known **(HH)**. If at a length of 3 times the excavation depth no equilibrium of moments has been found, the program gives an error statement, and stops. In the course of the analysis the shearing force **Q(I)** and the bending moment **M(I)** are determined, neglecting the anchor force, As soon as the length of the wall is known, the value of the anchor force can be determined, from the condition that at the toe of the wall the bending moment must be zero, see line 360. Then the distributions of the shear force and the bending moment can be corrected for the influence of the anchor force, and the program prints some output data. It also prints the shear force and the bending moment at the toe of the wall. These quantities should be zero. Usually this is not precisely the case, which is an indication of the accuracy.
In the example: **H=6.0**, **DA=0.5**, **CA=0.3333**, **CP=3.0**, **GD=16.0**, **GN=20.0**, **WL=6.0**, **WR=1.0**. The program then gives that the length of the wall should be 11.825 m. The anchor force is 162.710 kN/m , and the maximum bending moment is 544.263 kNm/m . The bending moment at the toe appears to be exactly zero, but the shear force is 0.043 kN . This is a small error, that can be accepted.

```
100 CLS:PRINT "Sheet pile wall in homogeneous soil":PRINT:NN=10000
110 DIM M(NN),Q(NN),F(NN)
120 INPUT "Depth of the excavation (m) ...... ";H
130 INPUT "Depth of the anchor (m) .......... ";DA
140 INPUT "Active stress coefficient ........ ";CA
150 INPUT "Passive stress coefficient ....... ";CP
160 INPUT "Dry weight (kN/m3) ............... ";GD
170 INPUT "Saturated weight (kN/m3) ......... ";GN
180 INPUT "Depth of groundwater left (m) .... ";WL
190 INPUT "Depth of groundwater right (m) ... ";WR
200 N=NN/3:HH=H:DZ=HH/N:DZ2=DZ/2:WW=10:A$="#####.###":PRINT
210 TLZ=0:PL=0:TRZ=0:PR=0:MT=0:Z=0:F(0)=0:Q(0)=0:M(0)=0
220 FOR I=1 TO N:Z=Z+DZ:G=WW:W=WW:IF Z-DZ2<WL THEN G=0:W=0
230 TLZ=TLZ+G*DZ:PL=PL+W*DZ:SLZ=TLZ-PL:SLX=SLZ:TLX=SLX+PL
240 G=GN:W=WW:IF Z-DZ2<WR THEN G=GD:W=0
250 TRZ=TRZ+G*DZ:PR=PR+W*DZ:SRZ=TRZ-PR:SRX=CA*SRZ:TRX=SRX+PR
260 F(I)=TRX-TLX:FF=(F(I)+F(I-1))*DZ2:Q(I)=Q(I-1)-FF
270 M(I)=M(I-1)+(Q(I)+Q(I-1))*DZ2:MT=MT+FF*(Z-DA-DZ2):NEXT I
280 WHILE MT>0:N=N+1:Z=Z+DZ:G=GN:W=WW:IF Z-DZ2<WL THEN G=GD:W=0
290 TLZ=TLZ+G*DZ:PL=PL+W*DZ:SLZ=TLZ-PL:SLX=CP*SLZ:TLX=SLX+PL
300 G=GN:W=WW:IF Z-DZ2<WR THEN G=GD:W=0
310 TRZ=TRZ+G*DZ:PR=PR+W*DZ:SRZ=TRZ-PR:SRX=CA*SRZ:TRX=SRX+PR
320 F(N)=TRX-TLX:FF=(F(N)+F(N-1))*DZ2:Q(N)=Q(N-1)-FF
330 M(N)=M(N-1)+(Q(N)+Q(N-1))*DZ2:MT=MT+FF*(Z-DA-DZ2)
340 IF N=NN THEN PRINT "No solution":STOP:END
350 WEND
360 HH=Z:FT=-M(N)/(HH-DA):Z=0:MM=0
370 FOR I=1 TO N:Z=Z+DZ:IF (Z>DA) THEN Q(I)=Q(I)+FT:M(I)=M(I)+FT*(Z-DA)
380 IF (M(I)>MM) THEN MM=M(I)
390 NEXT I
400 PRINT "Minimum length (m) ............... ";:PRINT USING A$;HH
410 PRINT "Anchor force (kN/m) .............. ";:PRINT USING A$;FT
420 PRINT "Maximum moment (kNm/m) ........... ";:PRINT USING A$;MM
430 PRINT "Shear force at the toe ........... ";:PRINT USING A$;Q(N)
440 PRINT "Moment at the toe ................ ";:PRINT USING A$;M(N)
450 STOP:END
```

Program 37-2: Sheet pile in homogeneous soil, with groundwater.

Again, the computer program has been kept as simple as possible. It can be used as a basis for a more advanced program, with more refined input and output data handling. The input data might be collected in a data file, that can be edited separately, and the output data might be presented in tables or graphs on the screen or on the printer.

Problems

37.1 Verify a number of values in Table 37-1 and Table 37-2, using a computer program.

37.2 Also verify the values from Table 37-3, using a computer program.

37.3 A sheet pile wall is used to retain a height of 5 m, in dry sand, with $\phi = 30°$. The depth of the anchor is 1 m. Determine the minimum embedment depth, according to Table 37-1, and using one of the computer programs.

37.4 Verify the output of the example of Program 37-2. In this case the length of the wall appears to be very large, almost twice the depth of the excavation. What should be the length of the wall if the anchor is located somewhat deeper, say at a depth of 2.0 m?

37.5 Modify the Program 37-2 such that it presents a table of the load, the shear force and the bending moment, as a function of depth.

38 Blum's method

In the previous chapter a procedure has been presented for the determination of the minimum length of a sheet pile wall, needed to ensure equilibrium. This method is such that whenever the wall is shorter than that minimum length, no equilibrium is possible, and the wall will certainly fail. This suggests that it is advisable to choose the length of the wall somewhat larger than the minimum length, as a total failure of the wall would be disastrous. If the length is taken somewhat larger than required, the bending moments may perhaps be somewhat reduced. A method of analysing the deformation and bending of the wall has been developed by Blum. This method is presented in this chapter, including a simple computer program.

38.1 Blum's schematisation

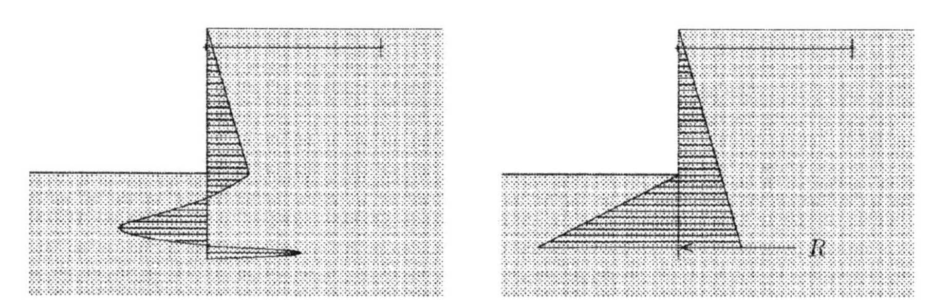

Figure 38-1. Blum's schematisation.

If the length of the sheet pile wall is somewhat larger than strictly necessary to ensure equilibrium, the passive earth pressure need not be developed over the entire length of the embedded part of the wall. It may be expected that the pressures against the wall will be of the form shown in Figure 38-1. Because of the extra length of the sheet pile wall the toe will act as a clamped edge, in which the lowest part may have a tendency to move to the right, building up a pressure towards the left. Together with the incomplete passive pressure towards the right this will constitute the clamping moment. Blum suggested to schematise the loads on the wall as shown in the right half of the figure. The force R (the *Ersatzkraft*) is equivalent to the pressure to the right at the extreme lower part of the wall. Its precise distribution is left undetermined. The toe of the sheet pile wall is now assumed to be a clamped edge, and it is also assumed that at the toe the bending moment is zero, but a shear force (of magnitude R) is allowed. In order that this force may indeed develop, and that there is enough material to form a clamped boundary, the actual length should be somewhat larger than assumed in the schematisation: usually the embedment depth is taken 20 % larger than calculated.

One of the ideas behind Blum's schematisation is that the clamping moment will probably lead to a reduction of the bending moments in the sheet pile wall, so that a lighter profile may be used. Thus the additional costs involved by taking a longer sheet pile wall is somewhat balanced by a lighter profile. That this is acceptable can be argued by noting that a failure by a wall that is too short is indeed disastrous, but that in case of failure by exceeding the maximum bending moment, some additional strength is available beyond the onset of plastic deformation of the steel. If a plastic deformation in bending is developed, the bending moment will at least be constant, and may even increase somewhat. Also, the soil pressures may be redistributed by the large deformations.

38.2 Blum's method

The basic principle of Blum's method of analysis is that the sheet pile wall is considered as fully clamped at its toe, with the additional condition that the bending moment at the toe is zero. The shear force, however, in general will be unequal to zero. This shear force is supposed to be the resultant force of the stresses in the vicinity of the toe, including some length below the toe. The clamping of the edge is supposed to be so strong that the displacement and the rotation (that is the first derivative of the displacement) are zero, and even the second derivative is zero, so that the bending moment is zero. There are three equations (horizontal equilibrium, displacement and moment equilibrium) with three unknowns (d, R and T). The length of the wall will be determined by the conditions of equilibrium, with active soil stresses on the high side and full passive stresses on the low side, and the condition that the horizontal displacement is zero at the level of the anchor. The procedure can best be illustrated by means of an elementary example.
The example refers to a sheet pile wall retaining a height h of homogeneous saturated soil, see Figure 38-2. To enable an analytical solution it is assumed that on the two sides of the wall the groundwater table coincides with the soil surface. To further simplify the problem the anchor is supposed to be acting at the top of the wall. The embedment depth d is unknown. This is one of the parameters that have to be determined by the analysis.
At the active side of the wall the vertical total stress is

$$\sigma_{zz} = \gamma z,$$

in which γ is the volumetric weight of the saturated soil. The pore pressures are

$$p = \gamma_w z,$$

so that the effective stresses are

$$\sigma'_{zz} = (\gamma - \gamma_w) z.$$

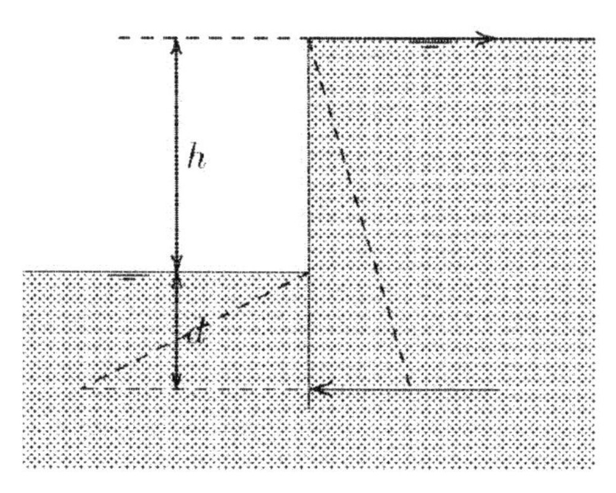

Figure 38-2. Example.

The horizontal effective stresses now are, for a cohesionless soil with $c = 0$,

$$\sigma'_{xx} = K_a(\gamma - \gamma_w)z.$$

The horizontal total stresses are obtained by adding the pore pressures,

$$\sigma_{xx} = [K_a(\gamma - \gamma_w) + \gamma_w]z.$$

This can also be written as

$$\sigma_{xx} = K_a^*\gamma z, \tag{38.1}$$

where

$$K_a^* = K_a(1 - \gamma_w/\gamma) + \gamma_w/\gamma. \tag{38.2}$$

If $K_a = 0.3333$ and $\gamma_w/\gamma = 0.5$, then $K_a^* = 0.6667$. It should be noted that the simple expression (38.1), linear in z, is valid only if the soil is homogeneous, with $c = 0$, and if the groundwater table coincides with the soil surface. In a more general case the computation of the horizontal total stresses proceeds in exactly the same way, but the result can not be expressed in the simple form of eq. (38.1).
In the same way the horizontal stresses at the passive side, for $z > h$, can be determined. The result is

$$\sigma_{xx} = K_p^*\gamma(z - h), \tag{38.3}$$

where

$$K_p^* = K_p(1 - \gamma_w/\gamma) + \gamma_w/\gamma. \tag{38.4}$$

If $K_p = 3.0$ and $\gamma_w/\gamma = 0.5$, then $K_p^* = 2.0$.
The resulting active and passive forces are

$$F_p = \tfrac{1}{2}K_p^*\gamma d^2.$$

The condition that the bending moment at the toe of the sheet pile wall must be zero, at the depth of the clamped edge, i.e. the point of application of the force R, give

$$T(h+d)=\tfrac{1}{6}K_a^*\gamma(h+d)^3-\tfrac{1}{6}K_p^*\gamma d^3. \tag{38.5}$$

For the computation of the horizontal displacement of the top of the sheet pile wall (which must be zero), the contribution of the three forms of loading can best be considered separately, see Figure 38-3.
The first loading case is the anchor force T, acting at the top of the sheet pile wall. This force leads to a displacement of the top of magnitude

$$u_1=\frac{T(h+d)^3}{3EI}. \tag{38.6}$$

This is a well known basic problem from applied mechanics.

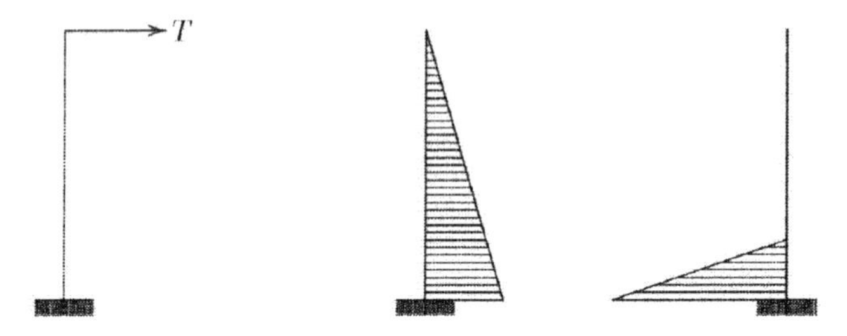

Figure 38-3. Loads on the clamped wall in Blum's schematisation.

For the case of a triangular load $f=az$ on a clamped beam of length l, the loading case in the central part of Figure 38-3, the displacements can be found using the classical theory of bending of beams, from applied mechanics. By integrating the differential equation $EId^4u/dz^4=f$, with the boundary conditions that at the top the bending moment and the shear force are zero, whereas at the toe the horizontal displacement u and its first derivative (the rotation) are zero, the displacement of the top can be obtained as

$$u_0=\frac{al^5}{30EI}. \tag{38.7}$$

The rotation of the top is found to be

$$\varphi_0=\frac{al^4}{24EI}. \tag{38.8}$$

Using these formulas the horizontal displacement of the top of the sheet pile wall caused by the active soil pressure on the right side is, with eq. (38.1) and (38.7),

$$u_2=-\frac{K_a^*\gamma(h+d)^5}{30EI}. \tag{38.9}$$

The minus sign indicates that this displacement is directed towards the left.
The displacement caused by the passive soil pressures at the left side of the sheet pile wall, as described by eq.(38.3), are found to be

$$u_3 = \frac{K_p^* \gamma d^5}{30EI} + \frac{K_p^* \gamma d^4 h}{24EI}. \tag{38.10}$$

The first term in this expression is the displacement at the top of the load, the second term is the additional displacement due to the rotation at the top of the load. Together these two quantities constitute the displacement at the top of the sheet pile wall. The upper, unloaded part of the wall, does not deform in this loading case.
The sum of the three displacements (38.6), (38.9) and (38.10) must be zero. This gives, with (38.5), and after multiplication by $EI/K_p^*\gamma$,

$$\frac{K_a^*}{K_p^*}\frac{(h+d)^5}{18} - \frac{d^3(h+d)^2}{18} - \frac{K_a^*}{K_p^*}\frac{(h+d)^5}{30} + \frac{d^5}{30} + \frac{d^4 h}{24} = 0,$$

or, after some rearranging of terms,

$$(\frac{d}{h})^3 = \frac{8(K_a^*/K_p^*)(1+d/h)^5}{20(1+d/h)^2 - 15d/h - 12(d/h)^2}. \tag{38.11}$$

From this equation the value of d/h can be solved iteratively, using an initial estimate, possibly simply $d/h = 0.0$.

```
100 CLS:PRINT "Sheet pile wall in homogeneous saturated soil"
110 PRINT "Blum":PRINT:A$="& ####.###"
120 INPUT "Volumetric weight of water ...... ";GW
130 INPUT "Volumetric weight of soil ....... ";GG
140 INPUT "Active stress coefficient ....... ";KA
150 INPUT "Passive stress coefficient ...... ";KP
160 KSA=KA*(1-GW/GG)+GW/GG:KSP=KP*(1-GW/GG)+GW/GG:D=0
170 C=8*(KSA/KSP)*(1+D)^5/(20*(1+D)^2-15*D-12*D*D)
180 IF C<0 THEN PRINT "No solution":END
190 C=C^(1/3):E=ABS(C-D):D=C:IF E>0.000001 THEN 170
200 PRINT USING A$;"d/h = ";D
210 T=(KSA*(1+D)^3-KSP*D^3)/(6*(1+D))
220 PRINT USING A$;"T/ghh = ";T
230 END
```

Program 38-1: Blum's method for saturated soil.

The computations can be made using the Program 38-1. The program only requests the input of the volumetric weights of water and (saturated) soil, and the values of the active and passive pressure coefficients, and then computes the values of d/h and $T/\gamma h^2$, using the equations (38.11) and (38.5). For the case that **GW=10**, **GG=20**, **CA=0.3333** and **CP=3.0** the result of the program is $d/h = 1.534$ and $T/\gamma h^2 = 0.239$. It appears that in this case the sheet pile wall needs a rather long embedment depth (more than 1.5 times the retaining height). This is the price that has to be paid for a more favourable distribution of the bending moments. The profile of the steel elements can be somewhat lighter, but the length is considerably larger than in the simple method of the previous chapter.

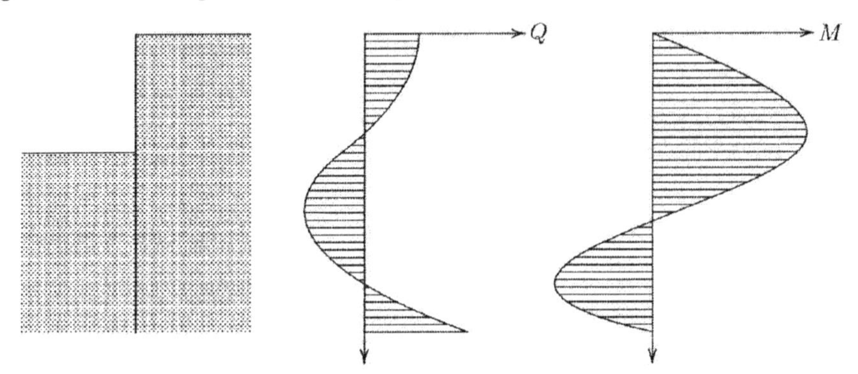

Figure 38-4. Shear force and bending moment.

The distribution of the shear force and the bending moment is shown in Figure 38-4. The shear force at the top is the anchor force. The value at the toe is Blum's concentrated force *R*. It appears that this force results in a reduction of the bending moments in the sheet pile wall, as mentioned before. For the determination of the profile of the wall it is favorable that the positive and negative bending moments are of the same order of magnitude.

The results of the computations for a number of values of the earth pressure coefficients K_a and K_p are given in Table 38-1. It has been assumed that the volumetric weight of the water is $\gamma_w = 10\ \text{kN/m}^3$, and that the volumetric weight of the saturated soil is $\gamma = 20\ \text{kN/m}^3$, a common value.

ϕ	K_a	K_p	d/h	$T/\gamma h^2$
10°	0.7041	1.4203	5.228	0.881
15°	0.5888	1.6984	3.406	0.554
20°	0.4903	2.0396	2.481	0.394
25°	0.4059	2.4639	1.917	0.300
30°	0.3333	3.0000	1.534	0.239
35°	0.2710	3.6902	1.255	0.196
40°	0.2174	4.5989	1.040	0.165
45°	0.1716	5.8284	0.868	0.141

Table 38-1: Blum's method for homogeneous soil.

The concentrated force *R* is an essential element in Blum's method. It should be remembered that this force actually represents the distributed load at the extreme toe of the sheet pile wall, which is produced by the deformation of the sheet pile wall. For the generation of this concentrated force the wall should be given some

additional length, by choosing the length of the wall somewhat larger than the theoretical value computed in the analysis. It is often assumed that the length of the embedment depth (the distance d in the example) should be taken 10 % or 20 % larger than computed. All this leads to a wall of considerable length. This is the price that has to be paid for the advantages of Blum's analysis: a lighter profile, and small displacements.
It may be noted that the example considered in this chapter is perhaps a very unfavorable case: the level of groundwater at the right side is very high, and on the left side it is very low. In the next chapter a more general method will be described. But also in more general cases it is observed that Blum's method leads to long sheet pile walls. The safety is large, but at a price.

Problems

38.1 Verify a number of values in Table 38-1 by substitution into eq. (38.11), or by a computation using Program 38-1.

38.2 A sheet pile wall is used to construct a building pit in a polder. The depth of the pit is 5 m, and on both sides the groundwater level coincides with the soil surface. The sheet pile wall is supported by a strut connecting to an identical wall at the other side of the building pit. Determine the necessary length of the sheet pile wall, assuming that $c=0$ and $\phi=30°$.

38.3 It has been found that the friction angle in the previous problem should be 40° instead of 30°. Determine the length of the sheet pile wall for this case.

38.4 Equation (38.11) applies to saturated soil, with the groundwater level coinciding with the soil surface. Derive a similar equation for homogeneous dry soil. Then compute the value of d/h for dry soil, with $\gamma=16\ \text{kN/m}^3$, $c=0$ and $\phi=30°$.

38.5 Verify the formulas (38.7) and (38.8) for the displacement and the rotation of the free end of a clamped beam loaded by a triangular stress.

39 Sheet pile wall in layered soil

For a sheet pile wall in a layered soil, the method of analysis is the same as for a wall in homogeneous soil, as considered in the previous chapter. The main difference is that the computation of the horizontal stresses against the wall is more complicated. The computation can best be performed using a computer program. In this chapter a simple program is presented, using Blum's method.
The complications are that the weight and the properties of the various layers may be different, and the zero level of the groundwater may also be different for each layer. The simplest approach is to consider the determination of the horizontal stresses against the wall as a separate problem, that precedes the analysis of the sheet pile wall. In principle, these stresses can easily be determined by analysing the stresses from the top of the soil downward, in each step adding the weight, and using the appropriate values of the lateral stress coefficients. The horizontal effective stress follows from $\sigma'_{zz} = \sigma_{zz} - p$, and the horizontal effective stress σ'_{xx} follows from the formula for active or passive earth pressure. The horizontal total stress finally is obtained by adding the pore pressure, $\sigma_{xx} = \sigma'_{xx} + p$. At the interfaces between succeeding layers the horizontal total stress may
be discontinuous, because the stress coefficients may be discontinuous.

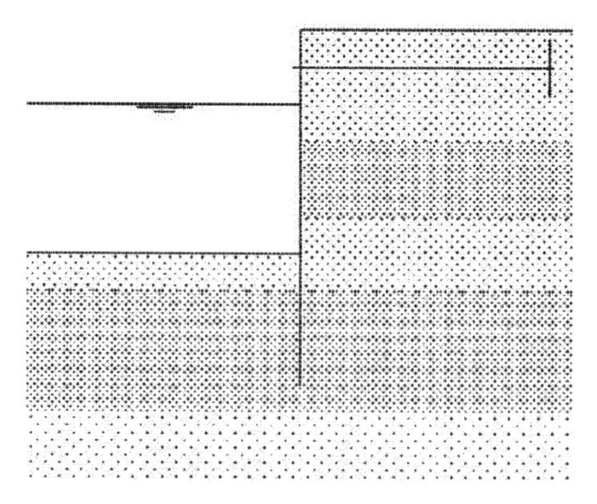

Figure 39-1. Layered soil.

39.1 Computer program

The computations can be executed by the Program 39-1. The input data are entered

interactively. After some general data the following data must be entered for each layer: the thickness of the layer, the cohesion, the coefficient of active earth pressure, the coefficient of passive earth pressure, the volumetric weight of the soil in dry condition (if there is no soil this is 0), the volumetric weight of the soil in saturated condition (if there is no soil this is 10, the volumetric weight of water), and the zero level of the groundwater, considered with respect to the top of the wall. These last three data must be given both for the left side and the right side of the wall. It is assumed that the excavation is on the left side, so that active pressures occur on the right side of the sheet pile wall, and passive stresses on the left side.
The program starts with an initial assumption of the length of the sheet pile wall, namely the sum of all layer thicknesses. The distribution of the loads on the two sides of the wall, active on the right side and passive on the left side, is then calculated for a large number of points (1000 in the program), from the data defining the layer thicknesses, their weight, the stress coefficients and the applicable depth of the groundwater. The stresses are cut off for tension, in line 370. The resulting load is indicated by **F(I)**. This is the difference of the soil pressures at the right side and the left side, see line 380. Next the shear force **Q(I)** and the building moment **M(I)** are calculated numerically, ignoring the anchor force, which is unknown at that stage.
The anchor force T is calculated from the condition that the bending moment is zero at the toe of the wall (in line 430). Next the rotation φ and the horizontal displacement u are computed by integrating the bending moment twice, using the boundary condition at the clamped toe of the wall, where $\varphi = 0$ and $u = 0$. Actually, this calculation should involve the bending stiffness EI, because the equations are $d\varphi/dz = M/EI$ and $du/dz = \varphi$, but as the conditions are that these variables must be zero, the value of the constant factor EI is irrelevant. In the program the variables φ and u are indicated by **P(I)** and **U(I)**. The displacement at the anchor point is **UA**. If that displacement is found to be positive, the length of the wall is shortened by a small amount, and the computation is repeated until **UA** turns negative. The procedure assumes that the initial estimate of the length of the wall is sufficiently large. Therefore the thickness of the deepest layer must be sufficiently large.
The output of the program consists of the computed length and anchor force only. The shear force and the bending moment are calculated, but not printed. Of course, it is very simple to modify the program so that it prints these quantities.

```
DA=0.0:N=2
D(1)=1.0:CA(1)=0.3333:CP(1)=3.0
GDL(1)=0.0:GNL(1)=10.0:WL(1)=1.0
GDR(1)=16.0:GNR(1)=20.0:WR(1)=0.0
D(2)=2.0:CA(2)=0.3333:CP(2)=3.0
GDL(2)=16.0:GNL(2)=20.0:WL(2)=1.0
GDR(2)=16.0:GNR(2)=20.0:WR(2)=0.0
```

The simple example of the previous chapter, for the case of a cohesionless soil, as shown in Figure 38-2, can be determined using the following data.

```
100 CLS:PRINT"Sheet pile wall in layered soil":NN=1000
110 PRINT"Blum":PRINT
120 DIM D(20),Z(20),CA(20),CP(20)
130 DIM GDL(20),GNL(20),WL(20),GDR(20),GNR(20),WR(20)
140 DIM M(NN),Q(NN),F(NN),P(NN),U(NN)
150 INPUT "Depth of anchor (m) .............. ";DA
160 INPUT "Number of layers ................. ";N
170 Z(0)=0:GW=10:FOR I=1 TO N:CLS:PRINT "Layer ";I:PRINT
180 INPUT "Thickness (m) .................... ";D(I)
190 INPUT "Cohesion (kN/m2) ................. ";CC(I)
200 INPUT "Active stress coefficient ........ ";CA(I)
210 INPUT "Passive stress coefficient ....... ";CP(I)
220 INPUT "Dry weight left (kN/m3) .......... ";GDL(I)
230 INPUT "Saturated weight left (kN/m3) .... ";GNL(I)
240 INPUT "Depth of groundwater left (m) .... ";WL(I)
250 INPUT "Dry weight left (kN/m3) .......... ";GDR(I)
260 INPUT "Saturated weight right (kN/m3) ... ";GNR(I)
270 INPUT "Depth of groundwater right (m) ... ";WR(I)
280 Z(I)=Z(I-1)+D(I):NEXT I
290 HH=Z(N):DZ=HH/NN:DZ2=DZ/2:TLZ=0:TRZ=0:J=1:ZZ=0
300 FOR I=1 TO NN:ZZ=ZZ+DZ:IF ZZ>Z(J) THEN J=J+1
310 IF ZZ<WL(J) THEN GL=GDL(J) ELSE GL=GNL(J)
320 IF ZZ<WR(J) THEN GR=GDR(J) ELSE GR=GNR(J)
330 IF ZZ<WL(J) THEN PL=0 ELSE PL=GW*(ZZ-WL(J))
340 IF ZZ<WR(J) THEN PR=0 ELSE PR=GW*(ZZ-WR(J))
350 TLZ=TLZ+DZ*GL:SLZ=TLZ-PL:TRZ=TRZ+DZ*GR:SRZ=TRZ-PR
360 SLX=CP(J)*SLZ+2*CC(J)*SQR(CP(J))
370 SRX=CA(J)*SRZ-2*CC(J)*SQR(CA(J)):IF SRX<0 THEN SRX=0
380 TLX=SLX+PL:TRX=SRX+PR:F(I)=TRX-TLX:NEXT I
390 F(0)=0:Q(0)=0:M(0)=0:JJ=0:FOR I=1 TO NN
400 FF=(F(I)+F(I-1))*DZ2:IF I*DZ<DA THEN JJ=I
410 Q(I)=Q(I-1)-FF:M(I)=M(I-1)+(Q(I)+Q(I-1))*DZ2
420 NEXT I:NH=NN+1:UA=1:WHILE UA>0:NH=NH-1
430 HT=NH*DZ:T=-M(NH)/(HT-DA):P(NH)=0:U(NH)=0
440 FOR I=NH-1 TO JJ STEP -1:M1=M(I)+T*(I*DZ-DA)
450 M2=M(I+1)+T*(I*DZ+DZ-DA):P(I)=P(I+1)+M1+M2
460 U(I)=U(I+1)-P(I)-P(I+1):NEXT I:UA=U(JJ)
470 WEND:NH=NH+1:HT=NH*DZ:T=-M(NH)/(HT-DA):A$="####.###"
480 FOR I=JJ TO NH:Q(I)=Q(I)+T:M(I)=M(I)+T*(I*DZ-DA):NEXT I
490 PRINT"Length : ";:PRINT USING A$;HT
500 PRINT"Anchor force : ";:PRINT USING A$;T
510 END
```

Program 39-1: Blum's method for a layered soil.

The depth of the excavation (the thickness of the first layer) has been taken as 1 m. The thickness of the second layer is 2 m. The initial estimate for the length then is 3 m. This will probably be sufficiently long. To simulate the excavation the dry volumetric weight of the soil at the left side is assumed to be 0.0, and its saturated

weight is assumed to be 10.0, the volumetric weight of water. Although there is no water there, but changing the groundwater level might change that.
Running the program shows that the length of the sheet pile wall must be 2.532 m, and the anchor force $T = 4.751$ kN/m. This means that $T/\gamma h^2 = 0.238$. These results are in agreement with the results of the analytical solution of the previous chapter.

39.2 Computation of anchor plate

The anchor is supposed to consist of a steel rod, connected to a plate. This plate should be capable of resisting the anchor force T. The maximum force of the anchor plate can be determined by an application of Coulomb's theory, see Figure 39-2. At the left side of the plate the stresses are supposed to be the passive earth pressure, and at the right side the active earth pressure is assumed to act. The maximum force then is, assuming that the plate is continuous,

$$T_{max} = \tfrac{1}{2}(K_p - K_a)\gamma b^2. \qquad (39.1)$$

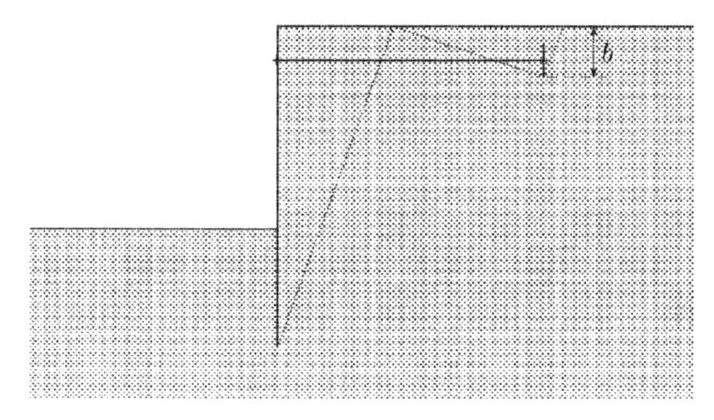

Figure 39-2. Anchor.

This value must be larger than the anchor force required for equilibrium of the sheet pile wall. A sufficiently large value of a safety factor must be taken into account (for instance 1.5). Of course, the anchor rod must also be strong enough to transfer this force. In engineering practice the anchor usually consists of a series of plates, at certain distances. The value (39.1) gives the force per unit length. This must be multiplied by the distance of the anchors to obtain the force in a single anchor.
The anchor plate need not run from its largest depth b to the soil surface. This might be an obstacle for other use of the soil. However, even when the anchor reaches only to a certain small depth, the sliding surface might be the same as for an anchor up to the surface. In practice an anchor is considered as fully embedded if its height is at least $\tfrac{1}{2}b$.
The distance of the anchor plate from the sheet pile wall should be sufficiently large to enable the necessary passive earth pressure to be developed. In principle, the distance must be so large that the active region of the sheet pile wall and the passive region of the anchor plate do not overlap, see Figure 39-2. The complete system, of

sheet pile wall, anchor and soil body, forms a retaining structure, that must be stable in itself. In order to verify that condition a stability analysis of the system must also be made, for instance using a circular slip plane passing below the sheet pile wall and the anchor plate. See Chapter 46.
In engineering practice some more advanced methods are often used for the analysis of a sheet pile wall. A familiar method is to consider the wall as a beam supported by a large number of elasto-plastic springs. The characteristics of these springs are defined such that the maximum soil stress is the passive earth pressure, and the minimum soil stress is the active earth pressure. The actual soil stress is supposed to depend upon the displacement, with the passive stress being developed if the displacement is towards the soil, and sufficiently large. The active state of stress is developed when the wall displaces away from the soil, and the displacement is sufficiently large. The actual displacements are calculated by considering the differential equations of equilibrium and deformation of the sheet pile wall. In such an analysis the excavation process can be modelled in a number of loading and unloading stages, with anchors being installed in different stages. The main advantages of this method is that the analysis includes a reasonably accurate computation of the deformations, and that it enables to analyse walls with multiple anchors.
Some fully numerical methods of analysis of soil structures based upon the finite element method also include a module for sheet pile walls. An example is the program PLAXIS, originally developed at the Delft University of Technology.

Problems

39.1 Calculate the case shown in Figure 38-2, with an anchor at depth $0.2h$. Determine the length of the sheet pile wall, if $h = 5$ m.

39.2 Also consider the case shown in Figure 38-2, with an anchor at a depth $0.2h$, and a water level of $0.2h$ below the top of the wall, at both sides. Determine the length of the sheet pile wall, if $h = 5$ m.

39.3 Extend the Program 39-1 with statements that print the load, the shear force and the bending moment. Let the program also calculate the largest positive and negative bending moments.

Part VII
Shallow and pile foundations

40 Limit analysis

Coulomb's method for the analysis of soil pressures considers extreme conditions, in which the soil is on the verge of failure. This type of analysis can be given a firm theoretical basis by the *theory of plasticity*. This also enables to generalize the method, and to investigate the possible limitations and the validity of the method.
In the analysis of stresses and strains in continuum mechanics three types of equations are needed : equilibrium conditions, constitutive relations, and compatibility equations. The general purpose is to determine the stresses and strains in a certain body, under the influence of given stresses and displacements on the surface of that body. Even for the simplest type of material, a linear elastic body, for which the constitutive relations are linear relations between stresses and strains (Hooke's law), this is a formidable task, which can be solved only for simple cases, such as a half space, a perfect sphere or a cylindrical body. Approximate solutions may be found for various materials, including linear elastic materials, using advanced numerical methods, such as the finite element method. Such numerical methods will not be considered in this book, however. An alternative may be formed by *limit analysis*, on the basis of plasticity theory, which aims not to give the complete field of actual stresses and deformations, but is restricted to give a possible upper or lower limit of the stresses or the deformations.

40.1 Basic theorems of plasticity theory

In considerations of limit analysis not all the details of the constitutive relations are taken into account, but one aspect is given priority, namely the failure criterion of the material. For soils this may be the Mohr-Coulomb criterion, described by a cohesion c and a friction angle ϕ. Also, not all the conditions of equilibrium and compatibility equations are taken into account, but only a subset of these equations. The purpose of limit analysis is not to determine the complete field of actual stresses and strains, but only to determine certain limiting values. The problem may be to determine a *lower bound* for the maximum allowable load on a soil body, or to determine an *upper bound* for this maximum load. If a lower bound for the failure load can be found, it is certain that no failure will occur as long as the real load remains below this lower bound. If an upper bound can be found it is certain that failure will occur if the real load is greater than this upper bound.
In its simplest form the theory of plasticity uses a single constant failure condition, which is a function of the stresses only. This condition expresses that for certain combinations of stresses in a point of the material the deformations increase without

bounds (this is called *plastic yielding*), and that for smaller stresses no plastic deformations will occur. A material with such a simple yield condition is called a *perfectly plastic material.* For soils a suitable yield condition is the Mohr-Coulomb criterion, although more complex yield conditions have also been studied.

In formulating the basic theorems of the theory of plasticity two types of fields are being used, which can be defined as follows.

1. An *equilibrium system*, or a *statically admissible field* of stresses is a distribution of stresses that satisfies the following conditions:
   - it satisfies the conditions of equilibrium in each point of the body,
   - it satisfies the boundary conditions for the stresses,
   - the yield condition is not exceeded in any point of the body.

2. A *mechanism*, or a *kinematically admissible field* of displacements is a distribution of displacements and deformations that satisfies the following conditions:
   - the displacement field is compatible, i.e. no gaps or overlaps are produced in the body (sliding of one part along another part is allowed),
   - it satisfies the boundary conditions for the displacements,
   - wherever deformations occur the stresses satisfy the yield condition.

The basic theorems of the plasticity theory are,

1. *Lower bound theorem.*
   The true failure load is larger than the load corresponding to an equilibrium system.

2. *Upper bound theorem.*
   The true failure load is smaller than the load corresponding to a mechanism, if that load is determined using the virtual work principle.

The first theorem states that if for a certain load an equilibrium system can be found (ignoring compatibility), then that load can certainly be carried. The second theorem states that if a mechanism can be found corresponding to a certain load (where equilibrium is taken into account only insofar as it corresponds to the chosen deformation), then this load can certainly not be carried.
It may be noted that in these theorems and in the definition of the statically or kinematically admissible fields, the constitutive relations are not mentioned, and therefore they play no role, except for the statement that the material will yield if the stresses satisfy the yield condition.
A proof of the two theorems is given in Appendix C. When studying these proofs it will appear that they have only a limited validity. The most important restriction is

that for a material with friction, such as a soil, for which the yield condition is the Mohr-Coulomb criterion, with a cohesion c and a friction angle ϕ, the theorems are valid only if during plastic deformation a continuing volume expansion occurs, of magnitude $\sin\phi$ times the rate of shear deformation. That seems to be an unrealistic behaviour, as it can be expected that in the case of continuing plastic deformations the volume will remain practically constant. This is also what has often been confirmed in experimental studies. An ever continuing plastic volume expansion would mean that the material expands without bounds, and that seems to be improbable. This means that the basic theorems of plasticity are not valid for soils, except for $\phi = 0$, i.e. for purely cohesive materials. For such a material the theory predicts that the volume is constant during plastic deformations, and that is in agreement with experimental evidence.

Because for $\phi = 0$ the theorems are valid, it follows that for such a material safe and unsafe predictions of the behaviour of a soil body can be made. For rapid loadings of saturated clays it can indeed be assumed that $\phi = 0$ and $c = s_u$, the undrained shear strength, see Chapter 26. For sands, for which it is essential that the friction angle $\phi > 0$, the theorems are not valid, at least in principle. In engineering practice they may nevertheless be used, often in a somewhat modified form. Great care should be taken in formulating conclusions from limit analysis for sands.

Actually, the limit theorems have been used in the Chapters 34 and 35. Rankine's considerations, see Chapter 34, are based upon equilibrium systems, choosing the horizontal stress such that the limit of yielding is reached. This means that the failure load is approached from below. In the analysis following Coulomb, see Chapter 35, the basis is a kinematic system, with sliding along a straight slip plane. Then the failure load is approached from above.

In the next chapters limiting states will be considered for a variety of structures, using limit analysis. These include the bearing capacity of a shallow footing, and the stability of slopes.

41 Strip footing

One of the simplest problems for which lower limits and upper limits can be determined is the case of an infinitely long strip load on a layer of homogeneous cohesive material ($\phi = 0$), see Figure 41-1. The weight of the material will be disregarded, at least in this

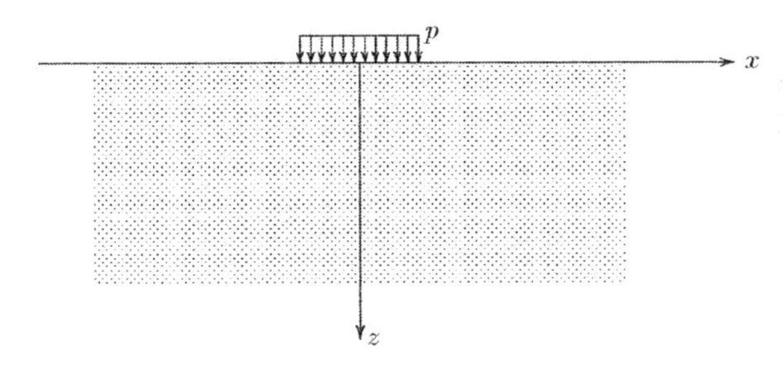

Figure 41-1. Strip footing.

chapter. That means that it is assumed that $\gamma = 0$. The problem is a first schematisation of the shallow foundation of a structure, using a long strip foundation, made of concrete,
for instance.
It will first be attempted to obtain a lower bound for the failure load, using an equilibrium system. Such a system should consist of a field of stresses that satisfies the conditions
of equilibrium in all points of the field, that agrees with the given stress distribution on the soil surface, and that does not violate the yield condition in any point.

41.1 Lower bound

An elementary solution of the conditions of equilibrium in a certain region is that the stresses in that region are constant, because then all
conditions are indeed satisfied. In a two-dimensional field these equilibrium conditions are, in the absence of gravity,

$$\frac{\partial \sigma_{xx}}{\partial x} + \frac{\partial \sigma_{zx}}{\partial z} = 0, \tag{41.1}$$

$$\frac{\partial \sigma_{xz}}{\partial x} + \frac{\partial \sigma_{zz}}{\partial z} = 0, \tag{41.2}$$

$$\sigma_{xz} = \sigma_{zx}. \tag{41.3}$$

The main difficulty is to satisfy the boundary condition, because the normal stress σ_{zz} is discontinuous along the surface, see Figure 41-1. This difficulty can be surmounted by noting that in a statically admissible field of stresses (an equilibrium system), not all stresses need be continuous.

Formally this can be recognized by inspection of the equations of equilibrium, eqs. (41.1) – (41.3). All partial derivatives in these equations must exist, which means that the stresses must at least be continuous in the directions in which they have to be differentiated. It follows that the shear stress σ_{xz} must be continuous in both directions, that the normal stress σ_{xx} must be continuous in x-direction, and the normal stress σ_{zz} must be continuous in z-direction. However, two of the partial derivatives, $\partial\sigma_{xx}/\partial z$ and $\partial\sigma_{zz}/\partial x$, do not appear in the equations of equilibrium, and therefore no conditions have to be imposed on the continuity of these two normal stresses in these directions. This means that σ_{xx} may be discontinuous in z-direction, and that σ_{zz} may be discontinuous in x-direction. Such a discontinuity is shown, for the vertical direction, in Figure 41-2.

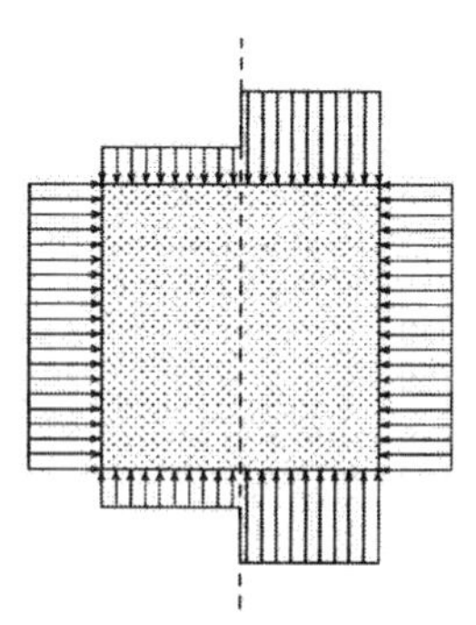

Figure 41-2. Stress discontinuity.

This figure shows a small element, with all the stresses acting upon its boundaries. The normal stress σ_{xx} must be continuous in x-direction, because of equilibrium, as can most easily be seen by letting the width of the element approach zero. Then the continuity of the stress σ_{xx} can be seen as a consequence of Newton's principle that the reaction must be equal to the action. The normal stress σ_{zz}, however, may jump across the vertical line, without disturbing equilibrium.

In Figure 41-2 the stress σ_{zz} is discontinuous in x-direction. The partial derivative $\partial\sigma_{zz}/\partial x$ is infinitely large at the location of the vertical axis, but the element, and all of its parts, are perfectly well in equilibrium.

This property of equilibrium systems has been applied by Drucker, one of the originators of the theory of plasticity, to construct equilibrium fields for practical problems. In this method the field is subdivided into regions of simple form, in each of which the stress is constant, so that the equations of equilibrium are automatically satisfied. The various subregions then are connected by requiring that all the stresses transferred on the boundary surfaces are continuous, allowing the normal stresses in the direction of these boundaries to be discontinuous. An example is shown in Figure 41-3, for the case of a strip footing. In a vertical strip below the load the stresses are supposed to be $\sigma_{xx} = 2c$, $\sigma_{zz} = 4c$, and $\sigma_{xz} = 0$. In the two regions to the left and right of this strip the stresses are $\sigma_{xx} = 2c$, $\sigma_{zz} = 0$, and $\sigma_{xz} = 0$. On the two vertical discontinuity lines only the vertical normal stress σ_{zz} is discontinuous. The other stresses are continuous, as required by equilibrium. This field of stresses satisfies all the conditions of equilibrium, and satisfies the boundary conditions on the upper surface. The shear stress $\sigma_{zx} = 0$, and the normal stress $\sigma_{zz} = 0$ if $|x| > a$, and $\sigma_{zz} = p$ if $|x| < a$, where $2a$ is the width of the loaded strip. The stress distribution should also satisfy the condition that the yield condition is never violated. This can be checked most conveniently by considering the Mohr circles for this case, as shown in the right half of Figure 41-3. In order that all circles remain within the yield envelope the value of the load p should be such that $p < 4c$. The stress distribution satisfies all the conditions for a statically admissible stress field, and it can be concluded that $p = 4c$ is a lower bound for the failure load. If the true failure load is denoted by p_c, it now has been shown that

$$p_c \geq 4c. \tag{41.4}$$

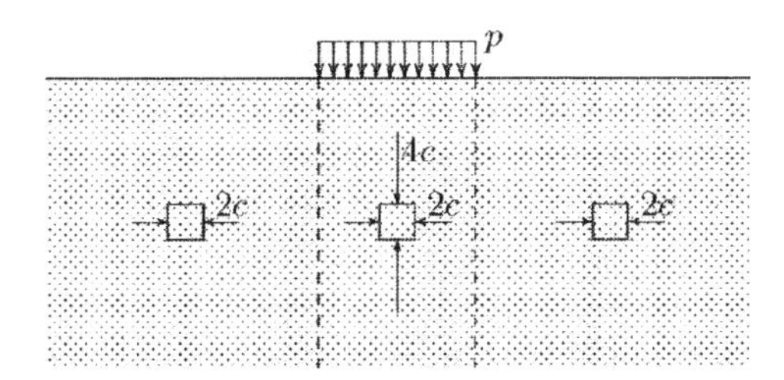

Figure 41-3. Drucker's equilibrium system.

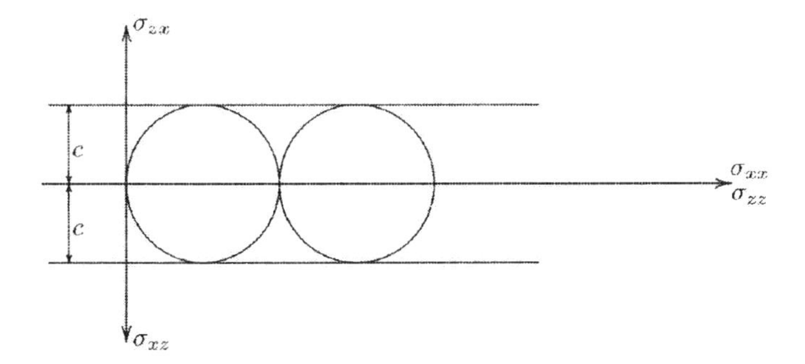

Figure 41-4. Circles of Morh belonging to Drucker's equilibrium system.

It is possible that by considering more than two discontinuity lines slightly higher lower bounds can be found. This will not be investigated here, however.
Another method to obtain a statically admissible stress field is to use an elastic solution, when available. Such a solution satisfies the equilibrium equations and the

boundary conditions. It also satisfies Hooke's law and the compatibility equations (which is not required for a statically admissible stress field, but not forbidden either). If the stress field is such that the maximum shear stress is not larger than the strength c, a lower bound of the failure load has been obtained. For the case of a strip load, see Figure 41-5, the elastic solution has been given in Chapter 31. It can be shown that the maximum shear stress is

$$\tau = \frac{p}{\pi} |\sin(\theta_1 - \theta_2)|. \tag{41.5}$$

This equation can be derived from the formulas (31.4)–(31.6) by noting that

$$\tau^2 = (\frac{\sigma_{xx} - \sigma_{zz}}{2})^2 + \sigma_{xz}^2. \tag{41.6}$$

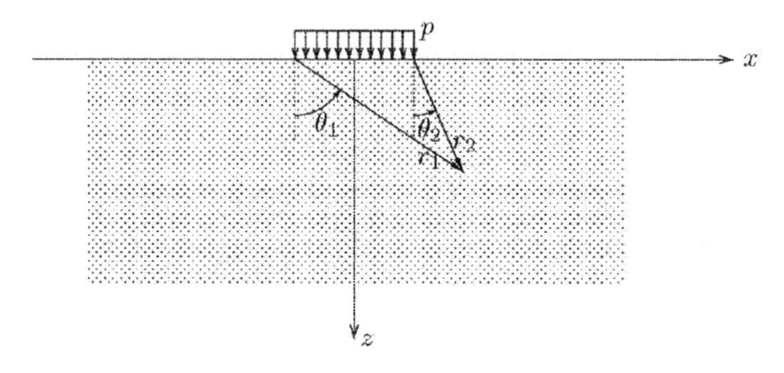

Figure 41-5. Elastic solution.

The maximum value of $|\sin(\theta_1 - \theta_2)|$ is 1, so that the maximum elastic shear stress is p/π. If this is taken equal to c, the load is $p = \pi c$. For this value of the load the elastic solution is a statically admissible stress field, and the corresponding load is a lower bound for the failure load, i.e.

$$p_c \geq 3.14c. \tag{41.7}$$

Unfortunately, this is a lower value than the value found before ($4c$), so that this elastic lower bound does not contribute to a better approximation of the failure load.

41.2 Upper bound

An upper bound for the failure load can be obtained by considering the mechanism shown in Figure 41-6. This mechanism consists of a displacement field in which half a circle, of radius a, rotates over a small angle, without internal deformations. This half circle slides along the remaining part of the body. The displacement field is compatible, and satisfies the boundary conditions on the displacements (that is very simple: these are only present on the circular slip surface). The load corresponding to this deformation can be determined by examining the moment equilibrium. If the circle rotates, a shear stress occurs at the periphery. If the shear stress is assumed to be maximal, so $\tau = c$, the moment with respect to the axis of rotation of the internal friction stresses at the perifery of the circle equals

$$\pi c a^2,$$

because the length of the circular arc is πa. The eccentricity of the external load is $\frac{1}{2}a$, so the exerted moment becomes

$$\tfrac{1}{2} p a^2.$$

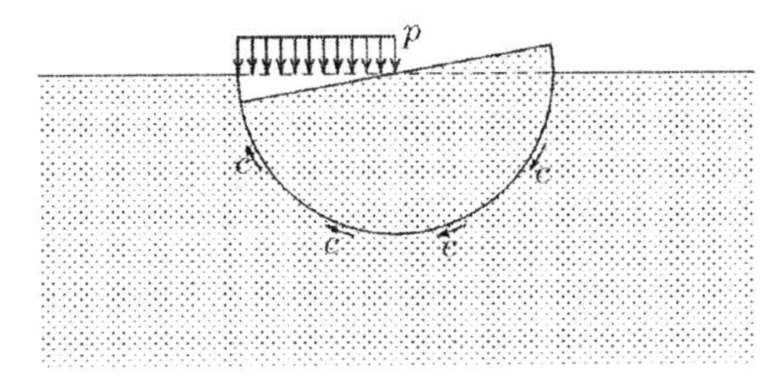

Figure 41-6. Mechanism 1.

Equating these two moments gives

$$p = 2\pi c.$$

This is an upper bound for the failure load p_c,

$$p_c \leq 6.28c. \tag{41.8}$$

A somewhat lower upper bound can be found by choosing the centre of the circle somewhat higher, see Figure 41-7. If the angle at the top is 2α, it follows

$$2cR^2\alpha = \tfrac{1}{2} p a^2,$$

and because $a = R\sin\alpha$, in which R is the radius of the circle and a the width of the load,

$$p = \frac{4c\alpha}{\sin^2\alpha}.$$

For $\alpha = \frac{1}{2}\pi$ the previous upper bound is recovered. The smallest value is obtained for $\alpha = 1.165562$, or $\alpha = 66.78^\circ$. The centre of the circle then is located at a height $0.429a$. The corresponding value of p is $5.52c$. This is an upper bound, hence

$$p_c \leq 5.52c. \tag{41.9}$$

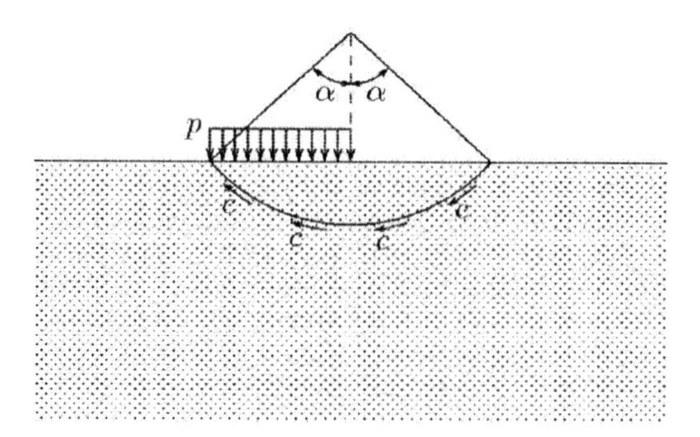

Figure 41-7. Mechanism 2.

It can be concluded at this stage that it has been shown that

$$4c \leq p_c \leq 5.52c. \tag{41.10}$$

In the next chapter the failure load will be approximated even closer.

42 Prandtl's solution

In 1920 Prandtl succeeded in finding a solution for the problem of a strip load on a half plane that is both statically admissible and kinematically admissible. This solution must therefore give the true failure load. The lower bound part of Prandtl's solution, with an equilibrium system of stresses, will be presented in this chapter. The proof that this solution is also kinematically admissible, which is much more difficult, will be omitted here. A complete proof can be found in textbooks on the theory of plasticity. As in the previous chapter, the material is considered to be weightless ($\gamma = 0$), and frictionless ($\phi = 0$), so that its only relevant property is the cohesive strength c. That is a great restriction, but it will be relaxed in later chapters. The stresses will be formulated using polar coordinates. In order to verify the equilibrium conditions, these must be expressed into polar coordinates first.

42.1 Equilibrium equations in polar coordinates

Figure 42-1 shows a small element of a two-dimensional region, in polar coordinates r and θ, with all the stresses acting upon it. Equilibrium in r-direction requires that

$$\sigma_{rr}^{*}(r+\Delta r)\Delta\theta - \sigma_{rr} r\Delta\theta + \sigma_{\theta r}^{*}\Delta r - \sigma_{\theta r}\Delta r - \sigma_{\theta\theta}\Delta r\Delta\theta = 0.$$

The last term in this equation is needed because the forces $\sigma_{\theta\theta}\Delta r$ and $\sigma_{\theta\theta}^{*}\Delta r$, which differ only infinitesimally, do not have precisely the same direction. Their directions differ by an amount $\Delta\theta$. Together they give a contribution to the forces in r-direction. By writing

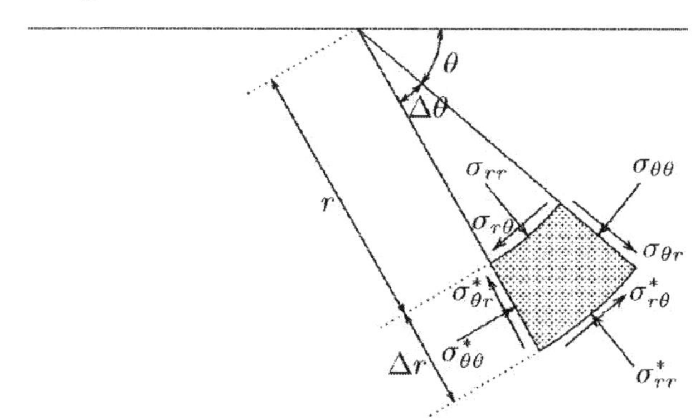

Figure 42-1. Polar coordinates.

$$\sigma^*_{rr} - \sigma_{rr} = \frac{\partial \sigma_{rr}}{\partial r}\Delta r,$$

$$\sigma^*_{\theta r} - \sigma_{\theta r} = \frac{\partial \sigma_{\theta r}}{\partial \theta}\Delta \theta,$$

the equilibrium equation becomes, after division by $r\Delta r\Delta\theta$,

$$\frac{\partial \sigma_{rr}}{\partial r} + \frac{1}{r}\frac{\partial \sigma_{\theta r}}{\partial \theta} + \frac{\sigma_{rr} - \sigma_{\theta\theta}}{r} = 0. \tag{42.1}$$

This is the equation of equilibrium in radial direction.
Equilibrium in tangential direction requires that

$$\sigma^*_{rr}\Delta r - \sigma_{\theta\theta}\Delta r + \sigma^*_{r\theta}(r+\Delta r)\Delta\theta - \sigma_{r\theta} r\Delta\theta + \sigma_{\theta r}\Delta r\Delta\theta = 0.$$

In this case the last term may deserve some explanation. This term is the result of the angle $\Delta\theta$ between the forces $\sigma_{\theta r}\Delta r$ and $\sigma^*_{\theta r}\Delta r$. Using the equality $\sigma_{\theta r} = \sigma_{r\theta}$ the following equation is obtained,

$$\frac{\partial \sigma_{r\theta}}{\partial r} + \frac{1}{r}\frac{\partial \sigma_{\theta\theta}}{\partial \theta} + \frac{2}{r}\sigma_{r\theta} = 0. \tag{42.2}$$

This is the equation of equilibrium in tangential direction.

42.2 Prandtl's solution

The basic principle of Prandtl's solution of the problem of the determination of the failure load of a half plane carrying a strip load on its surface is a subdivision of the region below the load into three zones, see Figure 42-2, two triangles and a wedge. In each of these three zones the stress state is assumed to be critical. The load can most simply be derived from a consideration of equilibrium.
In zone I the stresses are assumed to be

$$\text{I}: \quad \sigma_{xx} = 2c, \quad \sigma_{zz} = 0, \quad \sigma_{xz} = 0. \tag{42.3}$$

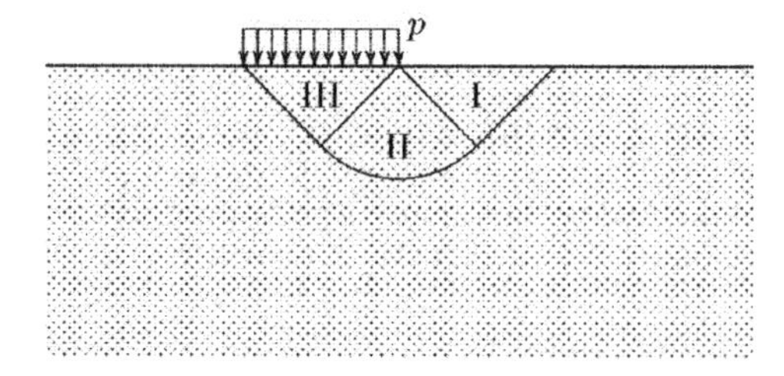

Figure 42-2. Prandtl's schematisation.

This stress state satisfies the equilibrium conditions and the boundary conditions on the upper surface (zero shear stress and zero normal stress), and it does not violate the yield condition in any point. Actually, in every point of this zone the yield condition is just reached. On a plane inclined at an angle of 45° the stresses are, see also Figure 42-3, $\sigma_{\theta\theta} = c$, and $\sigma_{\theta r} = -c$. The sign of these stresses can best be verified by comparison with the definitions of positive stress components, as illustrated in Figure 42-1, and the stress distribution shown in Figure 42-3. On the interface between zones I and II the normal stress in radial direction is $\sigma_{rr} = c$.

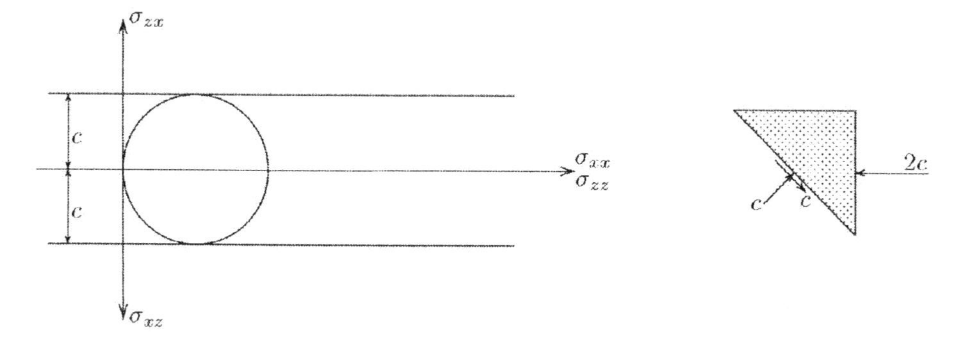

Figure 42-3. Stresses in zone I.

For zone II, the wedge, it is assumed that everywhere in this zone $\sigma_{rr} = \sigma_{\theta\theta}$ and $\sigma_{\theta r} = \sigma_{r\theta} = -c$. Throughout this zone the Mohr circle then just touches the envelope. The two equations of equilibrium, (42.1) and (42.2) reduce to

$$\frac{\partial \sigma_{rr}}{\partial r} = 0, \tag{42.4}$$

$$\frac{\partial \sigma_{\theta\theta}}{\partial \theta} = 2c. \tag{42.5}$$

These equations can be satisfied by the stress field

$$\text{II}: \quad \sigma_{rr} = \sigma_{\theta\theta} = c + 2c(\theta - \tfrac{1}{4}\pi), \quad \sigma_{\theta r} = \sigma_{r\theta} = -c, \tag{42.6}$$

where the integration constant has been chosen such that $\sigma_{\theta\theta}$ is continuous on the interface between zone I and zone II, where $\theta = \frac{1}{4}\pi$. On the interface between zone II and zone III the angle $\theta = \frac{3}{4}\pi$. Then

$$\theta = \tfrac{3}{4}\pi: \quad \sigma_{rr} = \sigma_{\theta\theta} = c + 2c(\pi + 1), \quad \sigma_{\theta r} = \sigma_{r\theta} = -c. \tag{42.7}$$

In zone III the stresses are again be assumed to be constant. A possible stress field is, see also Figure 42-4,

$$\text{III}: \quad \sigma_{xx} = \pi c, \quad \sigma_{zz} = (\pi + 2)c, \quad \sigma_{xz} = 0. \tag{42.8}$$

The boundary conditions for the stresses are satisfied if $p = (\pi + 2)c$. This solution satisfies all conditions for an equilibrium system. This means that the load in this solution is a lower bound. The failure load is at least equal to that lower bound,

$$p_c \geq (\pi + 2)c = 5.14c. \tag{42.9}$$

It can be shown that Prandtl's solution is also an upper bound, by considering a convenient deformation field, with the wedge being subdivided into a large number of small triangular wedges. In that case a complication arises in the top corner of the wedge, where the displacements are singular. The derivation for this case, which also happens to yield a load of $(\pi+2)c$, will not be considered here, see any textbook on the theory of plasticity.

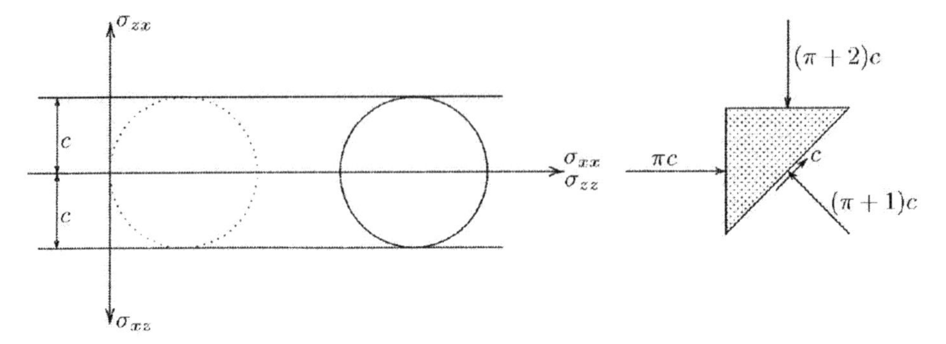

Figure 42-4. Stresses in Zone III.

Assuming that the proof that the value $(\pi+2)c$ is an upper bound can indeed be given, it follows that the true failure load in the case of a strip load is

$$p_c = (\pi+2)c = 5.14c. \tag{42.10}$$

This value is indeed higher than the lower bounds obtained in the previous chapter, and lower than the upper bounds obtained in that chapter. This confirms the validity of the upper and lower bound theorems.

43 Brinch Hansen

Although the computation of limit states for materials with internal friction does not lead to results that are certainly on the safe side or on the unsafe side of the failure load (as they are for cohesive materials), these computations can still be very valuable, because it can be expected that the results will at least give an approximation of the failure load. Engineering calculations are always based upon a number of assumptions and approximations, and engineers have to accept that their results at best are an approximation of reality. The circumstance that it can not be stated with certainty that a given load is above or below the failure load does not make the results useless. As long as the computations are based upon reasonable assumptions regarding the stresses and deformations, and if they incorporate certain experiences from the field or the laboratory, the results may be considered as giving a useful approximation of the real world. In this process there is also room for *engineering judgement*, which is a combination of intuition, experience and common sense. Finally, it may be mentioned that the applicability of a certain methodology may be enhanced by its possible agreement with known results for special cases.
In this chapter the case of a strip footing on cohesive material, considered in Chapters 41 and 42, is extended to a general type of shallow foundation, on a soil characterized by its cohesion c, friction angle ϕ and volumetric weight γ. The soil is assumed to be completely homogeneous. Although the formulas were originally intended to be applied to foundation strips of buildings, at a shallow depth below the soil surface, they are also applied to large caisson foundations used in offshore engineering for the foundation of huge oil production platforms.

43.1 Bearing capacity of strip foundation

An important problem of foundation engineering is the computation of the maximum load (the *bearing capacity*) of a strip foundation, i.e. a very long foundation, of constant width, at a certain depth below the soil surface. The influence of the depth of the foundation is accounted for by considering a surcharge at the foundation level, to the left and the right of the applied load. For the simplest case, of a strip of infinite length, on weightless soil, the first computations were made by Prandtl, see Figure 43-1, on the basis of the assumption that in a certain region at the soil surface the stresses satisfy the
equilibrium conditions and the Mohr-Coulomb failure criterion. In this entire region the soil then is on the verge of yielding. This analysis is a direct generalization of the problem considered in Chapter 42 of a strip load on the surface of a cohesive

material. The foundation pressure is denoted by p. The surcharge q, next to the foundation, is supposed to be given. It can be used to represent the effect of the depth of the foundation (d) below the soil surface. In that case $q=\gamma d$, where γ is the unit weight of the soil.

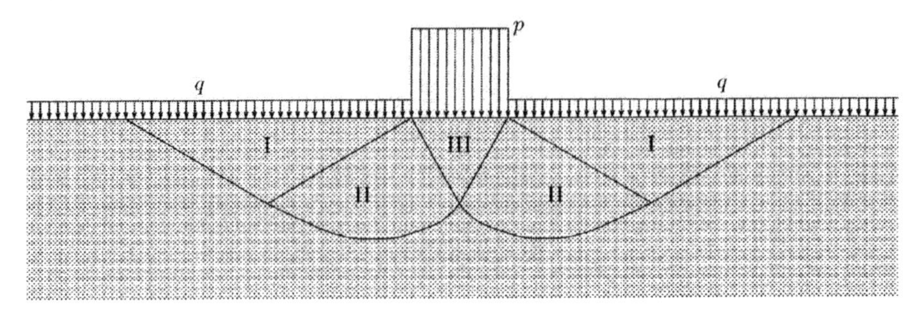

Figure 43-1. Strip foundation.

Prandtl's solution, which will not be derived in detail here, again uses a subdivision of the soil into three zones, see Figure 43-1. In zone I the horizontal stress is supposed to be larger than the vertical stress, which is equal to the surcharge q. This horizontal stress is then the passive lateral stress corresponding to the vertical stress q. In zone III the vertical normal stress is supposed to be the largest stress, and its value is equal to the unknown load p. The transition is formed by the wedge shaped zone II (*Prandtl's wedge*), which is bounded below by a logarithmic spiral. The results of the analysis can be written as

$$p=cN_c+qN_q, \tag{43.1}$$

where the coefficients N_c en N_q are dimensionless constants, for which Prandtl obtained the following expressions,

$$N_q=\frac{1+\sin\phi}{1-\sin\phi}\exp(\pi\tan\phi), \tag{43.2}$$

$$N_c=(N_q-1)\cot\phi. \tag{43.3}$$

In Table 43-1 the values of N_c and N_q are given, as a function of the friction angle ϕ. In the limiting case $\phi=0$ the value of $N_c=2+\pi$, as found in Chapter 42. If $c=0$ and $\phi=0$ the bearing capacity must be equal to the surcharge, i.e. $p=q$. Even a layer of mud can support a certain load, provided that it is the same all over its surface. This is expressed by the value $N_q=1$ for $\phi=0$.
Prandtl's formula (43.1) has been extended by Keverling Buisman, Caquot, Terzaghi and Brinch Hansen with various terms, including one for the unit weight of the soil. The complete formula is written in the form

$$p=cN_c+qN_q+\tfrac{1}{2}\gamma BN_\gamma, \tag{43.4}$$

ϕ	N_c	N_q	N_γ	ϕ	N_c	N_q	N_γ
0	5.142	1.000	0.000	20	14.835	6.399	3.930
1	5.379	1.094	0.003	21	15.815	7.071	4.661
2	5.632	1.197	0.014	22	16.833	7.821	5.512
3	5.900	1.309	0.032	23	18.049	8.661	6.504
4	6.185	1.433	0.060	24	19.324	9.603	7.661
5	6.489	1.568	0.099	25	20.721	10.662	9.011
6	6.813	1.716	0.151	26	22.254	11.854	10.558
7	7.158	1.879	0.216	27	23.942	13.199	12.432
8	7.527	2.058	0.297	28	25.803	14.720	14.590
9	7.922	2.255	0.397	29	27.860	16.443	17.121
10	8.345	2.471	0.519	30	30.140	18.401	20.093
11	8.798	2.710	0.665	31	32.671	20.631	23.591
12	9.285	2.974	0.839	32	35.490	23.177	27.715
13	9.807	3.264	1.045	33	38.638	26.092	32.590
14	10.370	3.586	1.289	34	42.164	29.440	38.366
15	10.977	3.941	1.576	35	46.124	33.296	45.228
16	11.631	4.335	1.913	36	50.586	37.753	53.404
17	12.338	4.772	2.307	37	55.630	42.920	63.178
18	13.104	5.258	2.767	38	61.352	48.933	74.899
19	13.934	5.798	3.304	39	67.867	55.957	89.007
20	14.835	6.399	3.930	40	75.313	64.195	106.054

Table 43-1: Bearing capacity coefficients.

where B is the total width of the loaded strip, and γ is the volumetric weight of the soil. That all effects may be superimposed, as has been assumed in eq. (43.4), has been confirmed by various investigations, but has never been proved rigorously. For the coefficient N_γ various suggestions have been made, on the basis of theoretical analysis or experimental evidence, for instance

$$N_\gamma = 2(N_q - 1)\tan\phi. \tag{43.5}$$

There appears to be general agreement on the character of this expression, but various researchers have proposed different values for the constant factor. Brinch Hansen used a factor $\frac{3}{2}$ rather than a factor 2, probably to avoid an overestimation, and therefore including some safety. In modern engineering it is considered that safety factors should be kept apart from the theoretical formulas, so that it was agreed that the best value of the multiplication factor is 2. A safety factor must be taken into account explicitly, in the design stage, by reducing the soil strength, or as a load factor.

Later the formula (43.4) has been further extended with various correction coefficients, in order to take into account the shape of the loaded area, the inclination of the load, a possible inclined soil surface, and a possible inclined loading area. Most of these effects were assembled into a single formula by Brinch Hansen,

$$p = i_c s_c c N_c + i_q s_q q N_q + i_\gamma s_\gamma \tfrac{1}{2}\gamma B N_\gamma. \tag{43.6}$$

In this equation the coefficients i_c and i_q are correction factors for a possible inclination of the load (inclination factors), and s_c and s_q are correction factors for the shape of the loaded area (shape factors). Some other factors may be used (for a sloping soil surface, or a sloping foundation foot), but these are not considered here.

43.2 Inclination factors

In case of an inclined load, i.e. loading by a vertical force and a horizontal load, see Figure 43-2, the bearing capacity is considerably reduced. This can be understood by noting that sliding would occur if the horizontal force approaches the maximum possible shear force on the foundation surface,

$$t \le c + p\tan\phi. \tag{43.7}$$

The formulas should be such that for this limiting value of the shear stress t (with respect to the constant value of the vertical stress p) the bearing capacity reduces to zero.

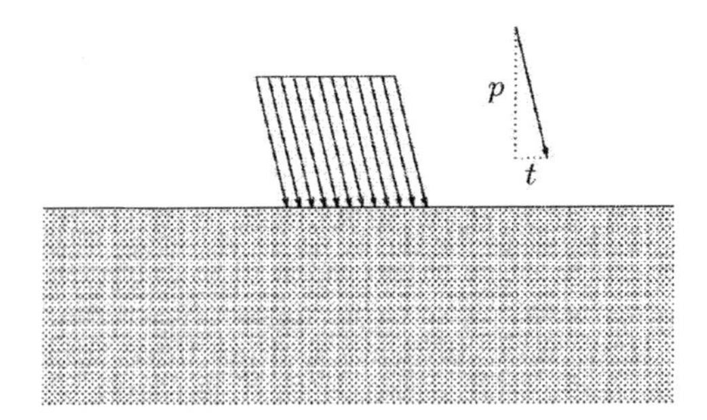

Figure 43-2. Inclined load.

For cases in which the shear force is smaller than its maximum possible value, the correction factors for the inclination of the load are usually expressed as

$$i_c = 1 - \frac{t}{c + p\tan\phi}, \tag{43.8}$$

$$i_q = i_c^2, \tag{43.9}$$

$$i_\gamma = i_c^3. \tag{43.10}$$

There is no general agreement on the precise value of these reduction factors, on an international level and even on a national level. Various researchers prefer slightly different values, and even the national standards may (e.g. NEN 6744) may give different values. The formulas given above at least are in agreement with certain special cases. The coefficients approach 0 if the shear stress approaches the

maximum value $\tau_{max} = c + \sigma_n \tan\phi$. The other extreme case is when the load is vertical ($t = 0$). Then all factors reduce to 1, as required.
Although it would be better to use the friction angle δ instead of the angle of internal friction ϕ (see Chapter 24), this is not customary.

43.3 Shape factors

If the shape of the foundation area is not an infinitely long strip, but a rectangular area, of width B and length L (where it is assumed, for definiteness, that the width is the shortest dimension, i,e, $L \geq B$), the usual correction factors are of the form

$$s_c = 1 + 0.2\frac{B}{L}, \tag{43.11}$$

$$s_q = 1 + \frac{B}{L}\sin\phi, \tag{43.12}$$

$$s_\gamma = 1 - 0.3\frac{B}{L}. \tag{43.13}$$

There is no international agreement on the precise values of these correction factors either. Some consultants prefer to take $s_q = 1$, for all values of ϕ, and some use coefficients with slightly different values for the factors 0.2 and 0.3. It may be noted that for $B/L = 0$, the formulas all give a factor 1, in agreement with the basic results for an infinite strip. It should also be remembered that $B/L \leq 1$, by definition.

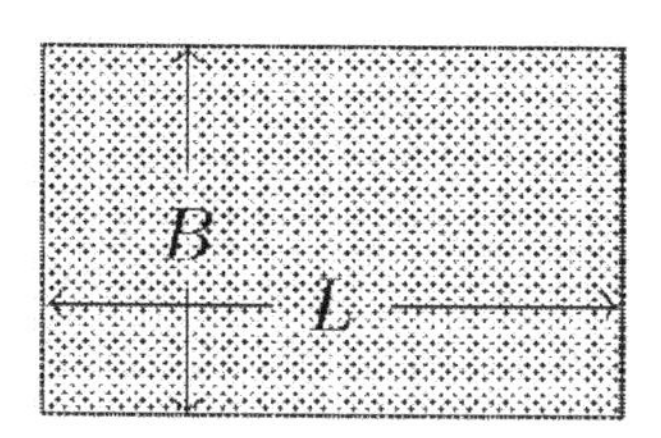

Figure 43-3. Rectangular area.

Some justification for values of the coefficients s_c and s_q larger than 1 is that when loading a rectangular plate, some of the soil surrounding the plate will also deform with the plate, so that a larger area contributes to the bearing capacity of the plate. The coefficient s is smaller than 1 because tests indicate that for a rectangular plate in sand a sliding surface may occur that is less deep than the sliding surface for a long strip. In case of an eccentric resultant force of the load, the width B and the length L may be reduced such that the resulting force does apply in the centre of the reduced area, see Figure 43-4. Part of the foundation plate then does not contribute to the bearing capacity, at least for this loading case. It may, of course, give a contribution to the bearing capacity of other loading cases.
As mentioned before, there is no general agreement about the values of many of the correction factors, because the results obtained by researchers in different countries,

from theoretical or experimental studies, appear to give different results. Great care is needed when using data from literature. When a certain value has been obtained by one single researcher, and deviates from the results of many others, that value may well be in error.

It is also very inconvenient that there is sometimes no agreement about the basic formula (43.4). In some older publications the factor $\frac{1}{2}$ is omitted. Then the values of N_γ are (approximately) half as large, so that the final result is the same, but it may give rise to some confusion when using a formula from one publication, and taking the coefficients from another publication. In this book Terzaghi's original formula has been used, as is common practice internationally.

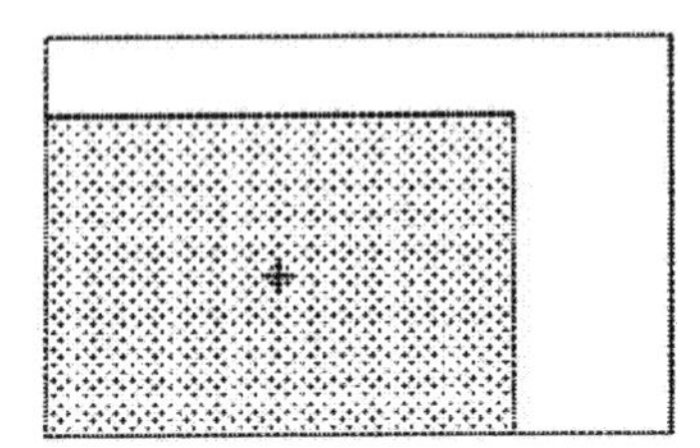

Figure 43-4. Eccentric load.

The formulas presented in this chapter have originally been derived for foundations on land, with relatively modest dimensions, say a few square meters. The third term in Brinch Hansen's formula (43.6) then is rather small, because of the small value of B, and it is often omitted. In offshore engineering the development of *gravity foundations* has meant that production platforms may be founded on huge concrete caissons that are placed on the sea bottom, in deep water. The surface area may be up to $80 \text{ m} \times 80 \text{ m}$. For the design of such structures the bearing capacity of the foundation is of great importance, and then the third term in the formulas (43.4) or (43.6), which describes the influence of the unit weight of the soil (i.e. the *gravity* term), is the most important term of all, giving the major contribution to the bearing capacity, especially in sand. This is why considerable attention has been paid to a more accurate definition of this term.

It must be emphasized that all the considerations of this chapter are restricted to dry soils, in which there is no difference between effective stresses and total stresses. For saturated soils the formulas should be expressed in terms of effective stresses. Usually this can be accomplished simply by replacing the volumetric weight γ by the effective volumetric weight $\gamma_s - \gamma_w$. That is a simple, but very fundamental adjustment.

43.4 CPT and undrained shear strength

In the Netherlands the cone penetration test is mainly used as a model test for pile foundations. In the Western parts of the Netherlands the soil usually consists of 10 – 20 meters of very soft soil layers (clay and peat), on a rather stiff sand layer. This soil structure is very well suited for a pile foundation, of wooden or concrete piles of

about 20 cm – 40 cm diameter, reaching just into the sand. The weight of the soft soil acts as a surcharge on the sand, which has a considerable cone resistance. The allowable stress on the sand depends upon its friction angle ϕ, its cohesion c (usually very small, or zero), and the surcharge q. The dimensions of the foundation pile have very little influence, because this parameter appears only in the third term of Brinch Hansen's formula, which is a small term if the width is less than, say, 1 meter. This means that the maximum pressure for a large pile and the thin pile of a cone penetrometer will be practically the same, so that the allowable pressure on a pile can be determined by simply measuring the cone resistance. This will be elaborated in Chapter 44.
The cone penetration test can also be used to determine physical parameters of the soil, especially the shear strength. It can be postulated, for instance, that in clays the cone resistance will be determined mainly by the undrained shear strength of the soil (s_u). In agreement with the analysis of Brinch Hansen the relation will be of the form

$$q_c - \sigma_v = N_c s_u, \tag{43.14}$$

where σ_v is the local vertical stress caused by the surcharge, and N_c is a dimensionless factor. For a circular cone in a cohesive material a cone factor N_c of the order of magnitude 15 – 18 is usually assumed, on the basis of plasticity calculations for the insertion of a cone into a cohesive material of infinite extent. layer. By measuring the cone resistance q_c the undrained shear strength s_u can be determined. The results are not very accurate, because of theoretical shortcomings and practical difficulties, but the measurement has the great advantage of being done in situ, on the least disturbed soil. The alternative would be taking a sample, bringing it to a laboratory, and then doing a laboratory test. This process includes many possible sources of disturbance, that are avoided by doing a test in situ.

Problems

43.1 Show that the expression (43.3) tends to $\pi + 2$ if $\phi \rightarrow 0$.

43.2 For sand it can often be assumed that the value of the friction angle is at least 30°. If there is some indication that the value is somewhat larger, say 10 % larger, then is it worthwhile, for the design of a foundation, to test the soil to determine the precise value of ϕ ?

43.3 Estimate the bearing capacity of a gravity foundation, with surface dimensions of $60 \text{ m} \times 60 \text{ m}$, on a sandy soil, under water.

43.4 Estimate the bearing capacity at the point of a pile, of cross section $40 \text{ cm} \times 40 \text{ cm}$, founded on a sand layer, at a depth of 20 m below a system of soft soil (clay) layers.

44 Pile foundations

In many areas of the Netherlands, especially in the Western part of the country, the soil consists of layers of soft soil (clay and peat), of $10-20$ meter thickness, on a stiff sand layer, of pleistocene origin. The bearing capacity of this sand layer is derived for a large part form its deep location, with the soft layers acting as a surcharge. And the properties of the sand itself, a relatively high density, and a high friction angle, also help to give this sand layer a good bearing capacity, of course. The system of soft soils and a deeper stiff sand layer is very suitable for a pile foundation. In this chapter a number of important soil mechanics aspects of such pile foundations is briefly discussed.

44.1 Bearing capacity of a pile

For the determination of the bearing capacity of a foundation pile it is possible to use a theoretical analysis, on the basis of Brinch Hansen's general bearing capacity formula. In this analysis the basic parameters are the shear strength of the sand layer (characterized by its cohesion c and its friction angle ϕ), and the weight of the soft layers, which can be taken into account as a surcharge q. In engineering practice a simpler, more practical and more reliable method has been developed, on the basis of a cone penetration test, considering this as a model test. It would be even better to perform a pile loading test on the pile, in which the pile is loaded, for instance by concrete blocks on a steel frame, with a test load approaching its maximum bearing capacity. This is very expensive, however, and the CPT is usually considered reliable enough.

In a homogeneous soil it can be assumed that under static conditions the failure load of a long pile is independent, or practically independent of the diameter of the pile. This means that the cone resistance measured in a CPT can be considered to be equal to the bearing capacity of the pile tip. A possible theoretical foundation behind this statement is that the failure is produced by shear deformations in a zone around the pile, the dimensions of which are determined by the only dimension in the problem, the diameter of the pile. If the pile diameter is taken twice as large, the dimensions of the failure zone around the pile will also be twice as large. The total force (stress times area) then is four times as large, see Figure 44-1. This is also in agreement with the theory behind Brinch Hansen's formula, provided that the third term (representing the weight of the soil below the foundation level, and the width of the foundation) is small. This will be the case if the pile diameter is small compared to its length.

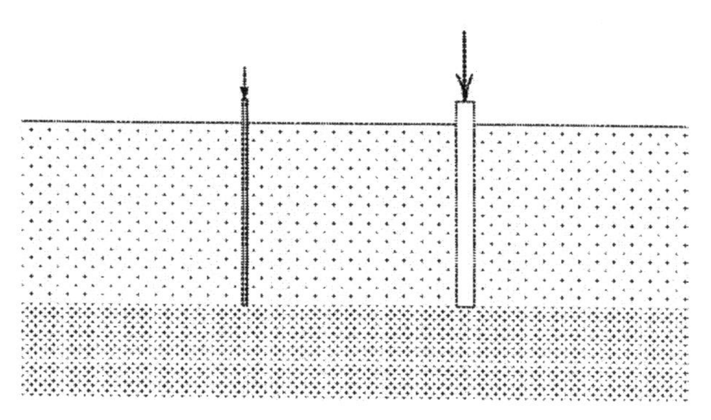

Figure 44-1. CPT and pile.

In reality the soil around the pile tip usually is not perfectly homogeneous. Very often the soil consists of layers having different properties. For this case practical design formulas have been developed, which take into account the different cone resistance below and above the level of the pile tip. Moreover, in these design formulas the possibility that the failure mode will prefer the weakest soil can be accounted for. In engineering practice the Koppejan formula is often used. In this formula the resistance of the pile is assumed to consist of three contributions,

$$p = \tfrac{1}{2}[\tfrac{1}{2}(p_1 + p_2) + p_3]. \tag{44.1}$$

In this equation p_1 is the smallest value of the cone resistance q_c below the pile tip, up to a depth of $4d$, where d is the diameter of the pile, p_2 is the average cone resistance in that zone, and p_3 is a representative low value of the cone resistance above the pile tip, in a zone up to $8d$ above the pile tip. In this way a representative average value of the cone resistance around the pile tip is obtained, in which engineering judgement is combined with experience. A pile may also have a bearing capacity due to friction along the length of the pile. This is very important for piles in sand layers. In applications in very soft soil (clay layers), the contribution of friction is generally very unreliable, because the soil may be subject to settlements, whereas the pile may be rigid, if it has been installed into a deep sand layer. It may even happen that the subsiding soil exerts a downward friction force on the pile, *negative skin friction*, which reduces the effective bearing capacity of the pile. Friction is of course very important for tension piles, for which it is the only contributing mechanism.

The maximum value of the skin friction can be determined very well using a friction cone, that is a penetration test in which the sleeve friction is also measured. The local values are often very small, however, so that the measured data are not very accurate. For sandy soils the friction therefore is often correlated to the cone resistance.

44.2 Statically determinate pile foundation

If the maximum allowable load on a single pile is known, from a theoretical analysis, or from the interpretation of a cone penetration test, or from a pile loading test, the number of piles in the foundation of a large structure can be determined from the total load of the structure, including its weight. In this process a sufficiently large safety coefficient (e.g. 1.5 or 1.8) must be taken into account, to avoid possible failure. If all the loads are vertical the piles may all be vertical as well. The installation is then also simplest, as driving a vertical pile is easier than driving a tilting pile. A small horizontal force may be transmitted by a vertical pile, by bending of the pile, but if large horizontal forces must be transferred to the soil (due to wind or waves), it is better to use some tilted piles, so that the pile forces can all be axial, and the deformations of the piles remain small. The analysis of the pile forces deserves some special attention.

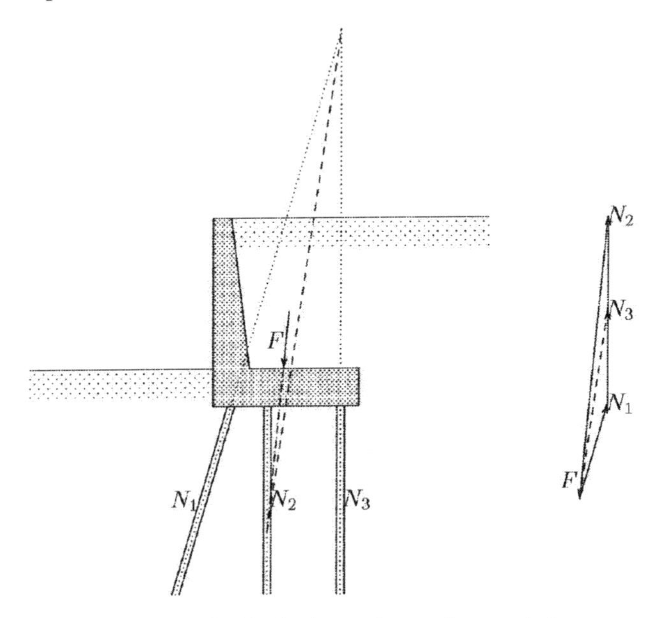

Figure 44-2. Statically determinate pile foundation.

As an example a retaining wall may be considered, see Figure 44-2. In this case there is a considerable horizontal force, which can most easily be transferred to the ground by using a tilted pile. For this foundation system it may be assumed that the force in each pile is directed along its axis. The reason for that is that a pile is much stiffer in axial loading than it is in lateral loading. In the case shown in Figure 44-2, with three rows of piles, the force in each row can be determined from the equilibrium equations alone. This is called a *statically determinate system*. The analysis can be performed graphically. The starting point is that the loading force F must be equilibrated by the sum of the forces in the piles, N_1, N_2 and N_3. Because N_2 and N_3 are vertical the force diagram shown in the right part of Figure 44-2 can be constructed. The precise contributions of N_2 and N_3 is still unknown in the first stage. However, because the resulting force of F and N_2 must be in equilibrium with the resulting force of N_1 and N_3 these two resultants must have the same line of action, and they must be of equal magnitude, in opposite direction. The resulting force of N_1

and N_3 should pass through the intersection point of these two forces, and similarly the resulting force of F and N_2 passes through the intersection of these two forces. Thereby the line of action of these resultants is known. In the force diagram this line of action can then be drawn as well, as its direction is known. The three pile forces have now been determined, and the problem is solved.

44.3 Statically indeterminate pile foundation

If there are more than three rows of piles, the problem of determining the individual pile forces is statically indeterminate. The solution then depends upon the flexibility of each of the piles, and of the superstructure. A well known procedure is to assume that the pile forces are directed along their axes (which means that the bending resistance of the piles is neglected with respect to their axial stiffness), and then to consider the piles as linear springs. For each pile one may write

$$N_i = k_i u_i. \tag{44.2}$$

in which N_i is the force in pile i, u_i the displacement of the pile top, and k_i the spring constant of the pile. This spring constant could be taken as the stiffness of the pile EA/l, but that would be valid only if the pile tip is fully fixed. In reality the soil surrounding the pile tip will also somewhat deform if the pile is loaded, so that the value of the spring constant k_i should be reduced. In the absence of further information about the stiffness of the soil it is sometimes assumed, as a first estimate, that the deformation of the pile top is twice as large as the deformation of the pile itself, leading to a value $k_i = \frac{1}{2}EA/l$. It can also be argued, however, that the pile force will not be constant along the pile, due to friction, which would lead to a larger value of k_i. In general, it is recommended to try to determine the spring constants by a careful analysis of the load transfer from the foundation to the soil.

If the superstructure can be considered as infinitely stiff, the computations can be performed using the displacement method. This will be illustrated by considering an example, see Figure 44-3. In this two-dimensional case there are three basic parameters to describe the displacement of the foundation : the horizontal and vertical displacements, and the rotation. It is assumed that all pile rows have the same stiffness (k). The load is supposed to consist of a vertical component of 2000 kN and a horizontal component of 200 kN. The line of action of this force is supposed to pass through the point $x = 1$ m, $y = 0$, see Figure 44-3. The slope of row 1 is 3:1 (vertical to horizontal).

The solution of the problem of determining the forces in each row of piles can be obtained by the standard procedure of the displacement method. This procedure is : first determine the basic displacement parameters (in this case the two displacements and the rotation), then express the internal forces into these parameters, and finally formulate the equations of equilibrium. In this case the procedure is as follows.

In case of a horizontal displacement u the forces in the pile rows are:

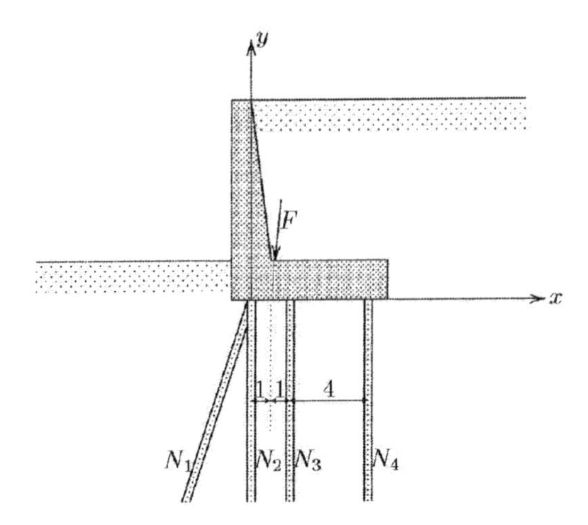

Figure 44-3. Statically indeterminate pile foundation.

$$N_1 = \frac{1}{\sqrt{10}} ku, \quad N_2 = 0, \quad N_3 = 0, \quad N_4 = 0.$$

For a vertical displacement v the forces are :

$$N_1 = \frac{3}{\sqrt{10}} kv, \quad N_2 = kv, \quad N_3 = kv, \quad N_4 = kv.$$

For a rotation around the origin, of magnitude $\theta = w/1$ m the forces are :

$$N_1 = 0, \quad N_2 = 0, \quad N_3 = 2kw, \quad N_4 = 6kw.$$

Addition of these forces gives

$$\begin{pmatrix} N_1 \\ N_2 \\ N_3 \\ N_4 \end{pmatrix} = \begin{pmatrix} \frac{1}{\sqrt{10}} & \frac{3}{\sqrt{10}} & 0 \\ 0 & 1 & 0 \\ 0 & 1 & 2 \\ 0 & 1 & 6 \end{pmatrix} \begin{pmatrix} ku \\ kv \\ kw \end{pmatrix} \tag{44.3}$$

The forces in the piles have been considered positive for pressure.
The equations of equilibrium of the foundation plate are that the sum of the horizontal forces should be 200 kN, the sum of the vertical forces should be 2000 kN, and the sum of the moments with respect to the origin should be 2000 kNm. These equations can be written as

$$\begin{pmatrix} \frac{1}{\sqrt{10}} & 0 & 0 & 0 \\ \frac{3}{\sqrt{10}} & 1 & 1 & 1 \\ 0 & 0 & 2 & 6 \end{pmatrix} \begin{pmatrix} N_1 \\ N_2 \\ N_3 \\ N_4 \end{pmatrix} = \begin{pmatrix} 200\,\text{kN} \\ 2000\,\text{kN} \\ 2000\,\text{kN} \end{pmatrix} \tag{44.4}$$

Substitution of (44.3) into (44.4) yields the equilibrium equations expressed into the displacements,

$$\begin{pmatrix} \frac{1}{10} & \frac{3}{10} & 0 \\ \frac{3}{10} & \frac{39}{10} & 8 \\ 0 & 8 & 40 \end{pmatrix} \begin{pmatrix} ku \\ kv \\ kw \end{pmatrix} = \begin{pmatrix} 200\,\text{kN} \\ 2000\,\text{kN} \\ 2000\,\text{kN} \end{pmatrix} \tag{44.5}$$

This is a system of three equations with three unknowns. The solution is a simple mathematical problem. The result is

$$\begin{aligned} ku &= -143\,\text{kN}, \\ kv &= 714\,\text{kN}, \\ kw &= -93\,\text{kN}. \end{aligned} \tag{44.6}$$

The pile forces then are

$$\begin{aligned} N_1 &= 632\,\text{kN}, \\ N_2 &= 714\,\text{kN}, \\ N_3 &= 529\,\text{kN}, \\ N_4 &= 157\,\text{kN}. \end{aligned} \tag{44.7}$$

The vertical component of the force in row 1 is 600 kN, and its horizontal component is 200 kN. That result could have been obtained immediately, as this is the only pile that can transfer a horizontal load.

The distribution of the pile forces appears not to be uniform. The force in row 4 appears to be considerably smaller than in the other rows. If this is the only load that the foundation must carry, it may be considered to place the piles in row 4 at larger mutual distances (in y-direction). This would mean that the stiffness in that row would be smaller, and the computations should be repeated for the new stiffness parameters.

The procedure illustrated here can easily be generalized to the three-dimensional case. Then there are six degrees of freedom (three displacements and three rotations), and six equations of equilibrium. The number of piles may be very large. The procedure is very well suited for numerical analysis, using a simple computer program.

Problems

44.1 Repeat the computation of the pile forces, see Figure 44-3, for the case that the stiffness of pile row 4 is half the stiffness of the other pile rows. Predict the pile force in row 1.

44.2 Can a computer program for the analysis of space frames be used for the computation of pile forces in a pile foundation?

Part VIII
Slope and stability

45 Vertical slope in cohesive material

A well known and important problem of soil mechanics is the case of a vertical cutoff in a purely cohesive material ($\phi = 0$), as occurs when making a vertical excavation, or a vertical slope, see Figure 45-1. The problem to be considered in this chapter is the determination of a lower bound or an upper bound for the maximum possible height h_c of the slope, for a material having a constant cohesive strength c, and a constant volumetric weight γ.

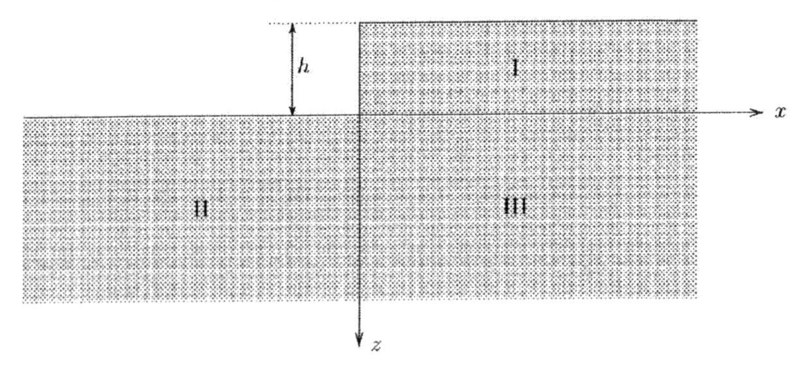

Figure 45-1. Vertical slope.

45.1 Lower bound

In this case it is essential that the weight of the material is taken into account. The equations of equilibrium now are

$$\frac{\partial \sigma_{xx}}{\partial x} + \frac{\partial \sigma_{zx}}{\partial z} = 0, \tag{45.1}$$

$$\frac{\partial \sigma_{xz}}{\partial x} + \frac{\partial \sigma_{zz}}{\partial z} - \gamma = 0. \tag{45.2}$$

A simple equilibrium system is shown in Figure 45-2, consisting of three zones. On the interfaces between the zones the normal stresses parallel to these interfaces may be discontinuous, without disturbing equilibrium, see Chapter 41. The boundary conditions for the stresses are that the normal stresses and the shear stresses are zero, all along the upper surface. These conditions are exactly satisfied by the stress fields indicated in Figure 45-2. This field can be constructed by starting to assume that in the entire field the shear stress $\sigma_{xz}=0$, because this shear stress must be zero on the two horizontal boundaries, and on the vertical slope. In order to satisfy the condition that on the vertical slope the horizontal stress $\sigma_{xx}=0$, it follows from the equation of horizontal equilibrium, eq. (45.1) that this stress must be zero throughout zone I. The expressions for the vertical normal stress σ_{zz} follow immediately from the equation of vertical equilibrium (45.2) by putting $\sigma_{zx}=0$, and using the boundary conditions at the top of the soil. The expressions for the horizontal stress σ_{xx} in zones II and III can be chosen arbitrarily, but they must be constant in x-direction (to satisfy horizontal equilibrium), and preferably as close to σ_{zz} as possible, to keep the maximum shear stress as small as possible. By choosing $\sigma_{xx}=\gamma z$ the Mohr circle in zone II, in the lower left part, reduces to a point. This seems to be very attractive, but the consequence is that in zone III, the lower right part, in zone II, the difference of the stresses σ_{xx} and σ_{zz} is rather large, $\sigma_{zz}-\sigma_{xx}=\gamma h$.

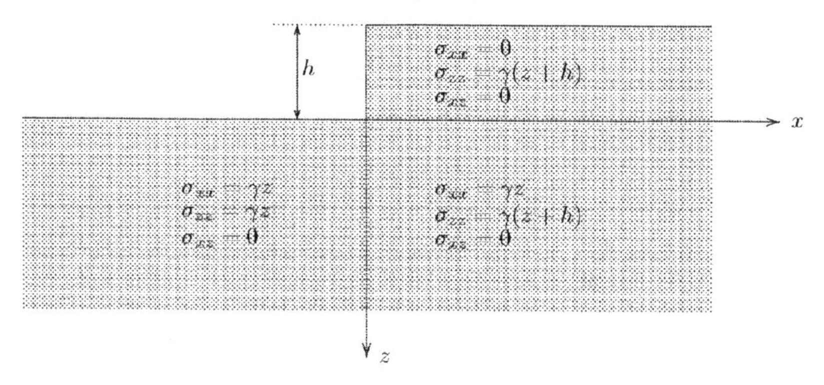

Figure 45-2. Equilibrium system.

The vertical and horizontal stresses are principal stresses in this case, because everywhere $\sigma_{xz}=0$. Therefore, the Mohr-Coulomb criterion now is $|\sigma_{xx}-\sigma_{zz}|\leq 2c$. As the largest value of $\sigma_{xx}-\sigma_{zz}$ occurs in the lower right part, it follows that the critical value of the height $h=2c/\gamma$.
This is a lower bound, i.e.

$$h_c \geq \frac{2c}{\gamma}. \tag{45.3}$$

In soil mechanics literature some higher values for a lower bound can be found, on the basis of more complex stress fields. De Josselin de Jong (1965) and Heyman (1973) obtained

$$h_c \geq \frac{2.82c}{\gamma}. \tag{45.4}$$

In the seventies of the 20th century De Josselin de Jong obtained gradually higher values, up to

$$h_c \geq \frac{3.39c}{\gamma}. \tag{45.5}$$

An even higher value was obtained by Pastor, in 1978,

$$h_c \geq \frac{3.64c}{\gamma}. \tag{45.6}$$

All these values are correct lower bounds. A higher value then Pastor's limit has not yet been found.

45.2 Upper bound

A simple upper bound can be found by considering a mechanism consisting of a single straight slip surface, at an angle α with the vertical direction, see Figure 45-3. The weight of the sliding wedge is $W = \frac{1}{2}\gamma h^2 \tan\alpha$, and it follows from the condition of equilibrium in the direction of sliding that

$$T = W\cos\alpha = \tfrac{1}{2}\gamma h^2 \sin\alpha.$$

Because the length of the slip plane is $h/\cos\alpha$ it follows that

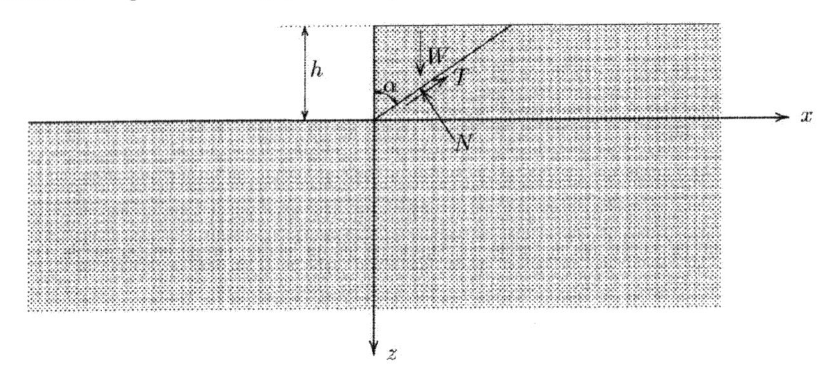

Figure 45-3. Mechanism with straight slip surface.

$$T = \frac{ch}{\cos\alpha}.$$

Combination of these two equations gives

$$h = \frac{4c}{\gamma}\frac{1}{\sin 2\alpha}. \tag{45.7}$$

The height of the excavation appears to depend upon the angle α. The critical sliding plane is the one for which h is a minimum. This minimum occurs if $\sin 2\alpha$ has its maximum value, i.e. $2\alpha = \frac{1}{2}\pi$, or $\alpha = 45°$. Because this is an upper bound it follows that

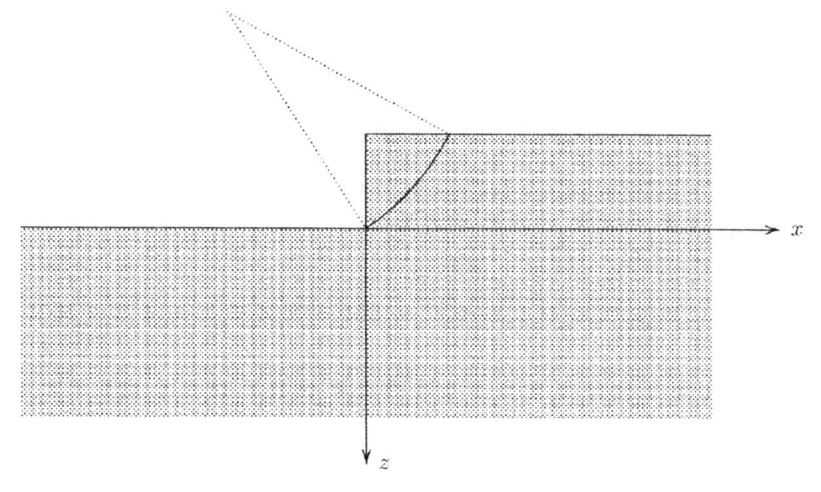

Figure 45-4. Fellenius' mechanism.

$$h_c \leq \frac{4c}{\gamma}. \tag{45.8}$$

This is the upper bound for straight slip surfaces.
Using circular slip planes Fellenius has found a lower upper bound, in 1927. Its value is

$$h_c \leq \frac{3.83c}{\gamma}. \tag{45.9}$$

see Figure 45-4. (The axis of rotation is located 1.41 h left and 2.21 h above the bottom corner.) There have been many attempts to find a lower value, but they have not been successful.
It can be concluded that for the problem of the critical height h_c of a vertical excavation or slope in a cohesive material, without internal friction ($\phi = 0$), it has been shown that

$$\frac{3.64c}{\gamma} \leq h_c \leq \frac{3.83c}{\gamma}. \tag{45.10}$$

This encloses the failure load between rather narrow bounds, but there still is some room for improvement. It may be surprising that even for this simple case, of homogeneous soil, with constant strength c, and zero friction, there is still some difference between the highest lower bound and the smallest upper bound. The reason for this lies probably in the fact that for the lower bound the Mohr-Coulomb criterion has been used, while for the upper bound the Coulomb criterion has been used, see Chapter 21.

Problems

45.1 In the equilibrium system of Figure 45-2 the maximum shear stress in the lower right part (zone III) can be made equal to zero by taking $\sigma_{xx} = \gamma(z+h)$. Does this mean that the stresses in the other two zones must be adjusted as well?

45.2 Determine the lower bound for the equilibrium system mentioned in the previous problem.

45.3 In a soil layer having a cohesion of 10 kPa a telephone cable must be placed at a depth of 2 m. Can this be done in a trench with vertical walls?

45.4 By mixing sand with long plastic fibres a material is obtained with cohesion and friction. The vendor claims that the cohesion may be 100 kPa. What can be the height of a vertical slope in this material?

46 Stability of infinite slope

The evaluation of the stability of a slope, of an embankment or a dike, is an important problem of applied soil mechanics. In the previous chapter this problem has been considered for a vertical slope in a purely cohesive material ($c>0, \phi=0$). As a preparation for the general case, which will be considered in the next chapter, this chapter will present some solutions for slopes of infinite extent, in a homogeneous frictional material, without cohesion ($c=0, \phi>0$).

46.1 Infinite slope in dry sand

Consider an infinitely long slope, in dry sand, at inclination α, see Figure 46-1. The equations of equilibrium can now best be expressed using coordinates parallel and perpendicular to the slope,

$$\frac{\partial \sigma_{\xi\xi}}{\partial \xi}+\frac{\partial \sigma_{\eta\xi}}{\partial \eta}+\gamma_d \sin\alpha=0, \tag{46.1}$$

$$\frac{\partial \sigma_{\xi\eta}}{\partial \xi}+\frac{\partial \sigma_{\eta\eta}}{\partial \eta}-\gamma_d \cos\alpha=0. \tag{46.2}$$

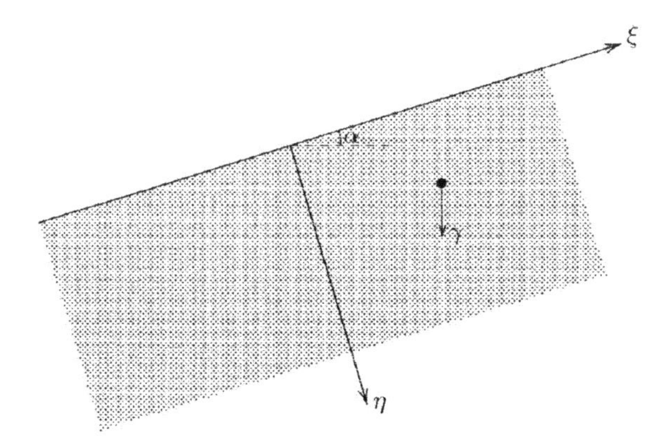

Figure 46-1. Infinite slope in dry sand.

The stresses in these equations are total stresses, but as there are no pore pressures, they are effective stresses as well, in this case of a dry soil. The state of stress is not uniquely determined by the equilibrium conditions. One of the possible solutions can be obtained by assuming that the state of stress is independent of ξ, the coordinate along the slope. That seems to be a reasonable assumption, because the slope extends towards infinity both in upward and in downward direction. There is no absolute need for the independence of ξ, however, and it is not more than an assumption. Using this assumption the equilibrium equations give, when expressed in effective stresses,

$$\sigma'_{\eta\xi} = -\gamma_s \eta \sin\alpha, \tag{46.3}$$

$$\sigma'_{\eta\eta} = +\gamma_s \eta \cos\alpha. \tag{46.4}$$

The integration constants have been taken as zero, because at the surface $\eta = 0$ the stresses $\sigma'_{\eta\eta}$ and $\sigma'_{\eta\xi}$ must be zero. It follows that

$$\frac{|\sigma'_{\eta\xi}|}{|\sigma'_{\eta\eta}|} = \tan\alpha. \tag{46.5}$$

A *stability factor* can be introduced as

$$F = \frac{\text{strength}}{\text{load}}. \tag{46.6}$$

The factor F may also be called the *safety factor*. For the infinite slope this is

$$F = \frac{|\sigma'_{\eta\xi}/\sigma'_{\eta\eta}|_{\max}}{|\sigma'_{\eta\xi}/\sigma'_{\eta\eta}|}. \tag{46.7}$$

The Coulomb failure criterion states that in a cohesionless material ($c = 0$) this ratio can not be larger than $\tan\delta$. This means that α can not be larger than δ, so: $\alpha < \delta$ and:

$$F = \frac{\tan\delta}{\tan\alpha}. \tag{46.8}$$

Although it is inaccurate and unsafe (see Chapter 24), the assumption $\delta \approx \phi$ is often made, which results in

$$F \approx \frac{\tan\phi}{\tan\alpha}. \tag{46.9}$$

If $\alpha < \phi$ then F is greater than 1, so the slope is stable. If $\alpha > \phi$, the value of F is smaller than one, so the slope is unstable.

It should be noted that the stability factor F appears to be independent of the volumetric weight γ. That is a characteristic of frictional materials. In case of loading by the weight of a frictional material the load is proportional to the volumetric weight, but so is the strength. The result is that the volumetric weight cancels in the ratio, so that the safety is independent of the volumetric weight.

It has been seen that the steepest possible slope in dry sand is ϕ. This property can be used as a simple method to determine the value of the friction angle ϕ of dry sand: it is the inclination of the steepest slope. It should be noted that this property holds only for a soil without cohesion, and a completely dry sand. A small amount of water can easily disturb it.

46.2 Infinite slope under water

For the case of a very long slope under water, see Figure 46-2, the critical slope can be determined as follows.

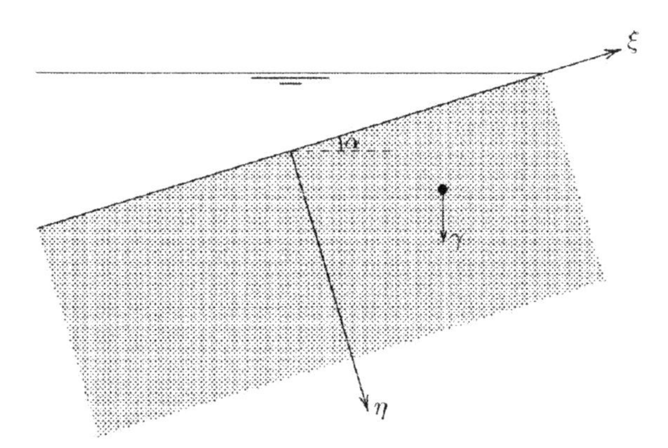

Figure 46-2. Infinite slope under water.

Equilibrium is again described by the equations (46.1) and (46.2), but in this case there is a certain pore pressure. Terzaghi's effective stress principle the total stresses can be expressed by $\sigma_{\xi\xi} = \sigma'_{\xi\xi} + p$ and $\sigma_{\eta\eta} = \sigma'_{\eta\eta} + p$, so that the equations of equilibrium can be expressed in terms of effective stresses as

$$\frac{\partial \sigma'_{\xi\xi}}{\partial \xi} + \frac{\partial \sigma'_{\eta\xi}}{\partial \eta} + \frac{\partial p}{\partial \xi} + \gamma_s \sin\alpha = 0, \tag{46.10}$$

$$\frac{\partial \sigma'_{\xi\eta}}{\partial \xi} + \frac{\partial \sigma'_{\eta\eta}}{\partial \eta} + \frac{\partial p}{\partial \eta} - \gamma_s \cos\alpha = 0. \tag{46.11}$$

If the groundwater is at rest, the pressure distribution is hydrostatic. If the z-axis is directed vertically upward, the pressures in the groundwater can be written as

$$p = p_0 - \gamma_w z = p_0 + \gamma_w \eta \cos\alpha - \gamma_w \xi \sin\alpha. \tag{46.12}$$

The reference pressure p_0 in this expression is the pressure at the level $z = 0$. If the entire slope is located under water, the phreatic surface (the level at which $p = 0$) must be located at an infinite height. The pore pressure at the level $z = 0$ then is infinitely large, $p_0 = \infty$. The present example is not completely realistic, which is a consequence of considering an *infinite* slope. At its best is the limiting form of a very long slope.

Substitution of (46.12) into (46.10) and (46.11) gives

$$\frac{\partial \sigma'_{\xi\xi}}{\partial \xi}+\frac{\partial \sigma'_{\eta\xi}}{\partial \eta}+(\gamma_s-\gamma_w)\sin\alpha=0, \tag{46.13}$$

$$\frac{\partial \sigma'_{\xi\eta}}{\partial \xi}+\frac{\partial \sigma'_{\eta\eta}}{\partial \eta}-(\gamma_s-\gamma_w)\cos\alpha=0. \tag{46.14}$$

These are precisely the same equations as in the dry case, except that γ_d has been replaced by $\gamma_s-\gamma_w$. Because it was found earlier that the stability factor F is independent of γ_d, see eq. (46.8), it follows that this is also valid in this case of a slope under water, i.e.

$$F\approx\frac{\tan\phi}{\tan\alpha}. \tag{46.15}$$

It appears that a slope under water can also be maintained at an inclination ϕ. This conclusion seems to be in contradiction with experimental evidence, which suggests that a slope under water usually is less steep than a slope above water, in the same material. A possible explanation is that under water other processes may disturb the stability of a slope, such as erosion by waves or by flowing groundwater. In a basin with water a slope at rest can indeed be as steep as a slope in dry sand.

46.3 Flow parallel to the slope

An interesting problem is the stability of an embankment or dam in which groundwater flows parallel to the slope, in downward direction, see Figure 46-3. This may occur in a dike that is just not high enough to retain the water in a river, so that water flows over the slope. This water penetrates into the dike material, and after some time a flow of groundwater parallel to the slope may be created, as shown in Figure 46-4.

If the flow is uniform the pressure distribution must be linear in ξ and η, i.e.

$$p=A\eta+B\xi+C.$$

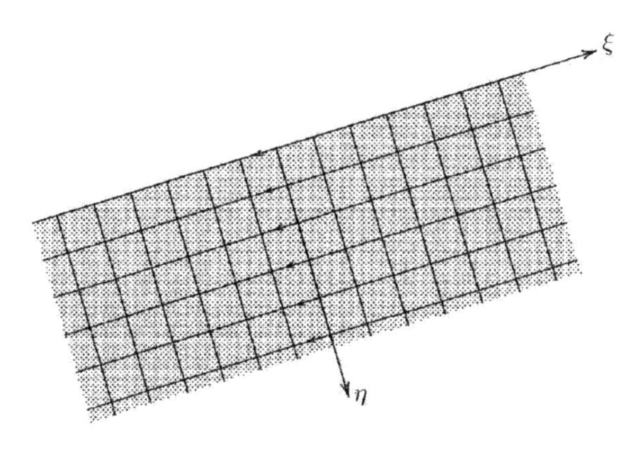

Figure 46-3. Parallel groundwater flow.

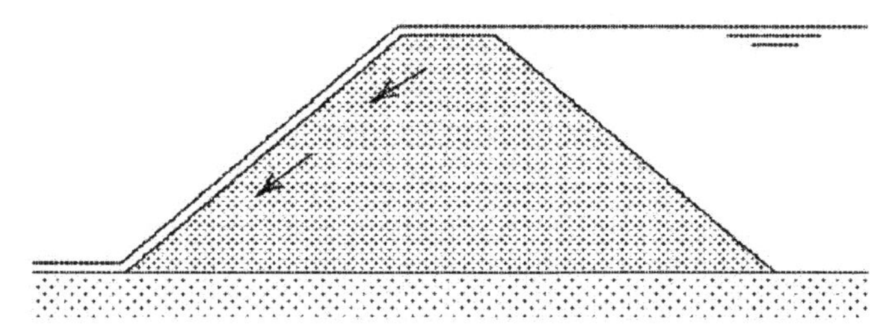

Figure 46-4. Parallel flow.

Along the surface the pressure must be zero (this will be the case if the soil is saturated, with merely a thin film of water flowing over the slope), i.e. $p=0$ for $\eta=0$. It then follows that $B=C=0$, so that the pressure distribution reduces to $p=A\eta$. This means that the groundwater head h is

$$h=z+\frac{p}{\gamma_w}=-\eta\cos\alpha+\xi\sin\alpha+A\frac{\eta}{\gamma_w}.$$

If the flow is parallel to the soil surface the component of the specific discharge vector perpendicular to the surface must be zero, $q_\eta=0$, and therefore $\partial h/\partial\eta=0$. It follows that $A=\gamma_w\cos\alpha$, so that the pressure p is

$$p=\gamma_w\eta\cos\alpha. \tag{46.16}$$

Substitution of this pressure distribution into the equations of equilibrium (46.10) and (46.11) gives

$$\frac{\partial\sigma'_{\xi\xi}}{\partial\xi}+\frac{\partial\sigma'_{\eta\xi}}{\partial\eta}+\gamma_s\sin\alpha=0, \tag{46.17}$$

$$\frac{\partial\sigma'_{\xi\eta}}{\partial\xi}+\frac{\partial\sigma'_{\eta\eta}}{\partial\eta}-(\gamma_s-\gamma_w)\cos\alpha=0. \tag{46.18}$$

A solution independent of ξ is

$$\sigma'_{\eta\xi}=-\gamma_s\eta\sin\alpha, \tag{46.19}$$

$$\sigma'_{\eta\eta}=(\gamma_s-\gamma_w)\eta\cos\alpha. \tag{46.20}$$

In this case the stability factor F is

$$F\approx\frac{\gamma_s-\gamma_w}{\gamma_s}\frac{\tan\phi}{\tan\alpha}. \tag{46.21}$$

Because $(\gamma_s-\gamma_w)/\gamma<1$ (usually about 0.5), it follows that the steepest possible slope in this case is much smaller than ϕ. The groundwater flow appears to have a negative influence on the stability of the slope.

It must be concluded that it is very unfavourable for the stability of the downstream slope of a dike if groundwater flows down the slope, parallel to the slope. This may occur in the case of groundwater exiting the slope along a seepage surface, or if the level of the free water at the upstream side of the dike is so high that it flows over the dike, and penetrates into the downstream slope. This mechanism is considered to have been responsible for the failure of many dikes in the 1953 flood in the South-West of the Netherlands.

46.4 Horizontal outflow

Another interesting example is a dike in which groundwater is flowing in horizontal direction, see Figure 46-5. If groundwater flows through the dike in horizontal direction, the groundwater head is independent of z, $\partial h/\partial z = 0$. Because $h = z + p/\gamma_w$ it then follows that $\partial p/\partial z = -\gamma_w$. Furthermore, along the surface, that is for $z = x\tan\alpha$, the pressure p must be zero. And if the flow is uniform the pressure distribution must be linear. The only pressure distribution that satisfies all these conditions is

$$p = \gamma_w x\tan\alpha - \gamma_w z. \tag{46.22}$$

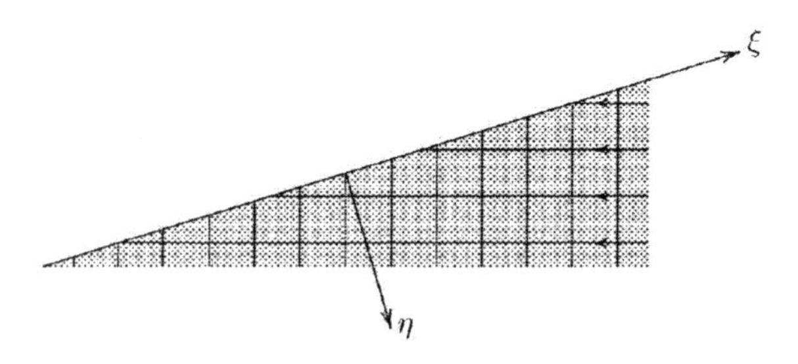

Figure 46-5. Horizontal groundwater flow.

This can be expressed into ξ and η using the transformation formulas for rotation of the coordinates,

$$x = \xi\cos\alpha + \eta\sin\alpha, \quad z = -\eta\cos\alpha + \xi\cos\alpha.$$

The result is

$$p = \gamma_w \eta/\cos\alpha. \tag{46.23}$$

Substitution into the equations of equilibrium (46.10) and (46.11) in this case gives

$$\frac{\partial\sigma'_{\xi\xi}}{\partial\xi} + \frac{\partial\sigma'_{\eta\xi}}{\partial\eta} + \gamma_s\sin\alpha = 0, \tag{46.24}$$

$$\frac{\partial\sigma'_{\xi\eta}}{\partial\xi} + \frac{\partial\sigma'_{\eta\eta}}{\partial\eta} - \gamma_s\cos\alpha - \gamma_w/\cos\alpha = 0. \tag{46.25}$$

A solution independent of ξ is

$$\sigma'_{\eta\xi} = -\gamma_s\eta\sin\alpha, \tag{46.26}$$

$$\sigma'_{\eta\eta} = (\gamma_s - \frac{\gamma_w}{\cos^2\alpha})\eta\cos\alpha. \tag{46.27}$$

The stability factor F now is

$$F \approx \frac{\gamma_s - (\gamma_w/\cos^2\alpha)}{\gamma_s}\frac{\tan\phi}{\tan\alpha}. \tag{46.28}$$

This value is even smaller than the value in the previous case, see (46.21), because the value of $\cos^2\alpha$ is always smaller than 1. It follows that a horizontal outflow of groundwater is even more dangerous than a flow parallel to the slope.
This case can be considered to occur, approximately, for a permeable dike or dam on an impermeable base. Large dams are often built on such an impermeable base, to prevent leakage from the lake through the subsoil. If the dam were built from homogeneous material, see Figure 46-6, groundwater will exit from the dam at the downstream slope, with a practically horizontal flow through the dam. This is a very unfavourable situation, and should be avoided.

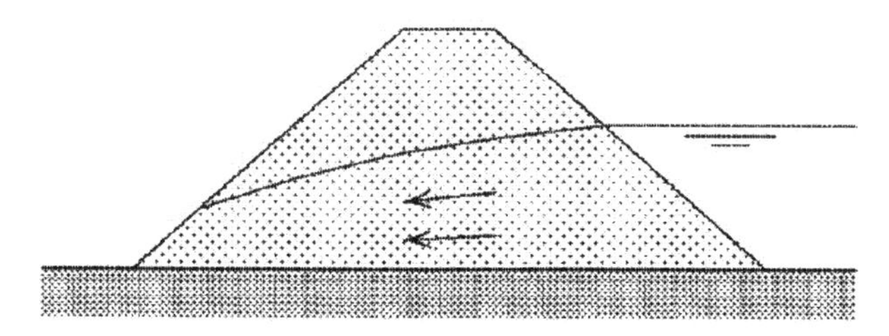

Figure 46-6. Flow through dike.

There are two good technical solutions. The first solution is to place a blanket of almost impermeable material (clay) on the upstream slope, or, even better, to construct a core of clay in the centre of the dam. This is better because it can not be damaged by poor maintenance or accidental damage. The second solution is to construct a filter at the toe of the dam or dike, consisting of very permeable material (for instance gravel). Such a filter will attract the groundwater and drain it away. Great care should to be taken to maintain the high permeability of the filter. Of course, the best solution is to apply the two solutions: a clay core in the centre, and a filter in the downstream toe. Failure of a large dam is such a catastrophe that it should be avoided at all cost.

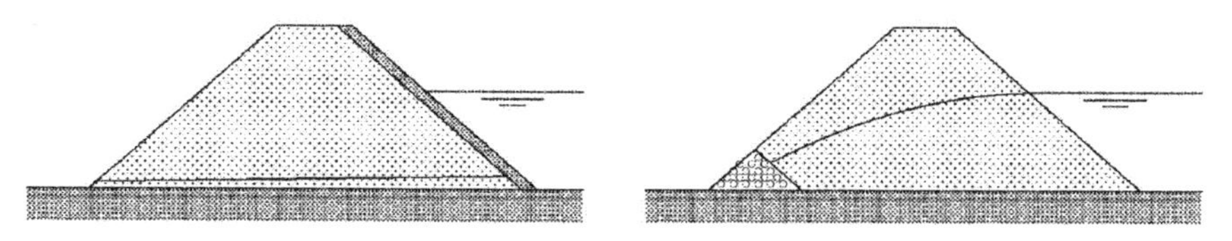

Figure 46-7. Dam with a clay blanket, or with a drain.

47 Slope stability

For the analysis of the stability of slopes of arbitrary shape and composition various approximate methods have been developed. Most of these assume a circular slip surface. Using a number of simplifying assumptions a value for the safety factor F, the ratio of strength and load, is determined. The circle giving the smallest of F is considered to be critical. The multitude of methods (developed by Fellenius, Taylor, Bishop, Morgenstern-Price, Spencer, Janbu, among others) in itself illustrates that none of them is exact. The results should always be handled with care. A value $F = 1.05$ gives no absolute certainty that the slope will stand. In this chapter two of the simplest methods will be presented.

47.1 Circular slip surface

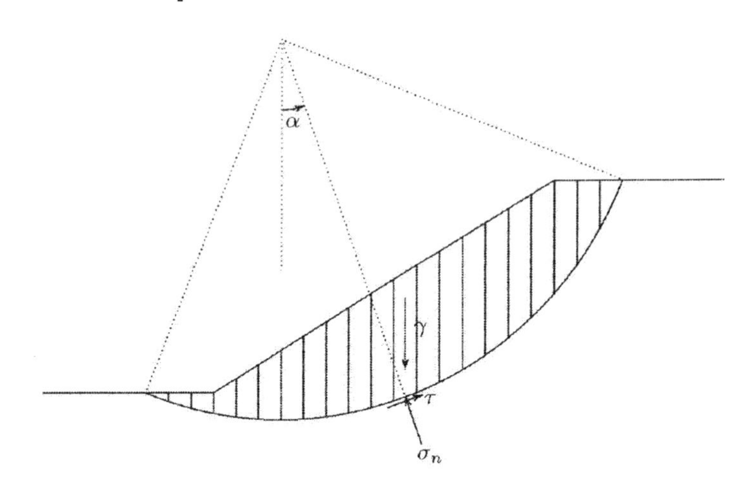

Figure 47-1. Circular slip surface.

Most methods assume that the soil fails along a circular slip surface, see Figure 47-1. The soil above the slip surface is subdivided into a number of *slices*, bounded by vertical interfaces. At the slip surface the shear stress is τ, which is assumed to be a factor F smaller than the maximum possible shear stress, i.e.

$$\tau = \frac{1}{F}(c + \sigma_n' \tan\phi). \tag{47.1}$$

The factor F is assumed to be the same for all slices, an assumption that is common to all methods. A second assumption is that $\phi = \delta$, which is an inaccurate estimation

according to Chapter 42.
The equilibrium equation to be used in conjunction with a circular slip surface is the equation of equilibrium of moments with respect to the centre of the circle. This equation gives

$$\sum \gamma h b R \sin\alpha = \sum \frac{\tau b R}{\cos\alpha}. \tag{47.2}$$

Here h is the height of a slice, b its width, γ the volumetric weight of the soil in the slice, and R is the radius of the circle. More generally it can be defined that γbh is the weight of the slice, possibly consisting of a sum of parts with different unit weight.
If all slices have the same width, it now follows from (47.1) and (47.2) that

$$F = \frac{\sum[(c+\sigma'_n \tan\phi)/\cos\alpha]}{\sum \gamma h \sin\alpha}. \tag{47.3}$$

This is the basic formula for many computation methods. The various methods usually differ in the method of calculating the normal effective stress σ'_n.

47.2 Fellenius

In Fellenius' method, the oldest method for the analysis of slope stability, it is assumed that there are no forces between the slices. The only remaining forces acting on a slice, see Figure 47-2, then are the weight γbh, a normal stress σ_n and a shear stress τ at the bottom of the slice. The normal stress σ_n can most conveniently be expressed into the known weight by considering the equilibrium of the slice in the direction perpendicular to the slip surface. This gives

$$\sigma_n = \gamma h \cos^2\alpha, \tag{47.4}$$

and, because $\sigma_n = \sigma'_n + p$,

$$\sigma'_n = \gamma h \cos^2\alpha - p. \tag{47.5}$$

Figure 47-2. Fellenius.

Substitution into (47.3) finally gives

$$F = \frac{\sum\{[c + (\gamma h \cos^2\alpha - p)\tan\phi]/\cos\alpha\}}{\sum \gamma h \sin\alpha}. \qquad (47.6)$$

This is the Fellenius formula.
For a slope in homogeneous soil the computation can be executed by assuming a certain location of the circle, and subdividing the sliding soil wedge into 10 or 20 slices. By measuring the values of α and h for each slice the value of the stability factor F can be determined. This must be repeated for a large number of circles, to determine the smallest value of F. In non-homogeneous soil the computation is somewhat more complicated because for each slice the value of γh must be determined as the sum of the contributions
of a number of layers in the slice.
Several objections can be made against this method. To begin with, a sound fundamental base lacks for all slip surface methods for materials with internal friction, as seen before. Therefore it is unknown what is the ratio between the stability factor of the most unfavourable circular slip surface and the real safety of the slope. But there are other objections as well. Disregarding the forces transmitted between the slices is a severe approximation, and vertical equilibrium is violated. Furthermore, there is an internal inconsistency in stating on the one hand that sliding occurs along the circle, and on the other hand stating that the horizontal and vertical directions are the directions of principal stress (as it is assumed that there are no shear stresses on vertical planes). This inconsistency can best be seen by considering the slice in the centre, for which $\alpha = 0$. At that slice $\sigma_n = \gamma h$, and it is assumed that there is a shear stress $(\sigma_n - p)/F$ on that slice. This violates the assumption that the vertical direction is a direction of principal stress. Horizontal equilibrium of that slice is also clearly violated. For other slices vertical equilibrium is violated, as only the condition of equilibrium perpendicular to the slip surface is taken into account.
Fellenius' method has the property that in a number of special cases it confirms certain limiting values. For instance, for an infinite slope in a dry frictional material without cohesion, one obtains from (47.6), assuming a straight slip surface at a depth d below the slope, and taking $p = c = 0$,

$$F = \frac{\sum \gamma d \cos\alpha \tan\phi}{\sum \gamma d \sin\alpha} = \frac{\tan\phi}{\tan\alpha}.$$

This is in perfect agreement with formula (46.9) in the previous chapter.
In the case of a slope under water, in the absence of groundwater flow, see Figure 46-2, the limiting value (46.15) is not immediately recovered. For such problems the Fellenius formula might be modified by using the volumetric weight under water, $(\gamma_s - \gamma_w)h$ rather than γh, and using the excess water pressure with respect to the hydrostatic water pressure for p. This is somewhat artificial, however, and for this reason the Fellenius method is rarely used.

47.3 Bishop

A method that is frequently used in engineering practice is Bishop's method. In this method the forces between the slices are not neglected, but it is assumed that the resultant force is horizontal, see Figure 47-3. By considering the vertical equilibrium of each slice only, the horizontal forces do not enter into the computations, however. The basic equation again is the equation of moment equilibrium, eq. (47.3). Vertical equilibrium of a slice now requires that

$$\gamma h = \sigma_n + \tau \frac{\sin\alpha}{\cos\alpha} = \sigma_n' + p + \tau \frac{\sin\alpha}{\cos\alpha}.$$

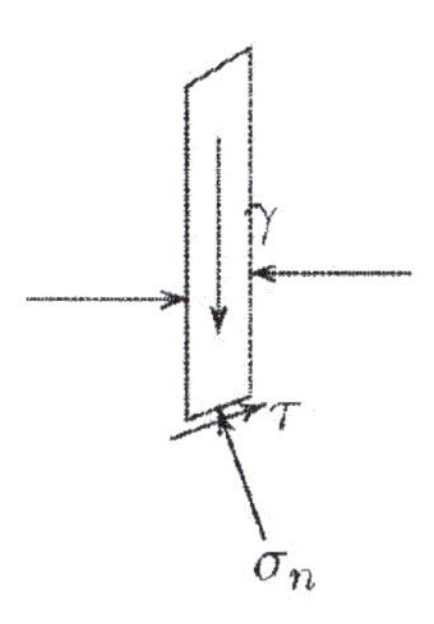

Figure 47-3. Bishop.

If in this equation the value of τ is written, in agreement with (47.1), as $\tau = (c + \sigma_n' \tan\phi)/F$, the result is

$$\sigma_n'(1 + \frac{\tan\alpha \tan\phi}{F}) = \gamma h - p - \frac{c}{F}\tan\alpha. \tag{47.7}$$

Substitution of σ_n' into (47.3) now leads to the final equation for Bishop's method,

$$F = \frac{\sum \frac{c + (\gamma h - p)\tan\phi}{\cos\alpha(1 + \tan\alpha \tan\phi / F)}}{\sum \gamma h \sin\alpha}. \tag{47.8}$$

Because the stability factor F also appears in the right hand side, it must be determined iteratively, by starting from an initial estimate (for instance $F = 1$), and then calculating an updated value using (47.8). This must be repeated until the value of F no longer changes. In general the procedure converges rather fast. As the computations must be executed by a computer program anyhow (many circles have to be investigated) the iterations can easily be incorporated into the program.

If $\phi = 0$ the Bishop and Fellenius methods are identical. If $\phi > 0$ Bishop's method usually gives somewhat smaller values. Because Bishop's method is more consistent (vertical equilibrium is satisfied), and it confirms known results for special cases, it is often used in geotechnical engineering. Various other methods have been developed, but the results often differ only slightly from those obtained by Bishop's method. That may explain its popularity.

Problems

47.1 Verify that Fellenius' method gives the correct limiting value for an infinite slope with a groundwater flow parallel to the slope.

47.2 Verify that Bishop's method gives the correct limiting values for the special cases considered in the previous chapter.

47.3 How can the effect of an earthquake be incorporated into Bishop's method?

Part IX
Appendices

A Stress analysis

In this appendix the main principles of stress analysis are presented. This includes the graphical method of representing the transformation formulas of the stress tensor by Mohr's circle. The considerations are restricted to two-dimensional states of stress, for reasons of simplicity.

A.1 Transformation formulas

Suppose that the state of stress in a certain point is described by the stresses σ_{xx}, σ_{xy}, σ_{yx} and σ_{yy}, see Figure A-1. Following the usual sign convention of applied mechanics a component of stress is considered positive when the force component on a plane whose outward normal vector is directed in a positive coordinate direction, acts in positive direction as well, or when the force component acts in negative direction on a plane whose outward normal vector is directed in negative coordinate direction. For normal stresses this means that tension is considered positive, and pressure is considered negative. In soil mechanics the usual sign convention is just the opposite. The difference is expressed in that in this appendix stresses are denoted by the symbol τ, whereas in the main text of the book stresses are denoted by σ. Formally, the relation is

$$\begin{aligned} \sigma_{xx} &= -\tau_{xx}, \\ \sigma_{xy} &= -\tau_{xy}, \\ \sigma_{yx} &= -\tau_{yx}, \\ \sigma_{yy} &= -\tau_{yy}. \end{aligned} \tag{A.1}$$

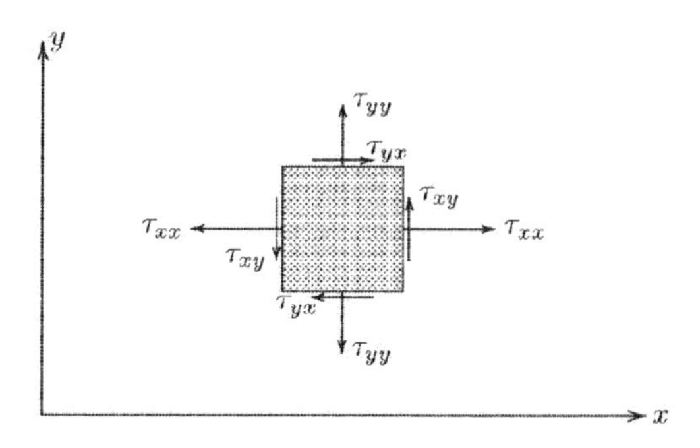

Figure A-1 Stress in two dimensions.

The state of stress in a certain point is completely defined by the four stress components τ_{xx}, τ_{xy}, τ_{yx} and τ_{yy}. Of these four stresses the shear stresses are equal, as can be shown by considering equilibrium of moments with respect to the centre of the element,

$$\tau_{xy} = \tau_{yx}. \tag{A.2}$$

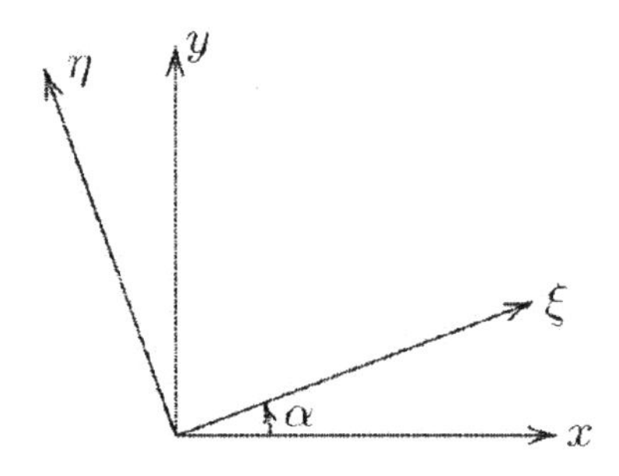

Figure A-2. Rotation of axis.

In many situations it is necessary to describe the stress transfer by also considering planes other than those in the directions of the Cartesian coordinates x and y. The stress state should then be described in a rotated set of coordinate axes, denoted by ξ and η, rotated with respect to the original coordinates over an angle α, see Figure A-2. The transformation formulas can be derived most conveniently by considering equilibrium of a suitably chosen elementary triangle, see Figure A-3.

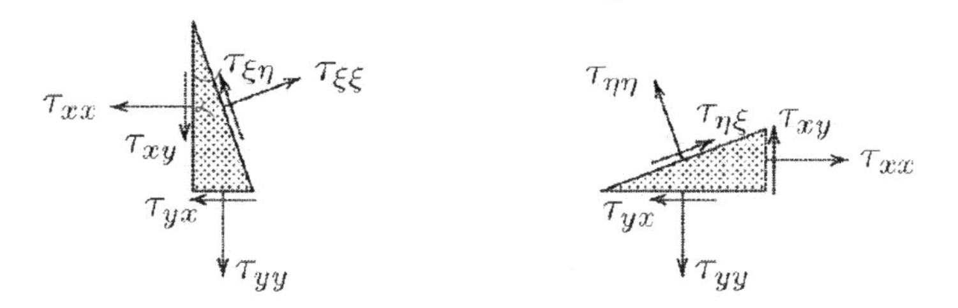

Figure A-3 Stress in a rotated system of coordinates.

In formulating the equilibrium conditions it should be remembered that the basis of this principle is equilibrium of forces, not stresses. This means that the magnitude of the various planes on which the various stress components act should be taken into account. By formulating the conditions of equilibrium in ξ-direction and in η-direction of a suitably chosen triangular element, with only two unknown stress components in the pair of equations, see Figure A-3, it follows that

$$\begin{aligned}
\tau_{\xi\xi} &= \tau_{xx}\cos^2\alpha + \tau_{yy}\sin^2\alpha + 2\tau_{xy}\sin\alpha\cos\alpha, \\
\tau_{\eta\eta} &= \tau_{xx}\sin^2\alpha + \tau_{yy}\cos^2\alpha - 2\tau_{xy}\sin\alpha\cos\alpha, \\
\tau_{\xi\eta} &= \tau_{\eta\xi} = \tau_{xy}(\cos^2\alpha - \sin^2\alpha) + (\tau_{yy} - \tau_{xx})\sin\alpha\cos\alpha.
\end{aligned} \tag{A.3}$$

These equations can be written in a more convenient form when using the parameter 2α,

$$\tau_{\xi\xi} = (\frac{\tau_{xx}+\tau_{yy}}{2}) + (\frac{\tau_{xx}-\tau_{yy}}{2})\cos 2\alpha + \tau_{xy}\sin 2\alpha,$$

$$\tau_{\eta\eta} = (\frac{\tau_{xx}+\tau_{yy}}{2}) - (\frac{\tau_{xx}-\tau_{yy}}{2})\cos 2\alpha - \tau_{xy}\sin 2\alpha, \qquad (A.4)$$

$$\tau_{\xi\eta} = \tau_{\eta\xi} = \tau_{xy}\cos 2\alpha - (\frac{\tau_{xx}-\tau_{yy}}{2})\sin 2\alpha.$$

A.2 Principal directions

For certain values of the rotation angle α the shear stresses $\tau_{\xi\eta}$ and $\tau_{\eta\xi}$ are zero. This means that there are certain planes on which only a normal stress is acting, and no shear stress. The directions normal to these planes are called the *principal directions* of the stress tensor. The value of α for which the shear stress is zero will be denoted by α_0. Its value can be determined by setting the last equation of (A.4) equal to zero. This gives

$$\tan 2\alpha_0 = \frac{\tau_{xy}}{\frac{1}{2}(\tau_{xx}-\tau_{yy})}. \qquad (A.5)$$

Because of the periodic property of the function $\tan 2\alpha_0$ it follows that there are two solutions, which differ by a factor $\frac{1}{2}\pi$. The corresponding values of the normal stresses can be found by substitution of this value of α into the first two equations of the system (A.4). These normal stresses are denoted by τ_1 and τ_2, the *principal stresses*. It is assumed that τ_1 is the largest of these two stresses, the *major* principal stress, and τ_2 is the smallest of the two stresses, the *minor* principal stress. Using some trigonometric relations, it can be shown that

$$\tau_{1,2} = (\frac{\tau_{xx}+\tau_{yy}}{2}) \pm \sqrt{(\frac{\tau_{xx}-\tau_{yy}}{2})^2 + \tau_{xy}^2}. \qquad (A.6)$$

The notions of principal stress and principal direction introduced here are special cases of the more general properties of *eigen value* and *eigen vector* of matrices and tensors.

A.3 Mohr's circle

The formulas derived above can be represented in a simple graphical form, using *Mohr's circle*. For this purpose it is most convenient to use the transformation formulas in the form (A.4), but expressed into the principal stresses. The orientation of the x-axis with respect to the direction of the major principal stress is denoted by γ, see Figure A-4. The directions of the major and the minor principal stresses are indicated by 1 and 2. The transformation formulas for the transition from the axes 1 and 2 to the axes x and y can easily be obtained from the formulas, by replacing x and y by 1 and 2 (with $\tau_{12}=0$), and replacing ξ and η by x and y, and the angle α by γ. The result is

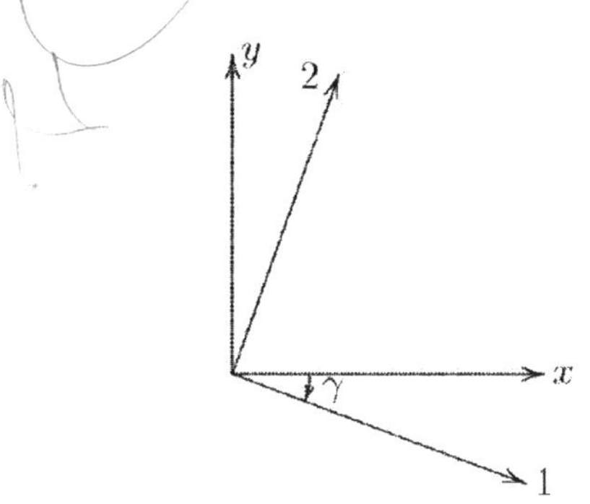

Figure A-4 Rotation of axis.

$$\begin{aligned}\tau_{xx} &= (\frac{\tau_1+\tau_2}{2})+(\frac{\tau_1-\tau_2}{2})\cos 2\gamma,\\ \tau_{yy} &= (\frac{\tau_1+\tau_2}{2})-(\frac{\tau_1-\tau_2}{2})\cos 2\gamma,\\ \tau_{xy} &= \tau_{yx} = -(\frac{\tau_1-\tau_2}{2})\sin 2\gamma.\end{aligned} \tag{A.7}$$

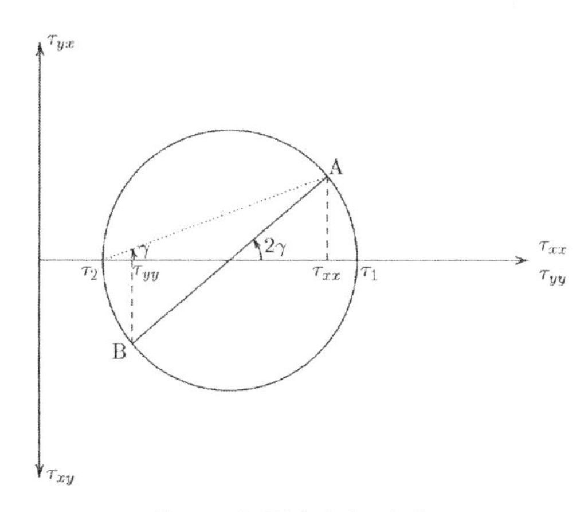

Figure A-5 Mohr's circle.

These formulas admit a simple graphical interpretation, see Figure A-5. In this figure, Mohr's diagram, the normal stresses τ_{xx} and τ_{yy} are plotted positive towards the right. The shear stress τ_{yx} is plotted positive in upward direction, and the shear stress τ_{xy} is plotted positive in downward direction. The pair of stresses τ_{xx} and τ_{xy} (i.e. the stresses acting on a plane with its normal in the x-direction) together constitute the point A in the diagram shown in Figure A-5. The stresses τ_{yy} and τ_{yx}

(i.e. the stresses acting on a plane with its normal in the *y*-direction) together constitute the point B in the figure. The formulas (A.7) indicate that these stress points describe a circle if the orientation angle varies. The centre of the circle is located in a point of the horizontal axis, at a distance $\frac{1}{2}(\tau_1+\tau_2)$ to the right of the origin, and the radius of the circle is $\frac{1}{2}(\tau_1-\tau_2)$. The location of the stress point on the circle is determined by the angle or, more precisely, by the central angle 2γ. If the angle increases, the stress points move along the circumference of the circle. In the case shown in the figure both τ_{xy} and τ_{yx} are negative.

A special point can be identified on the circle: the *pole*, the point from which the stresses in any direction can be found by a simple construction (Point P in Figure A-6).

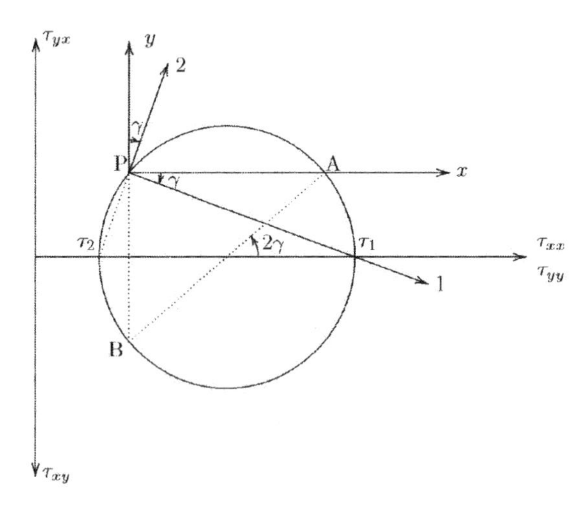

Figure A-6 Pole.

The pole can be found by drawing a line in *x*-direction from the stress point A, and intersecting this line with a line from the stress point B in the *y*-direction. The principal directions can now be found by drawing lines from the pole to the rightmost and leftmost points of the circle. The stresses on an arbitrary plane can be found by drawing a line in the direction of the normal vector to that plane, and intersecting it with the circle. The validity of the construction follows from the fact that an angle on the circumference of the circle, spanning a certain arc, is just one-half of the central angle on the same arc.

The graphical constructions described above are very useful in soil mechanics, to determine the directions of the major and the minor principal stresses, and also for the determination of the most critical planes, potential slip planes.

It may be mentioned that the considerations of this appendix apply to any symmetric second order tensor, strain as well as stress, for instance.

B Theory of elasticity

In this appendix the basic equations of the theory of elasticity are presented, together with some elementary solutions. The material is supposed to be isotropic, i.e. all properties are independent of the orientation.

B.1 Basic equations

The basic equations of the theory of elasticity describe the relations between stresses, strain and displacements in an isotropic linear elastic material.
The basic variables are the components of the displacement vector. In a cartesian coordinate system these can be denoted by u_x, u_y and u_z. The components of the strain tensor (or deformation tensor) can be derived from the displacements by differentiation,

$$\begin{aligned}
\varepsilon_{xx} &= \frac{\partial u_x}{\partial x}, \qquad \varepsilon_{xy} = \tfrac{1}{2}\left(\frac{\partial u_x}{\partial y} + \frac{\partial u_y}{\partial x}\right), \\
\varepsilon_{yy} &= \frac{\partial u_y}{\partial y}, \qquad \varepsilon_{yz} = \tfrac{1}{2}\left(\frac{\partial u_y}{\partial z} + \frac{\partial u_z}{\partial y}\right), \\
\varepsilon_{zz} &= \frac{\partial u_z}{\partial z}, \qquad \varepsilon_{zx} = \tfrac{1}{2}\left(\frac{\partial u_z}{\partial x} + \frac{\partial u_x}{\partial z}\right).
\end{aligned} \tag{B.1}$$

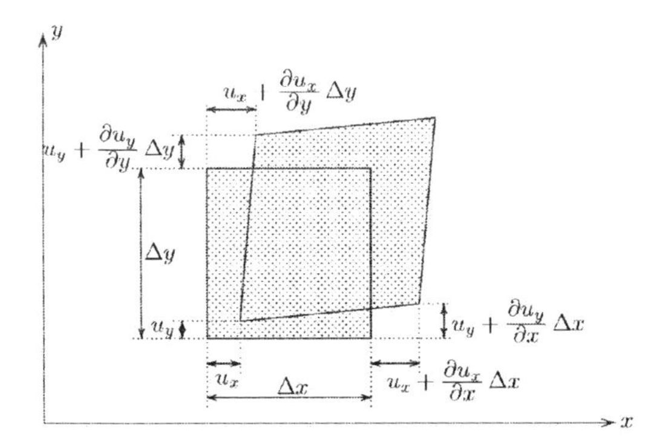

Figure B-1 Strains.

These expressions are illustrated in Figure B-1. It has been assumed that all the partial derivatives in the system of equations (B.1) are small. The strains ε_{xx}, ε_{yy} and ε_{zz} are a dimensionless measure for the relative change of length in the three coordinate directions. The shear strains ε_{xy}, ε_{yz} and ε_{zx} indicate the angular

deformations. The quantity ε_{xy}, for instance, is one half of the reduction of the right angle in the lower left corner of the element shown in Figure B-1.
The relative increase of the volume is the *volume strain*, and is denoted by the symbol ε_{vol},

$$\varepsilon_{vol} = \frac{\Delta V}{V}. \tag{B.2}$$

If the strains are small (compared to 1) this is the sum of the strains in the coordinate directions,

$$\varepsilon_{vol} = \varepsilon_{xx} + \varepsilon_{yy} + \varepsilon_{zz}. \tag{B.3}$$

For an isotropic linear elastic material the stresses can be expressed into the strains by Hooke's law,

$$\begin{aligned} \tau_{xx} &= \lambda\varepsilon_{vol} + 2\mu\varepsilon_{xx}, \quad & \tau_{xy} &= 2\mu\varepsilon_{xy}, \\ \tau_{yy} &= \lambda\varepsilon_{vol} + 2\mu\varepsilon_{yy}, \quad & \tau_{yz} &= 2\mu\varepsilon_{yz}, \\ \tau_{zz} &= \lambda\varepsilon_{vol} + 2\mu\varepsilon_{zz}, \quad & \tau_{zx} &= 2\mu\varepsilon_{zx}, \end{aligned} \tag{B.4}$$

The parameters λ and μ are Lamé's elastic constants. They are related to Young's modulus E and Poisson's ratio ν by

$$\lambda = \frac{\nu E}{(1+\nu)(1-2\nu)}, \qquad \mu = \frac{E}{2(1+\nu)}. \tag{B.5}$$

The sign convention for the stresses is that a stress component is positive when acting in positive coordinate direction on a plane having its outward normal in positive coordinate direction. This is the usual sign convention of continuum mechanics. It means that tensile stresses are positive, and compressive stresses are negative.
For a small element the stresses on the three visible faces are shown in Figure B-2. It may be noted that in soil mechanics the sign convention often is just the opposite, with compressive stresses being considered positive. Compressive stresses σ_{ij} can be related to the stresses τ_{ij} considered here, using the formula $\sigma_{ij} = -\tau_{ij}$.

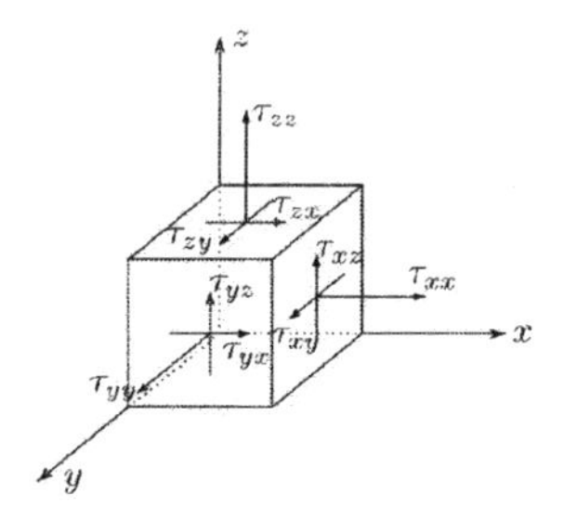

Figure B-2 Stresses on a small element.

The stresses should satisfy the equilibrium equations. In the absence of body forces these are

$$\begin{aligned}
\frac{\partial \tau_{xx}}{\partial x}+\frac{\partial \tau_{yx}}{\partial y}+\frac{\partial \tau_{zx}}{\partial z}=0, &\qquad \tau_{xy}=\tau_{yx},\\
\frac{\partial \tau_{xy}}{\partial x}+\frac{\partial \tau_{yy}}{\partial y}+\frac{\partial \tau_{zy}}{\partial z}=0, &\qquad \tau_{yz}=\tau_{zy},\\
\frac{\partial \tau_{xz}}{\partial x}+\frac{\partial \tau_{yz}}{\partial y}+\frac{\partial \tau_{zz}}{\partial z}=0, &\qquad \tau_{zx}=\tau_{xz}.
\end{aligned} \tag{B.6}$$

These equations can be derived by considering equilibrium of a small element, in the three coordinate directions, and equilibrium of moments about the three axes.

The stresses, strains and displacements in an isotropic linear elastic material should satisfy all the equations given above, and the appropriate boundary conditions at the surface of the body. Deriving solutions is not an easy matter. There are many books presenting techniques for the solution of elastic problems, for instance the book by S.P. Timoshenko & J.N. Goodier (*Theory of elasticity*, McGraw-Hill, 1970). In the next sections some special solutions will be presented.

For many solution methods it is convenient to express the equations of equilibrium into the displacement components. If the elastic coefficients λ and μ are constants (i.e. if the material is homogeneous), it follows from equations (B.1), (B.4) and (B.6) that

$$\begin{aligned}
(\lambda+\mu)\frac{\partial \varepsilon_{vol}}{\partial x}+\mu\nabla^2 u_x=0,\\
(\lambda+\mu)\frac{\partial \varepsilon_{vol}}{\partial y}+\mu\nabla^2 u_y=0,\\
(\lambda+\mu)\frac{\partial \varepsilon_{vol}}{\partial z}+\mu\nabla^2 u_z=0.
\end{aligned} \tag{B.7}$$

These equations form a system of three differential equations with three basic variables, the equations of Navier.

B.2 Boussinesq problems

For geotechnical engineering the class of problems of an elastic half space ($z>0$), bounded by the plane $z=0$, is of great importance. If the surface is loaded by normal stresses only, see Figure B-3, a solution can be found following methods developed by Boussinesq, in 1885.

Problems of this type, with given normal stresses on the boundary, and no shear stresses on the boundary, can be solved relatively easily by introducing a special potential function Φ. The displacements can be expressed into this potential by the equations

$$u_x = \frac{\partial \Phi}{\partial x} + \frac{\lambda+\mu}{\mu} z \frac{\partial^2 \Phi}{\partial x \partial z},$$

$$u_y = \frac{\partial \Phi}{\partial y} + \frac{\lambda+\mu}{\mu} y \frac{\partial^2 \Phi}{\partial y \partial z}, \tag{B.8}$$

$$u_z = -\frac{\lambda+2\mu}{\mu} \frac{\partial \Phi}{\partial z} + \frac{\lambda+\mu}{\mu} z \frac{\partial^2 \Phi}{\partial z^2}.$$

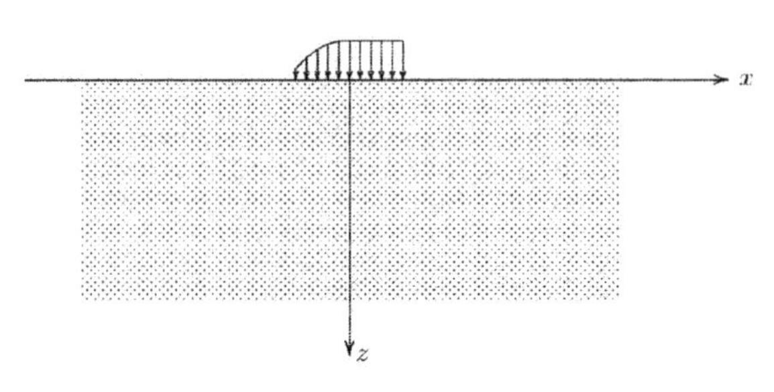

Figure B-3 Boussinesq problem.

Substitution into the equations (B.7) shows that all these equations are satisfied, provided that the function Φ satisfies Laplace's differential equation,

$$\nabla^2 \Phi = \frac{\partial^2 \Phi}{\partial x^2} + \frac{\partial^2 \Phi}{\partial y^2} + \frac{\partial^2 \Phi}{\partial z^2} = 0. \tag{B.9}$$

It follows that there is only a single unknown function, Φ, which should satisfy a rather simple differential equation, Laplace's equation. Many solutions of this equation are available.

The applicability of the potential Φ appears when the stresses are expressed in terms of this function. Using (B.1), (B.4) and (B.9), it follows that the normal stresses are

$$\frac{\tau_{xx}}{2\mu} = \frac{\partial^2 \Phi}{\partial x^2} + \frac{\lambda+\mu}{\mu} z \frac{\partial^3 \Phi}{\partial x^2 \partial z} - \frac{\lambda}{\mu} \frac{\partial^2 \Phi}{\partial z^2},$$

$$\frac{\tau_{yy}}{2\mu} = \frac{\partial^2 \Phi}{\partial y^2} + \frac{\lambda+\mu}{\mu} z \frac{\partial^3 \Phi}{\partial y^2 \partial z} - \frac{\lambda}{\mu} \frac{\partial^2 \Phi}{\partial z^2}, \tag{B.10}$$

$$\frac{\tau_{zz}}{2\mu} = -\frac{\lambda+\mu}{\mu} \frac{\partial^2 \Phi}{\partial z^2} + \frac{\lambda+\mu}{\mu} z \frac{\partial^3 \Phi}{\partial z^3}.$$

And the shear stresses are found to be

$$\frac{\tau_{xy}}{2\mu} = \frac{\partial^2 \Phi}{\partial x \partial y} + \frac{\lambda+\mu}{\mu} z \frac{\partial^3 \Phi}{\partial x \partial y \partial z},$$

$$\frac{\tau_{yz}}{2\mu} = \frac{\lambda+\mu}{\mu} z \frac{\partial^3 \Phi}{\partial y \partial z^2}, \tag{B.11}$$

$$\frac{\tau_{zx}}{2\mu} = \frac{\lambda+\mu}{\mu} z \frac{\partial^3 \Phi}{\partial x \partial z^2}.$$

The last two equations show that on the plane $z=0$ the shear stresses are automatically zero,

$$z=0: \quad \tau_{zx} = \tau_{zy} = 0, \tag{B.12}$$

whatever the function Φ is. This means that the potential Φ can be used only for problems in which the surface $z=0$ is free of shear stresses. That is an important restriction, which limits the use of this potential very severely. On the other hand, the class of problems of a half space loaded by normal stresses is an important class of problems for soil mechanics, and the differential equation is rather simple. On the surface $z=0$ the normal stress τ_{zz} may be prescribed, or the displacement u_z. Some examples will be given below.

B.3 Point load

A classical solution, described by Boussinesq, is the problem of a point load P on an elastic half space $z>0$, see Figure B-4.

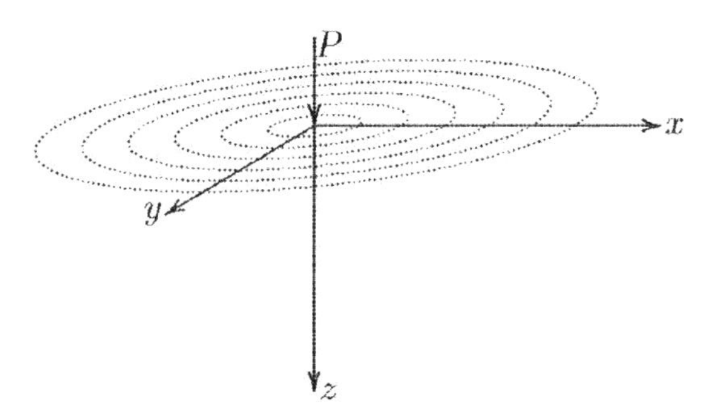

Figure B-4 Point load on half space.

The solution is assumed to be

$$\Phi = -\frac{P}{4\pi(\lambda+\mu)} \ln(z+R), \tag{B.13}$$

in which R is the spherical coordinate,

$$R = \sqrt{x^2+y^2+z^2}. \tag{B.14}$$

That this function satisfies the differential equation (B.9) can easily be verified by substitution into this equation. Next it must be checked that the boundary conditions are satisfied. The shear stresses on the surface $z=0$ are automatically zero, and the condition for the normal stresses can be verified as follows.
Differentiation of Φ with respect to z gives

$$\frac{\partial \Phi}{\partial z}=-\frac{P}{4\pi(\lambda+\mu)}\frac{1}{R}, \tag{B.15}$$

$$\frac{\partial^2 \Phi}{\partial z^2}=\frac{P}{4\pi(\lambda+\mu)}\frac{z}{R^3}, \tag{B.16}$$

$$\frac{\partial^3 \Phi}{\partial z^3}=\frac{P}{4\pi(\lambda+\mu)}(\frac{1}{R^3}-3\frac{z^2}{R^5}). \tag{B.17}$$

The vertical normal stress τ_{zz} now is, with (B.10),

$$\tau_{zz}=-\frac{3P}{2\pi}\frac{z^3}{R^5}. \tag{B.18}$$

On the surface $z=0$ this stress is zero, except in the origin, where the stress is infinitely large. The resultant force of the stress distribution can be obtained by integrating the vertical normal stress over an entire horizontal plane. This gives

$$\int_{-\infty}^{\infty}\int_{-\infty}^{\infty}\tau_{zz}dxdy=-P. \tag{B.19}$$

Every horizontal plane appears to transfer a force of magnitude P, as required. The solution (B.13) appears to satisfy all necessary conditions, and it can be concluded that it is the correct solution of the problem.
The vertical displacement is, with (B.8),

$$u_z=\frac{P}{4\pi\mu R}(\frac{\lambda+2\mu}{\lambda+\mu}+\frac{z^2}{R^2}). \tag{B.20}$$

The factor $(\lambda+2\mu)/(\lambda+\mu)$ can also be written as $2(1-\nu)$. The displacements of the surface $z=0$ is, when expressed in E and ν,

$$z=0:\quad u_z=\frac{P(1-\nu^2)}{\pi ER}. \tag{B.21}$$

This is singular in the origin, as might be expected for this case of a concentrated load.
All other stresses and displacements can easily be derived from the solution (B.13). That is left as an exercise for the reader.

B.4 Distributed load

On the basis of the elementary solution (B.13) many other interesting solutions can be derived. As an example the displacement in the centre of a circular area, carrying a uniform load will be derived, see Figure B-5.

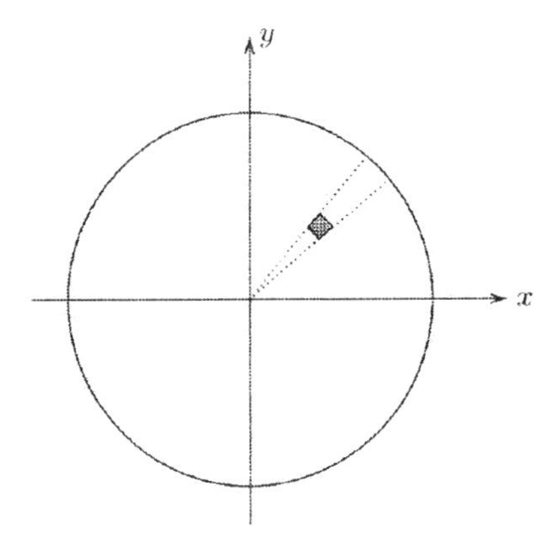

Figure B-5 Distributed load, on a circular area.

A load of magnitude pdA at a distance r from the origin leads to a displacement of the origin of magnitude

$$\frac{pdA(1-\nu^2)}{\pi E r},$$

which is in agreement with formula (B.21).
The displacement caused by a uniform load over a circular area, with radius a, can be found by integration over that area. Because $dA = r\,dr\,d\theta$, integration over θ from $\theta = 0$ to $\theta = 2\pi$, and integration over r from $r = 0$ to $r = a$ gives

$$r = 0,\ z = 0:\quad u = \frac{2pa(1-\nu^2)}{E}. \tag{B.22}$$

This is a well known and useful result.

B.5 Fourier transforms

A general class of solutions can be found by using Fourier transforms (I.N. Sneddon, Fourier Transforms, McGraw-Hill, 1951). As an example some problems of plane strain deformations (for which $u_y = 0$) will be considered here.
The solutions is assumed to be

$$\Phi = \int_0^\infty \{f(\alpha)\cos(\alpha x) + g(\alpha)\sin(\alpha x)\}\exp(-az)d\alpha, \tag{B.23}$$

in which $f(\alpha)$ and $g(\alpha)$ are undetermined functions in this stage.
That the expression (B.23) is indeed a solution follows immediately from substitution of the elementary solutions $\cos(\alpha x)\exp(-\alpha z)$ and $\sin(\alpha x)\exp(-\alpha z)$

into the differential equation (B.9). For $z \to \infty$ the solution tends towards zero, which suggests that these solutions may be
used for problems in which the stresses should vanish for $z \to \infty$.
The normal stress at the surface $z = 0$ is, with (B.10) and (B.23),

$$z = 0: \quad \frac{\tau_{zz}}{2\mu} = -\frac{\lambda + \mu}{\mu} \int_0^\infty \{\alpha^2 f(\alpha)\cos(\alpha x) + \alpha^2 g(\alpha)\sin(\alpha x)\} d\alpha. \tag{B.24}$$

Now suppose that the boundary condition is

$$z = 0, -\infty < x < \infty: \quad \tau_{zz} = q(x), \tag{B.25}$$

in which $q(x)$ is a given function. Then the condition is

$$\int_0^\infty \{A(\alpha)\cos(\alpha x) + B(\alpha)\sin(\alpha x)\} d\alpha = q(x), \tag{B.26}$$

in which

$$A(\alpha) = -2(\lambda + \mu)\alpha^2 f(\alpha), \tag{B.27}$$

and

$$B(\alpha) = -2(\lambda + \mu)\alpha^2 g(\alpha). \tag{B.28}$$

The problem of determining the functions $A(\alpha)$ and $B(\alpha)$ from (B.26) is the standard problem from the theory of Fourier transforms. The solution is provided by the *inversion theorem*. The derivation of this theorem will not be given here, see any book on Fourier analysis. The final result is

$$A(\alpha) = \frac{1}{\pi} \int_{-\infty}^\infty q(t)\cos(\alpha t) dt, \tag{B.29}$$

and

$$B(\alpha) = \frac{1}{\pi} \int_{-\infty}^\infty q(t)\sin(\alpha t) dt. \tag{B.30}$$

This is the solution of the problem, for an arbitrary load distribution $q(x)$ on the surface. The solution expresses that first the integrals (B.29) and (B.30) must be calculated, and then the results must be substituted into the general solution (B.23). The actual analysis may be quite complicated, depending upon the complexity of the load function $q(x)$. The procedure will be elaborated in the next section, for a simple example.

B.6 Line load

As an example the case of a line load will be elaborated, see Figure B-6. In this case the load can be described by the function

$$q(x) = \begin{cases} -F/2\varepsilon & \text{if } |x| < \varepsilon, \\ 0 & \text{if } |x| > \varepsilon, \end{cases} \tag{B.31}$$

where ε is a small length, with $\varepsilon \to 0$. From (B.29) and (B.30) it now follows that

$$A(\alpha) = -\frac{F}{\pi\varepsilon}\frac{\sin(\alpha\varepsilon)}{\alpha},$$

$$B(\alpha) = 0.$$

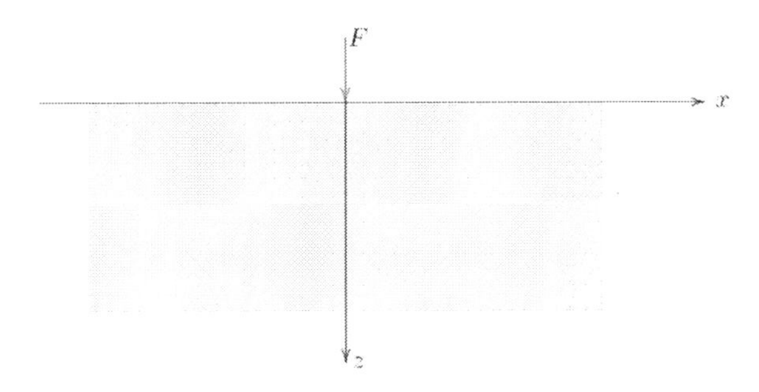

Figure B-6 Line load on half space.

If $\varepsilon \to 0$ this reduces to

$$A(\alpha) = -\frac{F}{\pi}, \tag{B.32}$$

$$B(\alpha) = 0. \tag{B.33}$$

With (B.27) and (B.28) the original functions are

$$f(\alpha) = \frac{F}{2\pi(\lambda+\mu)\alpha^2}, \tag{B.34}$$

$$g(\alpha) = 0. \tag{B.35}$$

The final solution of the problem is

$$\Phi = \frac{F}{2\pi(\lambda+\mu)}\int_0^\infty \frac{\cos(\alpha x)\exp(-\alpha z)}{\alpha^2}d\alpha. \tag{B.36}$$

Even though this integral does not converge, because of the behaviour of the factor α^2 in the denominator for $\alpha \to 0$, the result can be used to determine the stresses, for which the potential must be differentiated. For instance

$$\frac{\partial^2\Phi}{\partial x^2} = -\frac{F}{2\pi(\lambda+\mu)}\int_0^\infty \cos(\alpha x)\exp(-\alpha z)d\alpha,$$

and this integral converges. The result is

$$\frac{\partial^2\Phi}{\partial x^2} = -\frac{F}{2\pi(\lambda+\mu)}\frac{z}{x^2+z^2}. \tag{B.37}$$

In a similar way it can be shown that

$$\frac{\partial^2\Phi}{\partial z^2} = \frac{F}{2\pi(\lambda+\mu)}\frac{z}{x^2+z^2}. \tag{B.38}$$

Continuing the differentiation gives

$$\frac{\partial^3\Phi}{\partial z^3} = \frac{F}{2\pi(\lambda+\mu)}\frac{x^2-z^2}{(x^2+z^2)^2}, \tag{B.39}$$

$$\frac{\partial^3\Phi}{\partial x^2 \partial z} = -\frac{F}{2\pi(\lambda+\mu)}\frac{x^2-z^2}{(x^2+z^2)^2}. \tag{B.40}$$

The stresses finally are, with (B.10) and (B.11),

$$\tau_{xx} = -\frac{2F}{\pi}\frac{x^2 z}{(x^2+z^2)^2}, \tag{B.41}$$

$$\tau_{zz} = -\frac{2F}{\pi}\frac{z^3}{(x^2+z^2)^2}, \tag{B.42}$$

$$\tau_{xz} = -\frac{2F}{\pi}\frac{xz^2}{(x^2+z^2)^2}. \tag{B.43}$$

These formulas were first derived by Flamant, in 1892.
Many more solutions of elastic problems have been found, for instance for layered systems, and for bodies of more complex form than a half plane or a half space, for instance a plane with a row of circular holes (a problem of great interest to aeronautical engineers). Many of these solutions are very complex. A large number of solutions of interest for geotechnical engineering can be found in the book by H.G. Poulos & E.H. Davis (*Elastic Solutions for Soil and Rock Mechanics*, Wiley, 1974).

C Theory of plasticity

In this appendix the main theorems of plasticity theory are presented. These are the *limit theorems*, which enable to determine upper bounds and lower bounds of the failure load of a body.

C.1 Yield surface

The simplest description of plastic deformations is by considering a *perfectly plastic* material. This is a material that exhibits plastic deformations if (and only if) the stresses satisfy the *yield condition*. For a perfectly plastic material this yield condition is a function of the stresses only (and not of the deformations, or of the time). This yield condition is written in the form

$$f(\sigma_{ij})=0. \tag{C.1}$$

Plastic deformations can occur only if $f(\sigma_{ij})=0$. Stress states for which $f(\sigma_{ij})>0.$ are impossible, and if $f(\sigma_{ij})<0.$ there are no plastic deformations, but such states of stress are perfectly possible. The deformations then are elastic only.
The yield condition can be considered as a relation between the nine stresses σ_{ij}, with $i,j=1,2,3$, in a 9-dimensional space. In such a space the yield condition (C.1) is an 8-dimensional part of space. It is usually called the *yield surface*. If the state of stress can be described by three stresses (for instance the three principal stresses), the yield condition can be written as

$$f(\sigma_1,\sigma_2,\sigma_3)=0. \tag{C.2}$$

In the 3-dimensional space with axes σ_1, σ_2 and σ_3 this is a surface. For that reason the condition (C.1) in a higher dimensional space is also called the yield surface. In a 2-dimensional space, if there are only two parameters that determine yielding, the yield surface reduces to a (curved) line.
It is assumed that the origin $\sigma_{ij}=0$, that is the state of stress in which all stresses are zero, is located inside the yield surface. Furthermore, it is assumed that if a certain point σ_{ij}^e is located inside the yield surface, then $\alpha\sigma_{ij}^e$, with $\alpha<1$, is also inside the yield surface. In topology it is said that the yield surface is *star-shaped*. Later it will also be assumed that the yield surface is *convex*, see Figure C.1.
That is a more severe restriction than the assumption that it is star-shaped. These assumptions are essential for the derivations to be presented in this appendix.

To simplify the analysis it will be assumed that the material can deform only if $f(\sigma_{ij})=0$. This means that all elastic deformations are disregarded. Such a material is called *rigid plastic*.

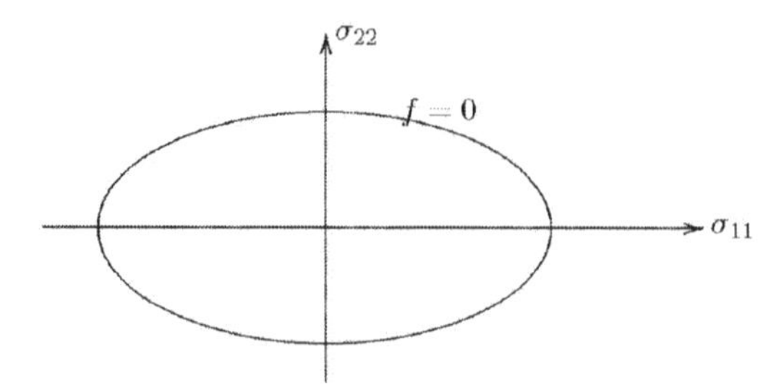

Figure C-1 Yield surface.

C.2 Some geometrical definitions

Before presenting the mechanics of plastic deformations it is useful to first derive some important geometrical relations, for the expression of a plane tangent to the yield surface, and for a line perpendicular to that surface. In a 9-dimensional space a plane tangent to the surface $f(\sigma_{ij})$ can be defined as

$$(\frac{\partial f}{\partial \sigma_{ij}})_1(\sigma_{ij}-\sigma_{ij}^1)=0. \tag{C.3}$$

Here $(\partial f/\partial \sigma_{ij})_1$ denotes the partial derivative of the function f with respect to the variable σ_{ij} in the point σ_{ij}^1. In equation (C.3) summation over the indices i and j is implied by the repetition of these indices. This is the summation convention of Einstein,

$$a_i b_i = \sum_{i=1}^{n} a_i b_i, \tag{C.4}$$

in which n is the dimension of space, usually 3, but in this case $n=9$. The definition (C.3) is a generalization to 9-dimensional space of the usual definition in 3-dimensional space.

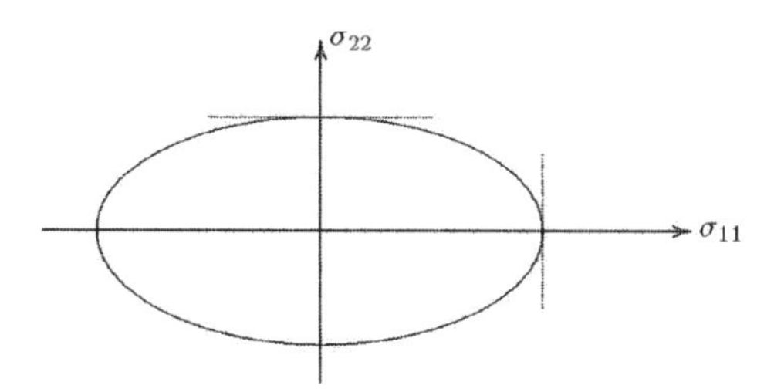

Figure C-2 Examples of tangents.

The significance of the definition (C.3) can be clarified as follows. On the yield surface the value of f is constant ($f = 0$). Suppose that σ_{ij}^1 is a point on that surface, and consider a small increment of the stress from that value, such that both σ_{ij}^1 and $\sigma_{ij}^1 + d\sigma_{ij}$ are located on the yield surface. The difference df of the functional values in these two points is zero, i.e.

$$df = (\frac{\partial f}{\partial \sigma_{ij}})_1 d\sigma_{ij} = 0. \tag{C.5}$$

Equation (C.3) is the generalization of (C.5) for arbitrary points, at an arbitrary distance form σ_{ij}^1, which is also linear in σ_{ij}. It follows that is indeed natural to denote (C.3) as the definition of the tangent plane.
Next the definition of a line perpendicular to the yield surface will be considered. For this purpose it may be noted that the general equation of a plane passing through the point σ_{ij}^1 is

$$A_{ij}(\sigma_{ij} - \sigma_{ij}^1) = 0, \tag{C.6}$$

in which the constants A_{ij} are given numbers, that define the slopes of the plane in the various directions. A straight line in this plane, through the point σ_{ij}^1, can be written as

$$\sigma_{ij} - \sigma_{ij}^1 = a(\sigma_{ij}^2 - \sigma_{ij}^1), \tag{C.7}$$

in which a is a variable parameter, and σ_{ij}^2 is a second point in the plane (C.6), which means that

$$A_{ij}(\sigma_{ij}^2 - \sigma_{ij}^1) = 0. \tag{C.8}$$

An arbitrary straight line through the point σ_{ij}^1, not necessarily in the plane considered, can be described by the equation

$$\sigma_{ij} - \sigma_{ij}^1 = cb_{ij}, \tag{C.9}$$

in which b_{ij} are constants, and c a variable parameter.
In general two straight lines

$$\sigma_{ij} - \sigma_{ij}^a = ac_{ij}, \tag{C.10}$$

$$\sigma_{ij} - \sigma_{ij}^b = bd_{ij}, \tag{C.11}$$

are considered to be perpendicular if the inner product of the directional vectors is zero,

$$c_{ij}d_{ij} = 0. \tag{C.12}$$

This is in agreement with the usual definition of orthogonality, by requiring that the inner product of two vectors is zero.
If equation (C.7) is now written as

$$\sigma_{ij} - \sigma_{ij}^1 = ac_{ij} = a(\sigma_{ij}^2 - \sigma_{ij}^1), \tag{C.13}$$

it follows that the line

$$\sigma_{ij} - \sigma_{ij}^0 = bA_{ij}, \tag{C.14}$$

is perpendicular to each line of the set (C.13), because $A_{ij}c_{ij}$ is always zero, see (C.8). The conclusion must be that the line (C.14) is perpendicular to the plane (C.6). The point σ_{ij}^0 needs not to be located on the yield surface, but this is not forbidden either, and the point may
even coincide with the point σ_{ij}^1. It follows that the line

$$\sigma_{ij} - \sigma_{ij}^1 = bA_{ij}, \tag{C.15}$$

passes through the point σ_{ij}^1, and is perpendicular to the plane (C.6).
If this property is applied to the tangent plane of the yield surface, as defined by equation (C.3), it follows that a line defined by

$$\sigma_{ij} - \sigma_{ij}^1 = b(\frac{\partial f}{\partial \sigma_{ij}})_1, \tag{C.16}$$

is perpendicular to the yield surface, in the point σ_{ij}^1.

As an example consider a yield surface in the form of an ellipse, see Figure C.1, with axes $2a$ and a,

$$f = \frac{\sigma_{11}^2}{4a^2} + \frac{\sigma_{11}^2}{a^2} = 0. \tag{C.17}$$

In this case the equation of the tangent plane (in this two-dimensional case this is a tangent line), is, following (C.3),

$$\sigma_{11}^1(\sigma_{11} - \sigma_{11}^1) + 4\sigma_{22}^1(\sigma_{22} - \sigma_{22}^1) = 0, \tag{C.18}$$

in which the superscript [1] indicates that the point is located on the yield surface.
In the rightmost point of the yield surface $\sigma_{11}^1 = 2a$ and $\sigma_{22}^1 = 0$. Equation (C.18) then defines the tangent as : $\sigma_{11} = 2a$. In the topmost point of the yield surface $\sigma_{11}^1 = 0$ and $\sigma_{22}^1 = a$. In that case equation (C.18) defines the tangent as : $\sigma_{22} = a$. These two tangents are shown as dotted lines in Figure C.2. These two lines are indeed tangent to the yield surface.

C.3 Convex yield surface

After the definition of some geometrical concepts we now return to the mechanics of plastic materials. As stated before, plastic deformation is governed by the location of the stress point σ_{ij} with respect to the yield surface $f(\sigma_{ij})=0$, in a 9-dimensional space. It is now assumed that the yield surface is convex. This is supposed to be defined by the requirement that

$$(\sigma_{ij}^1 - \sigma_{ij}^e)(\frac{\partial f}{\partial \sigma_{ij}})_1 > 0, \tag{C.19}$$

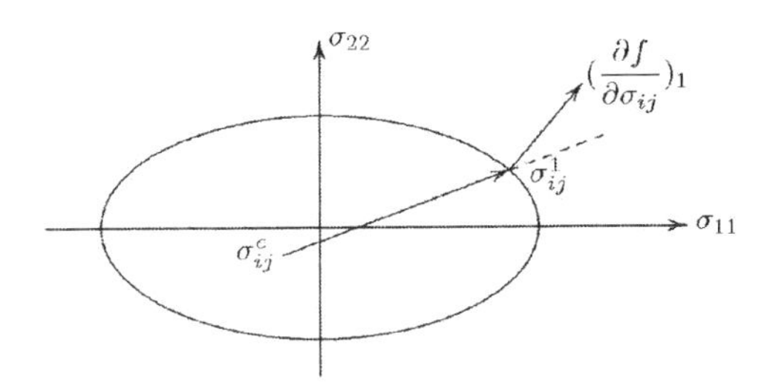

Figure C-3 Convex yield surface.

in which σ_{ij}^1 is a point of the yield surface, and σ_{ij}^e is an arbitrary point inside the yield surface. This means that $f(\sigma_{ij}^1)=0$ and $f(\sigma_{ij}^e)<0$. Equation (C.19) states that the inner product of the vector from σ_{ij}^e to σ_{ij}^1, and the vector $(\partial f/\partial \sigma_{ij})_1$, which is directed perpendicular to the yield surface, is positive. This means that the angle between these two vectors is smaller than $\pi/2$, which corresponds to the statement that the yield surface is convex, see Figure C.3. Only if the yield surface would have concave parts it would be possible that a vector from a point inside the yield surface to a point on that surface makes an angle greater than $\pi/2$ with the vector normal to the yield surface, in outward direction. This possibility is excluded here, by assuming that the yield surface is convex. This property will be used in later proofs.

C.4 Plastic deformations

It is assumed that the plastic deformations can be described by the *deformation rates* $\dot{\varepsilon}_{ij}$. It follows that

$$\begin{aligned} f(\sigma_{ij})<0 &: \dot{\varepsilon}_{ij}=0, \\ f(\sigma_{ij})=0 &: \dot{\varepsilon}_{ij}\neq 0. \end{aligned} \tag{C.20}$$

This means that plastic deformations, whenever they occur, will continue forever, at a certain rate. If time progresses, the deformations will increase indefinitely.
The plastic deformation rates $\dot{\varepsilon}_{ij}$ can also be plotted in a 9-dimensional space, and this can be done such that the axes coincide with the axes of stress space. The vectors σ_{ij} and $\dot{\varepsilon}_{ij}$ may then be represented in the same space.

C.5 Plastic potential

It is now postulated that the plastic strain rates can be derived from a *plastic potential* g, that depends on the stresses only, i.e. $g = g(\sigma_{ij})$, in such a way that the strain rates can be obtained by

$$f(\sigma_{ij}) < 0 \; : \; \dot{\varepsilon}_{ij} = 0, \tag{C.21}$$

$$f(\sigma_{ij}) = 0 \; : \; \dot{\varepsilon}_{ij} = \lambda \frac{\partial g}{\partial \sigma_{ij}}. \tag{C.22}$$

Here λ is an undetermined constant. The essential assumption is that such a function $g(\sigma_{ij})$, from which the strain rates can be determined by differentiation with respect to the corresponding stresses, see (C.22), exists.

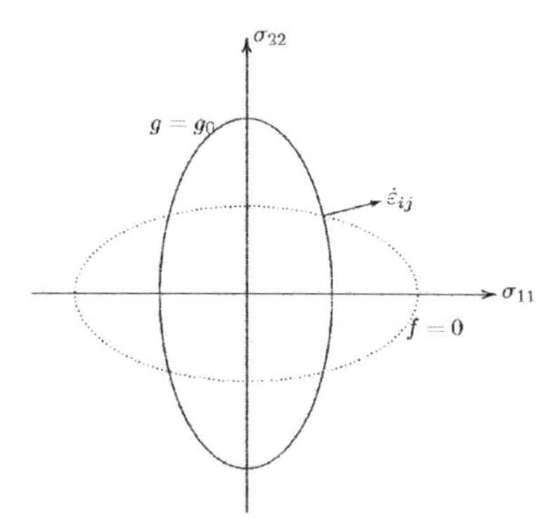

Figure C-4 Plastic potential.

From the geometrical considerations presented above, it follows that the vector of strain rates $\dot{\varepsilon}_{ij}$ in 9-dimensional space is perpendicular to the surface of the plastic potential g, see Figure C.4. In this figure the yield condition is shown by a dotted curve. The plastic potential passing through a certain point of the yield surface has been indicated by a fully drawn curve. The vector of strain rates is perpendicular to the plastic potential. Through each point of the yield surface a surface of constant values of g can be drawn, each with its own value of that constant. The shape of the plastic potential surfaces is unknown at this stage. It may be star-shaped, or convex, or perhaps not.

C.6 Drucker's postulate

It has been found, by comparing theoretical results with experimental data, that for metals very good agreement is obtained if the plastic potential g is identified with the yield function f. This is often called *Drucker's postulate*,

$$\text{Drucker} \; : \; f = g. \tag{C.23}$$

It has been attempted to find a theoretical derivation of this property, for instance on the basis of some thermodynamical principle. It has been found later, however, that there is no physical necessity for the validity of Drucker's postulate, other than that it provides a reasonable prediction for the plasticity behaviour of metals. For other materials, especially frictional materials such as sand, it is very unlikely that Drucker's postulate is valid, as it leads to unrealistic predictions. It is usually concluded that it may be applicable for materials without friction ($\phi = 0$), but is inapplicable if $\phi > 0$.

Notwithstanding the theoretical objections against Drucker's postulate, it may well be used for clays, especially in undrained conditions. For this reason its validity will be assumed in the sequel, until further notice. This will enable to derive limit theorems for clays. If the plastic potential is identified with the yield condition, equation (C.22) can be written as

$$f(\sigma_{ij}) < 0 \ : \ \dot{\varepsilon}_{ij} = 0, \tag{C.24}$$

$$f(\sigma_{ij}) = 0 \ : \ \dot{\varepsilon}_{ij} = \lambda \frac{\partial f}{\partial \sigma_{ij}}. \tag{C.25}$$

The direction of the vector of plastic deformations now is normal to the yield surface.

In the next sections the limit theorems will be derived, using the assumptions made before. The first step is the formulation and derivation of the virtual work principle.

C.7 Virtual work

Let there be considered a body in equilibrium. If the volume of the body is V the equilibrium conditions are that in the volume V the following equations are satisfied,

$$\sigma_{ij,i} + F_j = 0, \tag{C.26}$$

and

$$\sigma_{ij} = \sigma_{ji}, \tag{C.27}$$

where F_j is a given volume force. The comma indicates partial differentiation,

$$a_{,i} = \frac{\partial a}{\partial x_i}. \tag{C.28}$$

It is assumed that the boundary conditions are that on a part (S_1) of the boundary the stresses are prescribed, and that on the remaining part of the boundary (S_2) the displacements are prescribed,

$$\text{on } S_1 \ : \ \sigma_{ij} n_i = t_j, \tag{C.29}$$

$$\text{on } S_2 \ : \ u_i = f_i, \tag{C.30}$$

where t_j is given on S_1 and f_i is given on S_2.

In the sequel the following definitions are needed. A field of stresses that satisfies equations (C.26), (C.27) and (C.29) is a *statically admissible* stress field, or an *equilibrium system.* A field of displacements that satisfies certain regularity conditions (meaning that the material should retain its integrity, and that no overlaps or gaps may be created in the deformation, but that allows sliding of one part with respect to the rest of the body), and that satisfies equation (C.30), is a *kinematically admissible* displacement field, or a *mechanism.* To such a field a displacement field can be associated by

$$\varepsilon_{ij} = \tfrac{1}{2}(u_{i,j} + u_{j,i}). \tag{C.31}$$

Now consider an arbitrary statically admissible stress field σ_{ij}, and an arbitrary kinematically admissible displacement field u_i. These fields need not have any relation, except that they must be defined in the same volume V. In general one may write

$$\int_V \sigma_{ij,i} u_j dV = \int_V [(\sigma_{ij} u_j)_{,i} - \sigma_{ij} u_{j,i}] dV.$$

Using Gauss' divergence theorem and eq. (C.27) it follows that

$$\int_V \sigma_{ij,i} u_j dV = \int_S \sigma_{ij} u_j n_i dS - \int_V \tfrac{1}{2} \sigma_{ij} (u_{i,j} + u_{j,i}) dV.$$

With (C.31), (C.29), (C.30) and (C.27) it now follows that

$$\int_V \sigma_{ij} \varepsilon_{ij} dV = \int_{S_1} t_i u_i dS + \int_{S_2} \sigma_{ij} n_i f_j dS + \int_V F_i u_i dV. \tag{C.32}$$

Equation (C.32) is valid for any combination of an arbitrary statically admissible field and an arbitrary kinematically admissible displacement field, defined in the same body.

Equation (C.32) must also be valid for the combination of the statically admissible stress field σ_{ij} and the kinematically admissible displacement field $u_i + \dot{u}_i dt$ Because this field should also satisfy the boundary condition (C.30), in order to be kinematically admissible, it follows that

$$on\, S_2 \;:\; \dot{u}_i = 0. \tag{C.33}$$

The small increments of the displacement field $\dot{u}_i dt$, that satisfies (C.33) constitutes a *virtual displacement*. Similar to eq. (C.32) the following equation must be satisfied

$$\begin{aligned} \int_V \sigma_{ij} \varepsilon_{ij} dV + dt \int_V \sigma_{ij} \dot{\varepsilon}_{ij} dV = \int_{S_1} t_i u_i dS + dt \int_{S_1} t_i \dot{u}_i dS + \\ \int_{S_2} \sigma_{ij} n_i f_j dS + \int_V F_i u_i dV + dt \int_V F_i \dot{u}_i dV. \end{aligned} \tag{C.34}$$

If eq. (C.32) is subtracted from this equation, the result is, after division by dt,

$$\int_V \sigma_{ij} \dot{\varepsilon}_{ij} dV = \int_{S_1} t_i \dot{u}_i dS + \int_V F_i \dot{u}_i dV. \tag{C.35}$$

This is the *virtual work theorem.* It is valid for any combination of a statically admissible stress field, and a variation of a kinematically admissible displacement field. These fields need not be related at all.

The integral in the left hand side is the (virtual) work by the stresses on the given incremental deformations. The terms in the right hand side can be considered as the (virtual) work by the volume forces and the surface load during the virtual displacement. This virtual work appears to be equal to the work done by the stresses on the incremental strains.

C.8 Lower bound theorem

The *lower bound theorem* states that a lower bound for the failure load can be found by considering an equilibrium field. It can be proved in the following way.
Consider a body consisting of a perfectly plastic material, having a convex yield surface, and satisfying Drucker's postulate. Let the body be loaded by a surface load t_i on the part S_1 of the boundary, and by a volume force F_i. It is assumed that failure will occur for a certain combination of loads, say t_i^c and F_i^c. From now on only combinations of loads are considered that are proportional to the failure load, i.e.

$$t_i = \alpha t_i^c, \quad F_i = \alpha F_i^c, \tag{C.36}$$

where α is a constant.
The stresses at failure are assumed to be σ_{ij}^c, and the corresponding velocities are supposed to be $\dot{u}_i^c$. The virtual work theorem now gives

$$\int_V \sigma_{ij}^c \dot{\varepsilon}_{ij}^c dV = \int_{S_1} t_i^c \dot{u}_i^c dS + \int_V F_i^c \dot{u}_i^c dV. \tag{C.37}$$

Now assume that for a load $t_i^e = \alpha t_i^c$ and $F_i^e = \alpha F_i^c$ a statically admissible stress field σ_{ij}^e has been found, and that all these stresses are inside the yield criterion. Then this load is smaller than the failure load, i.e.

$$\alpha < 1. \tag{C.38}$$

The proof (ad absurdum) of this theorem can be given as follows. Let it be assumed that the theorem is false, i.e. assume that $\alpha > 1$. From the virtual work theorem it follows that

$$\int_V \sigma_{ij}^e \dot{\varepsilon}_{ij}^c dV = \int_{S_1} t_i^e \dot{u}_i^c dS + \int_V F_i^e \dot{u}_i^c dV,$$

or, with $t_i^e = \alpha t_i^c$ and $F_i^e = \alpha F_i^c$,

$$\int_V \frac{1}{\alpha} \sigma_{ij}^e \dot{\varepsilon}_{ij}^c dV = \int_{S_1} t_i^c \dot{u}_i^c dS + \int_V F_i^c \dot{u}_i^c dV. \tag{C.39}$$

From (C.37) and (C.39) it follows that

$$\int_V (\sigma_{ij}^c - \frac{1}{\alpha} \sigma_{ij}^e) \dot{\varepsilon}_{ij}^c dV = 0. \tag{C.40}$$

Using Drucker's postulate, which has been assumed to be valid, the strain rates at failure are

$$\dot{\varepsilon}_{ij}^{c} = \lambda(\frac{\partial f}{\partial \sigma_{ij}})_c. \tag{C.41}$$

Substitution into (C.40) gives

$$\lambda \int_V (\sigma_{ij}^c - \frac{1}{\alpha}\sigma_{ij}^e)(\frac{\partial f}{\partial \sigma_{ij}})_c dV = 0. \tag{C.42}$$

If $\alpha > 1$, and σ_{ij}^e is inside the yield surface (as had been assumed), then σ_{ij}^e/α is certainly inside the yield surface. Because of (C.19), i.e. because of the convexity of the yield surface, it now follows that

$$(\sigma_{ij}^c - \frac{1}{\alpha}\sigma_{ij}^e)(\frac{\partial f}{\partial \sigma_{ij}})_c > 0. \tag{C.43}$$

The integral of this quantity can not be zero, as equation (C.42) states. This means that the assumption $\alpha > 1$ must be false. Therefore $\alpha < 1$, and this is just what had to be proved.

The theorem means that a statically admissible stress field that does not violate the yield criterion, constitutes a lower bound for the failure load. The real failure load is always larger than the load for that equilibrium system. The load is on the safe side.

C.9 Upper bound theorem

The failure load can also be approached from above. This is expressed by the *upper bound theorem*, which can be derived as follows. Consider a body consisting of a perfectly plastic material, satisfying Drucker's postulate. The failure load again is t_i^c (on S_1) and F_i^c (in *V*). The corresponding stresses are σ_{ij}^c. These stresses are located on the yield surface, or partly inside it.

Suppose that a kinematically admissible velocity field $\dot{u}_i^k$ has been chosen, with the corresponding strain rates $\dot{\varepsilon}_{ij}^k$. The plastic strain rates can be derived from the yield function by the relations

$$\dot{\varepsilon}_{ij} = \lambda \frac{\partial f}{\partial \sigma_{ij}}.$$

Using these relations it is possible, at least in principle, to determine the stresses σ_{ij}^k in all points where $\dot{\varepsilon}_{ij}^k \neq 0$. Because the yield surface is convex, and the plastic strain rates are known, there is just one point where the vector of plastic strain rates is perpendicular to the yield surface. This point determines the stress state. Next the following integral can be calculated,

$$D = \int_V \sigma_{ij}^k \dot{\varepsilon}_{ij}^k dV. \tag{C.44}$$

This is the energy that would be dissipated by the assumed kinematic field, if it would occur. A load proportional to the failure load, $t_i^k = \beta t_i^c$ and $F_i^k = \beta F_i^c$, can now be calculated such that

$$\int_{S_1} t_i^k \dot{u}_i^k dS + \int_V F_i^k \dot{u}_i^k dV = D = \int_V \sigma_{ij}^k \dot{\varepsilon}_{ij}^k dV. \tag{C.45}$$

Although this formula has the same form as the virtual work principle, it does not follow from that theorem, because the stress field σ_{ij}^k in general is not an equilibrium system, and it need not satisfy the boundary condition for the stresses. Equation (C.45) is simply a procedure to determine the fictitious loads t_i^k and F_i^k.
The upper bound theorem is that the load t_i^k and F_i^k is larger than the failure load, or, in other words, that

$$\beta > 1. \tag{C.46}$$

The proof (ad absurdum) of this theorem is as follows. Let it be assumed that the theorem is false, i.e. assume that

$$\beta = t_i^k / t_i^c = F_i^k / F_i^c < 1.$$

From (C.45) it follows that

$$\int_V \sigma_{ij}^k \dot{\varepsilon}_{ij}^k dV = \beta \int_{S_1} t_i^c \dot{u}_i^k dS + \beta \int_V F_i^c \dot{u}_i^k dV. \tag{C.47}$$

Using the virtual work theorem the following equality can be formulated

$$\beta \int_V \sigma_{ij}^c \dot{\varepsilon}_{ij}^k dV = \beta \int_{S_1} t_i^c \dot{u}_i^k dS + \beta \int_V F_i^c \dot{u}_i^k dV. \tag{C.48}$$

From (C.47) and (C.48) it follows that

$$\int_V (\sigma_{ij}^k - \beta \sigma_{ij}^c) \dot{\varepsilon}_{ij}^k dV = 0. \tag{C.49}$$

In all points where $\dot{\varepsilon}_{ij}^k \neq 0$, so that there are contributions to the integral, the point σ_{ij}^k is located on the failure surface. The stress $\beta \sigma_{ij}^c$ is located inside the yield surface, because σ_{ij}^c is a point of the convex yield surface, and $\beta < 1$, by supposition. It then follows from (C.19) that

$$\dot{\varepsilon}_{ij}^k \neq 0: \ (\sigma_{ij}^k - \beta \sigma_{ij}^c)(\frac{\partial f}{\partial \sigma_{ij}})_k > 0.$$

The integral of this quantity can not be zero, as required by (C.49). This means that a contradiction has been obtained. The conclusion must be that the assumption that $\beta < 1$ must be false, at least if it is assumed that the other assumptions (validity of Drucker's postulate, convex yield surface) are true. Therefore $\beta > 1$, and this is what had to be proved.

The theorem means that a kinematically admissible velocity field, constitutes an upper bound for the failure load. The real failure load is always smaller than the load for that mechanism. The load is on the unsafe side.

C.10 Frictional materials

For a frictional material, such as most soils, in particular sands, the Mohr-Coulomb criterion is a good representation of the yield condition. For the case that the cohesion $c=0$ this is shown in Figure C.5. It is assumed that yielding of the material is determined by the stresses σ_{xx}, σ_{yy}, and $\sigma_{xy}=\sigma_{yx}$ only. The stresses are effective stresses, but as there are no pore pressures (by assumption) they are total stresses as well. The yield condition is that the radius of Mohr's circle equals $\sin\phi$ times the distance of the centre of the circle to the origin. This can be expressed as

$$\tfrac{1}{2}(\sigma_1-\sigma_3)=\tfrac{1}{2}(\sigma_1+\sigma_3)\sin\phi, \tag{C.50}$$

or, if the principal stresses are expressed in terms of the stress components in an arbitrary coordinate system of axes x and y,

$$f=(\frac{\sigma_{xx}-\sigma_{yy}}{2})^2+\tfrac{1}{2}\sigma_{xy}^2+\tfrac{1}{2}\sigma_{yx}^2-(\frac{\sigma_{xx}+\sigma_{yy}}{2})^2\sin^2\phi=0. \tag{C.51}$$

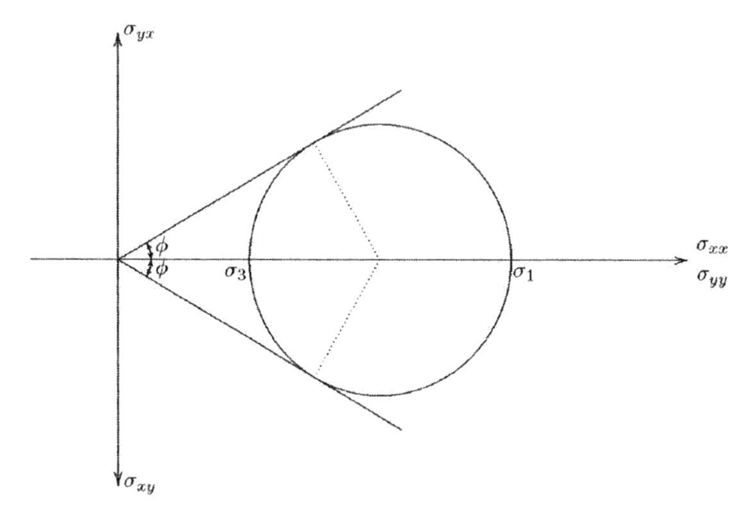

Figure C-5 Mohr-Coulomb criterion.

The circumstance that this yield condition depends upon the isotropic stress implies that Drucker's postulate will automatically lead to a deformation corresponding to that stress, i.e. a volume strain. This can be seen formally by calculating the strain rates using Drucker's postulate. This gives

$$\dot{\varepsilon}_{xx} = \lambda(\frac{\partial f}{\sigma_{xx}}) = \lambda\{(\frac{\sigma_{xx} - \sigma_{yy}}{2}) - (\frac{\sigma_{xx} + \sigma_{yy}}{2})\sin^2 \phi\}, \tag{C.52}$$

$$\dot{\varepsilon}_{yy} = \lambda(\frac{\partial f}{\sigma_{yy}}) = \lambda\{(\frac{\sigma_{yy} - \sigma_{xx}}{2}) - (\frac{\sigma_{xx} + \sigma_{yy}}{2})\sin^2 \phi\}, \tag{C.53}$$

$$\dot{\varepsilon}_{xy} = \lambda(\frac{\partial f}{\sigma_{xy}}) = \lambda \sigma_{xy}. \tag{C.54}$$

These strain rates can also be represented graphically in a Mohr diagram. If the radius of that circle is denoted by $\frac{1}{2}\dot{\gamma}$, it follows that

$$(\frac{\dot{\gamma}}{2})^2 = (\frac{\dot{\varepsilon}_{xx} - \dot{\varepsilon}_{yy}}{2})^2 + \dot{\varepsilon}_{xy}^2. \tag{C.55}$$

Using the expressions (C.52), (C.53) and (C.54) this can also be written as

$$(\frac{\dot{\gamma}}{2})^2 = \lambda^2 \{(\frac{\sigma_{xx} - \sigma_{yy}}{2})^2 + \sigma_{xy}^2\}, \tag{C.56}$$

or, because these stresses satisfy the yield criterion (C.51),

$$(\frac{\dot{\gamma}}{2})^2 = \lambda^2 (\frac{\sigma_{xx} + \sigma_{yy}}{2})^2 \sin \phi. \tag{C.57}$$

It follows that

$$\frac{\dot{\gamma}}{2} = \lambda(\frac{\sigma_{xx} + \sigma_{yy}}{2}) \sin \phi. \tag{C.58}$$

On the other hand the volume strain rate is

$$\dot{\varepsilon}_{vol} = \dot{\varepsilon}_{xx} + \dot{\varepsilon}_{yy} = -2\lambda(\frac{\sigma_{xx} + \sigma_{yy}}{2}) \sin^2 \phi. \tag{C.59}$$

From (C.58) and (C.59) it follows that

$$\dot{\varepsilon}_{vol} = -\dot{\gamma} \sin \phi. \tag{C.60}$$

Any plastic shear strain γ will be accompanied by a simultaneous volume strain ε_{vol}, in a ratio of $\sin\phi$. The minus sign indicates that this is a volume expansion. That the shear strains in a sand that is failing are accompanied by a continuous volume increase is not what is observed in experiments. It can also not be imagined very well that a sand in failure would continuously increase in volume, as long as it shears. The conclusion must be that Drucker's postulate is not valid for frictional materials. Plasticity theory for such materials must be considerably more complicated, and the proofs of the limit theorems, which heavily rely on the validity of Drucker's postulate, do not apply to frictional materials.

D Model tests

A useful tool in engineering is the analysis of the behaviour of a structure by doing a model test, at a reduced scale. The purpose of the test maybe just to investigate a phenomenon in a qualitative way, but more often its purpose is to obtain quantitative information. In that case the scale rules must be known. For a soil a special difficulty is that the mechanical properties often depend upon the state of stress, which is determined to a large extent by the weight of the soil itself. This means that in a scale model the soil properties are not well represented, because in the model the stresses are much smaller than in reality (the *prototype*).
An ingenious way to simulate the stresses in a model is to increase gravity, by placing the scale model in a centrifuge, in which the model is rotated at high speed. The principles of this method are briefly presented in this appendix. Some attention is also paid to $1g$-testing, the testing of a model without scaling gravity. It will appear that in some situations this can be useful method of model testing.
The scale rules of a certain area of physics can be derived by considering the basic equations that fully describe a certain process, and then taking care that all relevant terms in each of the equations are scaled by the same factor. The equations describing the process may be partly symbolic, if a detailed description can not be given, but the character of the relations is known. It is essential that all important factors are taken into account. Less important factors may be disregarded, if their small influence can be demonstrated.

D.1 Simple scale models

One of the most important properties of soils is that it may shear, possibly up to very large deformations, and that this shear is caused by the relative magnitude of the shear stress, compared to the normal stress. In Coulomb's failure criterion

$$\tau_{max} = c + \sigma' \tan\phi, \tag{D.1}$$

this appears if the first term, the cohesion c, is very small. This is the case for sand. In that case one may write

$$c = 0 : \quad \frac{\tau_{max}}{\sigma'} = \tan\phi. \tag{D.2}$$

It appears that failure is determined only by a ratio of the stresses, not by their magnitude. This does not necessarily mean that the ratio of shear stress to normal stress determines the soil behaviour throughout the entire range from zero deformation to failure. For very small deformations the behaviour is more or less elastic, and it is not certain that in that range the ratio τ/σ is the only parameter that governs the deformations. However, there is much evidence that the stiffness of soils increases with the stress level, both in shear as in compression (compare Terzaghi's logarithmic compression formula). Thus, it is not unreasonable to assume, at least for sandy soils, that the deformations can be described by a formula of the character

$$\varepsilon_{ij} = f(\frac{\sigma'_{ij}}{\sigma'_0}), \tag{D.3}$$

where σ'_0 is an invariant of the stress tensor, say the isotropic stress.

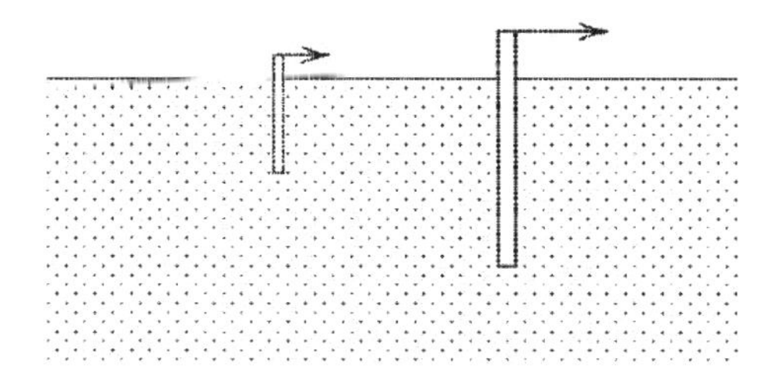

Figure D-1 Model test.

This means that the deformations are determined only by the ratio of the shear stresses and a characteristic normal stress, say the isotropic stress. For sands this is a useful approximation. It may be noted that in compression the deformation is also determined by a stress ratio, in this case the ratio of the stress to the initial stress. The assumption excludes effects as consolidation, creep and dilatancy. These must be small compared to shear and primary compression for the assumption (D.3) to be valid. Examples of problems for which the assumption is valid are a laterally loaded pile, or a caisson loaded
by cyclic forces.
If all the spatial dimensions are scaled down by a factor n_L, i.e.

$$x_{i-m} = x_{i-p}/n_L, \tag{D.4}$$

the equations of equilibrium, including the term representing the weight of the material, are satisfied if the scale factor for the stresses is also n_L,

$$\sigma_{ij-m} = \sigma_{ij-p}/n_L. \tag{D.5}$$

This can be verified by noting that the equations of equilibrium consist of terms of type $\partial\sigma_{xx}/\partial x$, and the gravity term γ. All these terms now are identical in the model and in the prototype.
If the relation between stresses and strain is of the form (D.3), the deformations are represented at scale 1,

$$\varepsilon_{ij-m} = \varepsilon_{ij-p}. \tag{D.6}$$

Because the deformations are related to the displacements by derivatives with respect to the spatial coordinates (for example $\varepsilon_{xx} = \partial u_x / \partial x$), the displacements are at the same scale as a length,

$$u_{i-m} = u_{i-p} / n_L , \tag{D.7}$$

In each of the relevant equations (equilibrium, compatibility and constitutive equations) the ratio of all terms in the model is the same as the corresponding terms in the prototype. This means that it is indeed possible to study the behaviour of the prototype in a scale model. The boundary values of stress and deformations must also be applied using the scale n_L.

A problem that can be studied in this way is a laterally loaded pile, see Figure D-1. Compression is not important in this case, so that it is unlikely that pore water pressures will be generated. The determining factor for the deformations is the ratio of shear stress to normal stress. In the model these ratios will be the same as in the prototype if the material is the same. The deformations then are at scale 1. Similarly, problems of sheet pile walls, or retaining walls, can be studied by $1g$-models, if the material is non-cohesive, i.e. sand.

Even dynamic problems may be studied by such a scale model, by noting that in that case the equations of motion contain terms of the type $\rho \partial^2 u_i / \partial t^2$. These terms will be the same in the model and in the prototype if the time is scaled according to the square root of the length scale,

$$t_m = t_p / \sqrt{n_L} . \tag{D.8}$$

Here it has been assumed that the density ρ is the same in the model as in the prototype, which is easy to accomplish, by using the same material. It may be noted that dynamic effects are important only for special problems, such as earthquakes and high speed trains. In the standard engineering problems dynamic effects usually play a minor role. Even in the cyclic loading of an offshore platform the dynamic effects are small because the period of the cycles (about 10 seconds) is so large.

Problems of consolidation can also be studied in $1g$-models, at least in a first approximation. In the consolidation equation,

$$\frac{\partial e}{\partial t} = -n\beta \frac{\partial p}{\partial t} + \frac{k}{\gamma_w} \left(\frac{\partial^2 p}{\partial x^2} + \frac{\partial^2 p}{\partial y^2} + \frac{\partial^2 p}{\partial z^2} \right), \tag{D.9}$$

all terms should then be scaled by the same factor. If all the stresses are scaled on the same scale (n_L) as a length, in order to model equilibrium, and the deformations on scale 1, the term in the left hand side of the equation can be in agreement with the other terms only if time is scaled on the length scale,

$$t_m = t_p / n_L . \tag{D.10}$$

The first term in the right hand side of the equation then is not scaled correctly, because this term consists of a ratio of two factors at length scale. But in many cases

this is a small term anyway, as the compressibility of the water (β) is very small. This means that the error in scaling the consolidation process will be very small.
It follows from the considerations given above that it is impossible to take both consolidation and dynamic effects into account, as these two phenomena lead to different requirements for the time scale. An ingenious way to solve this difficulty is to scale the permeability, without changing the porous material, by using a different fluid in the model, having a different viscosity, such that the two terms scale in the same way.
As mentioned before, all this does not apply if the material behaviour is more complex than is indicated by eq. (D.3). This will be so in the majority of problems, for instance in case of simultaneous elastic and plastic deformations, or in case of a cohesive material. This means that simple scale tests on clays are not representative for the behaviour in the prototype. They can be used only if friction is the dominant property in the mechanical behaviour, and the plastic deformations are relatively large.

D.2 Centrifuge testing

A general way of describing the relation between stresses and strains in a soil is

$$\Delta\varepsilon_{ij} = f(\sigma'_{ij}, \Delta\sigma'_{ij}, h_k), \tag{D.11}$$

where f is an arbitrary function, and h_k indicates that there may be some other physical parameters involved in the functional relationship, such as the cohesion c, or the stiffness parameters K and G. Equation (D.11) states that the incremental strains are determined by the stresses and the incremental stresses, in a not yet specified manner. Various types of behaviour can be described by relations of the type (D.11), such as elastic and plastic deformations. Of particular importance is that the incremental strains depend upon the actual stresses. This means that the stiffness may depend upon the stresses, which is a typical property of many soils. Dilatancy and contractancy can also be described by the general relation (D.11). And elastic deformations, in which the incremental strains are fully determined by the incremental stresses can also be described by (D.11), of course.
Assuming the validity of the general relationship (D.11), model testing is possible only if the stresses and the strains are all modelled at scale 1, and that the same soil is used, to ensure that the properties are the same. This implies that the stresses caused by the weight of the material must also be modelled at scale 1. In the equations of equilibrium terms of the type $\partial u_{xx}/\partial x$ appear with a term $\gamma = \rho g$. In order to model both these terms at the same scale, the volumetric weight γ must be inversely proportional to the length scale,

$$\gamma_m = \gamma_p \times n_L. \tag{D.12}$$

This can be realized by rotating the model very fast, in a *centrifuge*. Gravity then appears to be magnified, see Figure D-2. The facility consists of an arm that can be rotated around a central axis. At the two ends of the arm containers are placed, one containing the model, and the other containing a counter weight (or another model), to balance the arm. If the arm rotates a centrifugal force acts on the material in the two containers, which will rotate around a hinge. If the rotation is very fast the bottom of the two containers will be practically vertical.

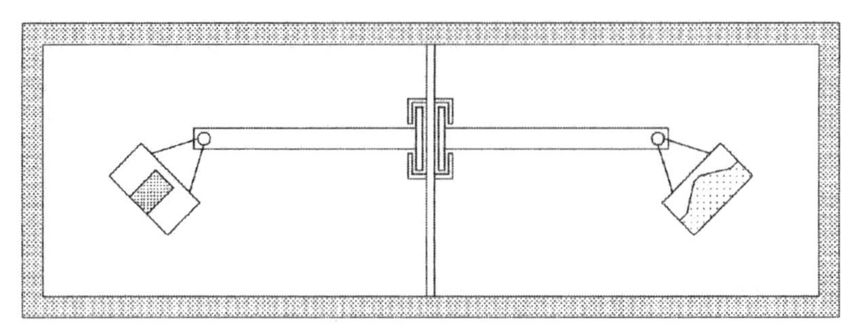

Figure D-2 Geotechnical centrifuge.

For safety, the centrifuge must be protected by heavy steel plates and concrete walls, to prevent damage in case of failure of a part of the system. For this reason the centrifuge is often located in the basement of a geotechnical laboratory.
An elementary consideration of the motion of a body moving along a circular path, of radius R, indicates that an acceleration perpendicular to the path occurs, of magnitude

$$a = \frac{v^2}{R}. \tag{D.13}$$

This is called the *centripetal acceleration.* In the case of a container filled with soil that rotates in a centrifuge this acceleration is caused by the force from the container on the soil, and transmitted through the soil, in upward direction. If the soil were not contained by the container, it would fly on, in a straight path, but it is retained in its circular path by the container. This requires a very large force, and this force is larger if the velocity is larger, or the radius smaller (at the same velocity). The stress state would be the same if the container were at rest, and a volumetric force would act upon the soil. If this volumetric force is denoted by g_m, we have

$$g_m = \frac{v^2}{R} = \omega^2 R, \tag{D.14}$$

in which ω is the angular velocity (or the frequency) of the centrifuge. Many geotechnical centrifuges have an arm length of about 5 m. This means that an acceleration of 100 g = 1000 m/s^2 is achieved if the velocity of the container is 71 m/s, or 254 km/h. The angular velocity then is 14.14 rad/s, which means that the container flies by every 0.444 s. This corresponds to 2.25 revolutions per second, or 135 revolutions per minute.
The major principle of centrifuge testing is that all stresses in the model are the same as the stresses in the prototype, so that it is practically guaranteed that soil will

behave in the same way as in reality. A geotechnical centrifuge is a reasonably complex machine, however, and it generates large forces in its parts. Furthermore, observing deformations and measuring stresses is not a simple matter. Electronic measuring devices may be built in, but these should be very small, and the measuring signals must be transmitted to the outside world, through the central axis. An alternative registration method is to record the measurements on a data recorder that is attached to the arm itself, and to read the data later. A video signal can be used to observe deformations in flight. Preparation of the samples also requires much attention, as the sample must be a good representation of the prototype, at a small scale. A small disturbance in the model corresponds to a large disturbance in reality. Consolidation problems, in which time is important, can be studied in a centrifuge if the terms $\partial e/\partial t$ and $\partial^2 p/\partial x^2$ are scaled in the same way. Because stresses and strains are at scale 1, it follows that the time scale must be the square of the length scale,

$$t_m = t_p / n_L^2. \tag{D.15}$$

If the time scale is determined by inertia effects (in dynamic problems) the terms $\rho\partial^2 u/\partial t^2$ must be scaled by the same factor as the derivatives of the stresses, $\partial\sigma_{xx}/\partial x$. That will be the case if the time scale equals the length scale,

$$t_m = t_p / n_L. \tag{D.16}$$

Again it is not easily possible to scale consolidation in combination with dynamic effects, except by using a fluid of different viscosity.

Problems

D.1 Is it possible to study the problem of slope stability of a sand dike in a $1g$-model test? And in a centrifuge?

D.2 Is it possible to study the problem of slope stability of a clay dike in a $1g$-model test? And in a centrifuge?

D.3 The arm of the centrifuge at GeoDelft is of 6 m length, capable of testing at maximum 300 g. What is then its number of rotations per minute? And the velocity of the container?

D.4 At a fair one may see a large rotating cylinder, in which people remain hanging against the wall if the bottom moves down. If it is supposed that the friction coefficient between man and steel wall is about 0.2, the radial acceleration must be about $0.2g$. If the radius of the cylinder is 4.5 m, then what is the velocity of the people, in km/h?

Answers to problems

1.1 Yes.
1.2 Outer slope.
1.3 Small.
1.4 Preloading by ice.
1.5 At the lower side.
1.6 At the higher side.
1.7 Tower close to channel.
4.1 Mass: 3000 kg, Volumetric weight: 15 kN/m^3.
4.2 $n = 0.42, e = 0.73$.
4.3 0.846 $\text{m}^3, \gamma = 1923\ \text{kg/m}^3$.
4.4 Settlement: 0.83 m.
4.5 No influence.
4.6 $n = 0.42$.
4.7 $\rho_k = 2636\ \text{kg/m}^3$.
5.1 Total stress unchanged, effective stress increase 5 Pa.
5.2 In the ship artificial air pressure. Effective stress equals air pressure. On the moon there is no atmospheric pressure. Effective stress zero.
5.3 Yes, if it sinks.
5.4 No, effective stresses unchanged.
5.5 No.
6.1 After reclamation, at 2 m depth: $\sigma = 36\,\text{kPa}, p = 0, \sigma' = 36$ kPa.
At 10 m depth: $\sigma = 180\,\text{kPa}, p = 80, \sigma' = 100$ kPa.
6.2 $\sigma = 125\,\text{kPa}, \sigma' = 125$ kPa.
6.3 $\sigma = 125\,\text{kPa}, p = 50, \sigma' = 75$ kPa.
6.4 Water level 10 m: $\sigma = 125\,\text{kPa}, p = 100, \sigma' = 25$ kPa.
Water level 150 m: $\sigma' = 25$ kPa.
6.5 $\sigma' = 86.6$ kPa.
6.6 $\sigma' = 62$ kPa.
6.7 $\Delta\sigma' = 32$ kPa.
7.1 1 $\text{m/d} = 1.16 \times 10^{-5}$ m/s. Normal: 1 m/d.
7.2 1 $\text{gpd/sqft} = 0.5 \times 10^{-6}$ m/s. Normal: 20 gpd/sqft.
7.3 $k = 3.33$ m/d.
8.1 $k = 1.48 \times 10^{-4}$ m/s.

8.2 $Q = 0.0628\,\text{cm}^3/\text{s}$.
8.3 To prevent leakage along the top of the sample.
8.4 $k = 0.5$ m/d.
9.1 $\sigma = 152\,\text{kPa}, p = 100, \sigma' = 52$ kPa.
9.2 $\sigma = 144\,\text{kPa}, p = 90, \sigma' = 54$ kPa.
9.3 $\sigma = 184\,\text{kPa}, p = 90, \sigma' = 94$ kPa.
9.4 5 m.
10.1 0.10 kN.
10.2 0.12 kN.
10.3 6.25 m.
10.4 1.75 m.
11.1 $Q = 0.4kHB$.
11.2 $i = 0.17$.
11.3 Yes, in case of holes in the clay layer.
12.1 No.
12.2 0.50 m.
12.3 $h \to -\infty$.
12.4 Not forever if there is no supply (rain).
13.1 Smaller.
13.2 More than 2 cm.
13.3 Dilatancy. Yes.
13.4 To the waist.
14.1 3300 kPa.
14.2 Very small, $\nu \approx 0.5$.
15.1 $C_{10} = 53$.
15.2 25 mm, 24 kPa.
15.3 2.5 cm.
15.4 E = 50 to 100 MPa.
15.5 $C_{10} = 4$. Just OK.
17.1 379 s.
17.2 Factor 4 larger.
17.3 650 d.
17.4 0.04 mm.
17.5 0.004.
17.6 Stop if JJ>100.
18.1 Smaller than 20,000 s.
18.2 Time step a factor 4 smaller.
18.3 Computation time a factor 8 larger.
19.1 $c_v = 1.25 \times 10^{-7}\,\text{m}^2/\text{s}$.
19.2 $m_v = 1.15\,\text{m}^2/\text{MN}, k = 1.44 \times 10^{-9}$ m/s.
20.1 First clay layer: 65 kPa, second clay layer: 141 kPa, load: 34 kPa.
20.2 27 cm, 33 cm, 39 cm.

20.3 70 days.

21.1 $\phi \geq 28^\circ$.

21.2 30 kPa.

21.3 $\sigma_{xx} = 2p, \sigma_{xy} = p$.

21.4 $\sigma_{nn} = 1.500p, \sigma_{nt} = 0.867p, \alpha = 30^\circ$.

22.1 $c = 0.12$ kPa, $\phi = 29.6^\circ$.

22.2 $F = 340$ N.

22.3 Yes.

23.1 $\phi = 29^\circ$.

23.2 $F = 153$ N.

24.1 41.22 kPa

24.2 $\phi = 22.4^\circ$

25.1 $c = 5$ kPa, $\phi = 30^\circ, A \times B = 0.2$.

25.2 $p = 40$ kPa.

25.3 Relatively dense.

25.4 $p = 40$ kPa.

26.1 $s_u = 85$ kPa.

26.2 $s_u = 53$ kPa.

26.3 $s_u = 69$ kPa.

29.1 $\sigma_{zz} = p/(1 + z/a)^2$.

29.2 $u_z = pa/E$.

29.3 $c \approx E/a$.

30.1 $\sigma_{zz} = 1.23$ kPa.

30.2 $\sigma_{zz} = 3.75$ kPa, in A: $\sigma_{zz} = 0$, at 8000 m depth: $\sigma_{zz} = 0$.

30.3 3.40 kPa, 1.72 kPa, 2.32 kPa.

30.4 Underestimated.

30.5 Yes, if ν is constant.

31.1 No.

31.2 $\sigma_{rr} = (2P/\pi r)\cos\theta$, $\sigma_{r\theta} = 0$, $\sigma_{\theta\theta} = 0$.

32.2 0.213 m.

32.3 0.070 m.

34.1 $\phi = 30^\circ : K_a = 0.333, K_p = 3.000$, etc.

34.2 $h = 2c/\gamma\sqrt{K_a}$.

34.3 Cambridge K_0 meter.

34.4 96 kN.

34.5 67 kN/m.

34.6 315 kN/m.

35.1 No.

35.2 $K_p = 1/K_a$.

35.3 45.3 kN/m.

35.4 11.4 % smaller.

35.5 408 kN/m.

36.1 OK.

36.2 OK.

36.4 Slope too steep for stability.

36.4 57.6 kN/m.

36.5 71.8 kN/m.

36.6 192 kN.

37.1 OK.

37.2 OK.

37.3 1.90 m.

37.4 11.507 m.

38.1 OK.

38.2 12.67 m.

38.3 10.20 m.

38.4 $d/h = 0.650$.

39.1 8.02 m.

39.2 8.22 m.

39.4 $F = T \times a$.

43.1 OK.

43.2 Yes.

43.3 15,120 MN.

43.4 700 kN

43.5 No, q_c is also total stress.

43.6 $q_c \approx 8$ MPa.

44.1 632 kN.

44.2 Yes.

45.1 Yes, σ_{xx} in the lower left region.

45.2 $h_c \geq 2c/\gamma$.

45.3 Yes.

45.4 20 m or more.

47.1 No.

47.2 Yes.

47.3 Introduction of horizontal force in equilibrium of moments.

App. D.1 Yes. Yes.

App. D.2 No. Yes.

App. D.3 3.56 Revolutions per second, $v = 134$ m/s.

App. D.4 $v = 3$ m/s.

Literature

R.F. Craig, *Soil Mechanics*, Van Nostrand Reinhold, New York, 1978.

Construeren met grond, CUR-publicatie no. 162, 1992.

G. Gudehus, *Bodenmechanik*, Enke, Stuttgart, 1981.

M.E. Harr, *Foundations of Theoretical Soil Mechanics*, McGraw-Hill, New York, 1966.

T.K. Huizinga, *Grondmechanica*, Waltman, Delft, 1969.

A.S. Keverling Buisman, *Grondmechanica*, Waltman, Delft, 1941.

T.W. Lambe and R.V. Whitman, *Soil Mechanics*, Wiley, New York, 1969.

G.W.E. Milligan and G.T. Houlsby, *BASIC SoilMechanics*, Butterworths, London, 1984.

C.R. Scott, *SoilMechanics and Foundations*, Applied Science Publishers, London, 1978.

R.F. Scott, *Principles of Soil Mechanics*, Addison-Wesley, Reading MA, 1963.

G.N. Smith, *Elements of Soil Mechanics*, Granada, London, 1978.

U. Smoltczyk (ed.), *Grundbau Taschenbuch*, Wilhelm Ernst, Berlin, 1980, 1982, 1986.

I.N. Sneddon, *Fourier Transforms*, McGraw-Hill, New York, 1951.

K. Terzaghi, *Theoretical Soil Mechanics*, Wiley, New York, 1940.

K. Terzaghi and R.B. Peck, *Soil Mechanics in Engineering Practice*, Wiley, New York, 1948.

S.P. Timoshenko and J.N. Goodier, *Theory of Elasticity*, 2nd ed., McGraw-Hill, New York, 1951.

C. van der Veen, E. Horvat en C.H. van Kooperen, *Grondmechanica met beginselen van de Funderingstechniek*, Waltman, Delft, 1981.

A.F. van Weele, *Moderne Funderingstechnieken*, Waltman, Delft, 1981.

Index

Coëfficient of active earth pressure:

$$K_a = \frac{\sin^2(\alpha+\phi)}{\sin^2\alpha\sin(\alpha-\delta)\left[1+\sqrt{\{\sin(\phi+\delta)\sin(\phi-\beta)\}/\{\sin(\alpha-\delta)\sin(\alpha+\beta)\}}\right]^2}$$

α=90°
β=0°

δ\φ	10°	15°	20°	25°	30°	35°	40°	45°
0°	0.704	0.589	0.49	0.406	0.333	0.271	0.217	0.172
5°	0.662	0.556	0.465	0.387	0.319	0.26	0.21	0.166
10°	0.635	0.533	0.447	0.373	0.308	0.253	0.204	0.163
15°	0.617	0.518	0.434	0.363	0.301	0.248	0.201	0.160
20°	0.607	0.508	0.427	0.357	0.297	0.245	0.199	0.160
25°	0.604	0.505	0.424	0.355	0.296	0.244	0.199	0.160
30°	0.606	0.506	0.424	0.356	0.297	0.246	0.201	0.162

α=80°
β=0°

δ\φ	10°	15°	20°	25°	30°	35°	40°	45°
0°	0.757	0.652	0.559	0.478	0.407	0.343	0.287	0.238
5°	0.720	0.622	0.536	0.460	0.393	0.333	0.280	0.233
10°	0.699	0.603	0.520	0.448	0.384	0.326	0.275	0.229
15°	0.687	0.592	0.511	0.441	0.378	0.323	0.273	0.228
20°	0.684	0.588	0.508	0.438	0.377	0.322	0.273	0.229
25°	0.689	0.591	0.510	0.440	0.379	0.325	0.276	0.232
30°	0.702	0.600	0.517	0.446	0.385	0.330	0.281	0.237

α=90°
β=10°

δ\φ	10°	15°	20°	25°	30°	35°	40°	45°
0°	0.970	0.704	0.569	0.462	0.374	0.300	0.238	0.186
5°	0.974	0.679	0.547	0.444	0.359	0.289	0.230	0.180
10°	0.985	0.664	0.531	0.431	0.350	0.282	0.225	0.177
15°	1.004	0.655	0.522	0.423	0.343	0.277	0.221	0.174
20°	1.032	0.654	0.518	0.419	0.340	0.275	0.220	0.174
25°	1.070	0.658	0.518	0.419	0.340	0.275	0.221	0.175
30°	1.120	0.669	0.524	0.422	0.343	0.278	0.223	0.177

α=80°
β=10°

δ\φ	10°	15°	20°	25°	30°	35°	40°	45°
0°	1.047	0.784	0.654	0.550	0.461	0.384	0.318	0.261
5°	1.067	0.766	0.636	0.534	0.448	0.374	0.311	0.255
10°	1.097	0.759	0.626	0.524	0.440	0.368	0.307	0.253
15°	1.138	0.759	0.622	0.520	0.437	0.366	0.305	0.252
20°	1.191	0.768	0.625	0.521	0.438	0.367	0.306	0.254
25°	1.259	0.785	0.634	0.528	0.443	0.371	0.310	0.257
30°	1.346	0.811	0.650	0.539	0.452	0.379	0.317	0.264

α=90°
β=20°

δ\φ	10°	15°	20°	25°	30°	35°	40°	45°
0°			0.883	0.572	0.441	0.344	0.267	0.204
5°			0.886	0.558	0.428	0.333	0.259	0.199
10°			0.897	0.549	0.42	0.326	0.254	0.195
15°			0.914	0.546	0.415	0.323	0.251	0.194
20°			0.940	0.547	0.414	0.322	0.250	0.193
25°			0.974	0.553	0.417	0.323	0.252	0.195
30°			1.020	0.565	0.424	0.328	0.256	0.198

α=80°
β=20°

δ\φ	10°	15°	20°	25°	30°	35°	40°	45°
0°			1.015	0.684	0.548	0.444	0.360	0.291
5°			1.035	0.676	0.538	0.436	0.354	0.286
10°			1.064	0.674	0.534	0.432	0.351	0.283
15°			1.103	0.679	0.535	0.432	0.350	0.284
20°			1.155	0.690	0.540	0.435	0.354	0.286
25°			1.221	0.708	0.551	0.443	0.360	0.292
30°			1.305	0.734	0.568	0.456	0.370	0.300

α=90°
β=30°

δ\φ	10°	15°	20°	25°	30°	35°	40°	45°
0°					0.750	0.436	0.318	0.235
5°					0.753	0.428	0.311	0.229
10°					0.762	0.423	0.306	0.226
15°					0.776	0.422	0.305	0.225
20°					0.798	0.425	0.305	0.225
25°					0.828	0.431	0.309	0.228
30°					0.866	0.442	0.315	0.232

α=80°
β=30°

δ\φ	10°	15°	20°	25°	30°	35°	40°	45°
0°					0.925	0.566	0.433	0.337
5°					0.943	0.563	0.428	0.333
10°					0.969	0.564	0.427	0.332
15°					1.005	0.570	0.430	0.333
20°					1.051	0.582	0.437	0.338
25°					1.111	0.600	0.448	0.346
30°					1.189	0.624	0.463	0.358

Coëfficient of passive earth pressure:

$$K_p = \frac{\sin^2(\alpha-\phi)}{\sin^2\alpha\sin(\alpha-\delta)\left[1-\sqrt{\{\sin(\phi-\delta)\sin(\phi+\beta)\}/\{\sin(\alpha-\delta)\sin(\alpha+\beta)\}}\right]^2}$$

α=90°
β=0°

δ\φ	10°	15°	20°	25°	30°	35°	40°	45°
0°	1.420	1.698	2.040	2.464	3.000	3.690	4.599	5.828
-5°	1.569	1.901	2.313	2.833	3.505	4.391	5.593	7.278
-10°	1.730	2.131	2.635	3.285	4.143	5.309	6.946	9.345
-15°	1.914	2.403	3.029	3.855	4.976	6.555	8.872	12.466
-20°	2.130	2.735	3.525	4.597	6.105	8.324	11.771	17.539
-25°	2.395	3.151	4.169	5.599	7.704	10.980	16.473	26.696
-30°	2.726	3.691	5.036	7.013	10.095	15.273	24.933	46.087

α=90°
β=10°

δ\φ	10°	15°	20°	25°	30°	35°	40°	45°
0°		2.099	2.595	3.235	4.080	5.228	6.841	9.204
-5°		2.467	3.086	3.908	5.028	6.605	8.923	12.518
-10°		2.907	3.700	4.783	6.314	8.569	12.076	17.944
-15°		3.456	4.496	5.969	8.145	11.536	17.225	27.812
-20°		4.166	5.572	7.652	10.903	16.370	26.569	48.891
-25°		5.122	7.093	10.181	15.384	25.117	46.474	108.431
-30°		6.470	9.371	14.274	23.468	43.697	102.545	426.159

α=80°
β=0°

δ\φ	10°	15°	20°	25°	30°	35°	40°	45°
0°	1.363	1.582	1.843	2.156	2.535	3.002	3.587	4.332
-5°	1.480	1.737	2.045	2.418	2.879	3.456	4.193	5.158
-10°	1.600	1.905	2.273	2.725	3.292	4.017	4.966	6.244
-15°	1.732	2.096	2.540	3.094	3.802	4.730	5.981	7.726
-20°	1.883	2.321	2.861	3.549	4.450	5.666	7.363	9.838
-25°	2.060	2.590	3.257	4.127	5.299	6.937	9.329	13.021
-30°	2.274	2.923	3.759	4.881	6.450	8.742	12.286	18.184

α=80°
β=10°

δ/φ	10°	15°	20°	25°	30°	35°	40°	45°
0°		1.935	2.308	2.767	3.343	4.079	5.043	6.340
-5°		2.218	2.668	3.233	3.96	4.914	6.201	7.998
-10°		2.541	3.093	3.805	4.742	6.010	7.783	10.372
-15°		2.922	3.614	4.528	5.767	7.504	10.045	13.969
-20°		3.387	4.272	5.474	7.162	9.636	13.465	19.844
-25°		3.975	5.131	6.759	9.148	12.854	19.039	30.500
-30°		4.740	6.295	8.583	12.137	18.084	29.127	53.188

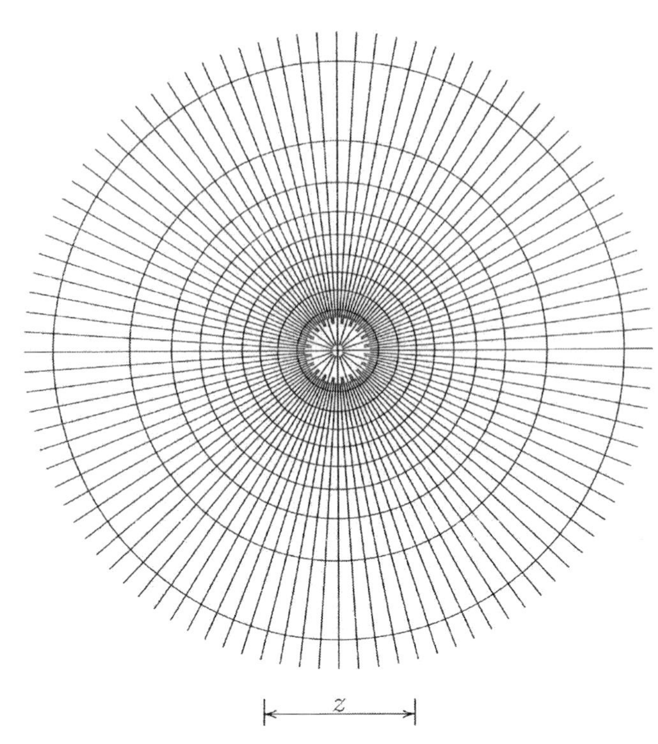

Equations Soil Mechanics

Particles: $n=\dfrac{V_p}{V_g};\quad e=\dfrac{V_p}{V_k};\quad S=\dfrac{V_w}{V_p};\quad w=\dfrac{W_w}{W_k};\quad RD=\dfrac{e_{\max}-e}{e_{\max}-e_{\min}};\quad w=Se\dfrac{\rho_w}{\rho_k};$

Stress: $\sigma_{xx}=\sigma'_{xx}+p;\quad \sigma_{xy}=\sigma'_{xy};\quad \sigma_0=\frac{1}{3}\left(\sigma'_{xx}+\sigma'_{yy}+\sigma'_{zz}\right)$

Plow: $q_x=-k\,i_x;\quad q_x=\dfrac{Q_x}{A_x};\quad i_x=\dfrac{\partial h}{\partial x};\quad h=z+\dfrac{p}{\gamma_w};\quad k=\dfrac{\kappa\,\gamma_w}{\mu};\quad \kappa=cd^2\dfrac{n^3}{(1-n)^2};\quad i_{kr}=\dfrac{\gamma_n-\gamma_w}{\gamma_w}$

Falling Head Test: $h=h_0\exp\left(-\dfrac{k}{a}\dfrac{At}{L}\right)$; (of $k=c_v\gamma_w m_v$ voor: $\beta\rightarrow\infty$);

Flow net: $\dfrac{\Delta\Phi}{\Delta s}=\dfrac{\Delta\Psi}{\Delta n}(\Phi=kh)$; $Q=\dfrac{n_{\text{flow}}}{n_{\text{potential}}}k\Delta h\,B$

Well and source (confined aquifer): $h_0-h=-\dfrac{Q}{2\pi kD}\ln(\dfrac{r}{R});$

Well and source (unconfined aquifer): $h_0^2-h^2=-\dfrac{Q}{\pi k}\ln(\dfrac{r}{R});$

Continuum:

$$\sigma_0=K\varepsilon_{vol};\quad \tau_{xy}=2G\varepsilon_{xy};\quad \varepsilon_{xx}=\frac{1}{E}\left[\sigma'_{xx}-\nu\left(\sigma'_{yy}+\sigma'_{zz}\right)\right];\quad \frac{\sigma'_v}{\varepsilon_v}=E_{oed}=\frac{1}{m_v}=K+\frac{4}{3}G=\frac{E(1-\nu)}{(1+\nu)(1-2\nu)};$$

$$K=\frac{E}{3(1-2\nu)};\ G=\frac{E}{2(1+\nu)};$$

Koppejan: $\varepsilon=U\left[\dfrac{1}{C_p}+\dfrac{1}{C_s}\log(\dfrac{t}{t_1})\right]\ln\dfrac{\sigma'}{\sigma'_1};$

Bjerrum: $-\Delta e=e_1-e=C_r\log(\dfrac{\sigma_{\text{pop}}}{\sigma_1})+C_c\log(\dfrac{\sigma}{\sigma_{\text{pop}}})+C_\alpha\log(\dfrac{t}{t_1});\quad \varepsilon=\dfrac{1}{C_{10}}\log(\dfrac{\sigma}{\sigma_1})\ (\sigma>\sigma_1);\quad \varepsilon=\dfrac{1}{A_{10}}\text{lo}$

$(\sigma<\sigma_1)$; ISO: $-\Delta e=C_c\log(\dfrac{\sigma}{\sigma_1});\quad \varepsilon=\dfrac{-\Delta e}{1+e};\quad U=\dfrac{\Delta h-\Delta h_0}{\Delta h_\infty-\Delta h_0}=1-\dfrac{8}{\pi^2}\sum_{j=1}^{\infty}\dfrac{1}{(2j-1)^2}\exp[-(2j-1)^2\dfrac{\pi^2 c_v t}{4h^2}$

$$U\approx\frac{2}{\sqrt{\pi}}\sqrt{\frac{c_v t}{h^2}}\ \text{if}\ U<0.7;\quad U\approx1-\frac{8}{\pi^2}\exp\left(-\frac{\pi^2}{4}\cdot\frac{c_v t}{h^2}\right)\ \text{if}\ U>0.5$$

idation:

$= c_v \frac{\partial^2 p}{\partial t^2}$; $c_v = \frac{k}{\gamma_w(m_v + n\beta)}$; $\frac{c_v t}{h^2} >> 0.1$: $\frac{p}{p_0} = \frac{4}{\pi}\cos[\frac{\pi}{2}(\frac{h-z}{h})]\exp(-\frac{\pi^2}{4}\frac{c_v t}{h^2})$; $\frac{c_v t_{99\%}}{h^2} = 1.784$;

$= 0.848$; $\frac{c_v t_{50\%}}{h^2} = 0.197$; $\frac{c_v t_{1\%}}{h^2} = 10^{-4}\frac{\pi}{4}$; $p_i(1+\Delta t) = p_i(t) + \alpha[p_{i+1}(t) - 2p_i(t) + p_{i-1}(t)]$; $\alpha = \frac{c_v \Delta t}{(\Delta z)^2}$;

oulomb: $(\frac{\sigma_1' - \sigma_3'}{2}) - (\frac{\sigma_1' + \sigma_3'}{2})\sin\phi - c\cos\phi = 0$; $\sin\phi = \frac{\frac{1}{2}(\sigma_1' - \sigma_3')}{c \cdot \cot\phi + \frac{1}{2}(\sigma_1' + \sigma_3')}$; $\sigma_1' = \sigma_3' \frac{1+\sin\phi}{1-\sin\phi} + 2c\frac{\cos\phi}{1-\sin\phi}$;

ned strength: $c_u = s_u \approx \frac{\sigma_1' - \sigma_3'}{2} = \frac{c\cos\phi + \sigma_0' \sin\phi}{1 - \frac{1}{3}\sin\phi}$; Coulomb: $\tau_f = c + \sigma_n' \tan\delta$;

on / Henkel: $\Delta p = B[\Delta\sigma_3 + A(\Delta\sigma_1 - \Delta\sigma_3)]$; $B = \frac{1}{1 + n\beta K}$; Triaxial test: $A \approx \frac{1}{3}$ (neutral): $A < \frac{1}{3}$ (dilatant)

(contractant);

esq: Point load: $r = 0$: $\sigma_{zz} = \frac{3}{2}\frac{P}{\pi z^2}$; Rigid plate: $z = 0$: $\frac{\sigma_{zz}}{p} = \frac{1}{2\sqrt{1-(r/a)^2}}$; $z = 0$:

$(1-\nu^2)\frac{\bar{p}a}{E} = \frac{\pi}{4}(1-\nu)\frac{\bar{p}a}{G}$;

e plate: $r = 0$: $\frac{\sigma_{zz}}{p} = 1 - \frac{z^3}{\left(\sqrt{a^2+z^2}\right)^3}$; $u_z = 2(1-\nu)^2\frac{pa}{E} = (1-\nu)\frac{pa}{G}$; Layer thickness

$f \approx 1 + \frac{h}{2a} - \frac{1 + \frac{1}{2}(h/a)^2}{\sqrt{1+(h/a)^2}}$;

t: Line load ($r = \sqrt{x^2+z^2}$): $\sigma_{zz} = \frac{2Fz^3}{\pi r^4}$; $\sigma_{xx} = \frac{2Fx^2 z}{\pi r^4}$; $\sigma_{xz} = \frac{2Fxz^2}{\pi r^4}$; Next to smooth wall:

: $Q_h = \frac{2}{\pi}\frac{F}{1+(a/h)^2}$ Strip: $x = 0$: $\sigma_{zz} = \frac{2P}{\pi}[\arctan(\frac{a}{z}) + \frac{az}{a^2+z^2}]$; $\sigma_{xx} = \frac{2P}{\pi}[\arctan(\frac{a}{z}) - \frac{az}{a^2+z^2}]$;

t to smooth wall: $Q_h = \frac{2}{\pi}ph\arctan(\frac{a}{h})$;

tal stress: $K_e = \frac{\nu}{1-\nu}$; $K_0 \approx 1 - \sin\phi$;

$= \frac{1-\sin\phi}{1+\sin\phi}$; $K_p = \frac{1+\sin\phi}{1-\sin\phi}$; $\sigma'_{h-\min} = K_a\sigma_v' - 2c\sqrt{K_a}$; $\sigma'_{h-\max} = K_p\sigma_v' + 2c\sqrt{K_p}$; $Q = \frac{1}{2}K\gamma h^2$; $Q_h = Q$

$\delta)$

undation:

dtl: ($\varphi = 0$): $p_c = (\pi+2)c$; Brinch Hansen: $p_c = cN_c i_c s_c + qN_q i_q s_q + \frac{1}{2}\gamma BN_\gamma i_\gamma s_\gamma$; $N_q = \frac{1+\sin\phi}{1-\sin\phi}\exp(\pi\tan\phi)$;

$N_q - 1)\cot\phi$; $N_\gamma = 2(N_q - 1)\tan\phi$;

$\frac{t}{c + p\tan\phi}$; $i_q = i_c^2$; $i_\gamma = i_c^3$; $s_c = 1 + 0.2\frac{B}{L}$; $s_q = 1 + \frac{B}{L}\sin\phi$; $s_\gamma = 1 - 0.3\frac{B}{L}$ $(L \geq B)$

slope: ($c = 0$): $F \approx \frac{\tan\phi}{\tan\alpha}$;

parallel surface: $F \approx \frac{\gamma_n - \gamma_w}{\gamma_n}\frac{\tan\phi}{\tan\alpha}$; Horizontal flow: $F \approx \frac{\gamma_n - (\gamma_w/\cos^2\alpha)}{\gamma_n}\frac{\tan\phi}{\tan\alpha}$

ope: Fellenius: $F \approx \frac{\sum[(c + (\gamma h\cos^2\alpha - p)\tan\phi)/\cos\alpha]}{\sum\gamma h\sin\alpha}$; Bishop: $F \approx \frac{\sum\frac{c + (\gamma h - p)\tan\phi}{\cos\alpha(1 + \tan\alpha\tan\phi/F)}}{\sum\gamma h\sin\alpha}$

9 789065 620583